ENGINEERING OF POLYMERS AND CHEMICAL COMPLEXITY

Volume I: Current State of the Art and Perspectives

ENGINEERING OF POLYMERS AND CHEMICAL COMPLEXITY

Volume I: Current State of the Art and Perspectives

Edited by

LinShu Liu, PhD, and Antonio Ballada, PhD

Gennady E. Zaikov, DSc, and A. K. Haghi, PhD
Reviewers and Advisory Board Members

Apple Academic Press

TORONTO   NEW JERSEY

Apple Academic Press Inc. | Apple Academic Press Inc.
3333 Mistwell Crescent | 9 Spinnaker Way
Oakville, ON L6L 0A2 | Waretown, NJ 08758
Canada | USA

©2014 by Apple Academic Press, Inc.

First issued in paperback 2021

Exclusive worldwide distribution by CRC Press, a member of Taylor & Francis Group
No claim to original U.S. Government works

ISBN 13: 978-1-77463-095-2 (pbk)
ISBN 13: 978-1-926895-86-4 (hbk)

Library of Congress Control Number: 2014933152

Library and Archives Canada Cataloguing in Publication

Engineering of polymers and chemical complexity.

Includes bibliographical references and index.
Contents: Volume I. Current state of the art and perspectives/edited by LinShu Liu, PhD, and Antonio Ballada, PhD; Gennady E. Zaikov, DSc, and A. K. Haghi, PhD, Reviewers and Advisory Board Members -- Volume II. New approaches, limitations and control / edited by Walter W Focke, PhD and Prof. Hans-Joachim Radusch.

ISBN 978-1-926895-86-4 (v. 1: bound).--ISBN 978-1-926895-87-1 (v. 2: bound)
1. Polymers. 2. Polymerization. 3. Chemical engineering. 4. Nanocomposites (Materials). I. Liu, LinShu, editor of compilation II. Ballada, Antonio, editor of compilation III. Focke, W. W. (Walter Wilhelm), editor of compilation IV. Radusch, Hans-Joachim, editor of compilation V. Title: Current state of the art and perspectives. VI. Title: New approaches, limitations and control.

TP156.P6E54 2014   668.9   C2014-901112-1

Apple Academic Press also publishes its books in a variety of electronic formats. Some content that appears in print may not be available in electronic format. For information about Apple Academic Press products, visit our website at **www.appleacademicpress.com** and the CRC Press website at **www.crcpress.com**

ABOUT THE EDITORS

LinShu Liu, PhD

LinShu Liu, PhD, is a Research Chemist of the Eastern Regional Research Center, Agricultural Research Service, US Department of Agriculture. He obtained his PhD from Kyoto University, Japan, in 1990, and he was a Postdoctoral Associate at MIT (Massachusetts Institute of Technology), Cambridge, Massachusetts, USA, from 1990–1992. His research interests fall in the areas of delivery systems from naturally occurring polymers for the controlled release of bioactive substances, functional composites from biomass or biobased materials, smart packaging technology and material, and biomedical devices for tissue regeneration, pharmaceutical, and cosmetic applications.

Antonio Ballada, PhD

Antonio Ballada, PhD, received degrees in industrial chemistry from Milano University and in business administration from Bocconi University, both in Milano, Italy. He has held many top positions in the chemical and pharmaceutical business in Italy, USA, Taiwan, and China. He was CEO of Antibioticos SpA, and he was also responsible for the Catalyst and Additives Business Division at Basell, Inc., where he developed important relationships with the academic world involved in related R&D. He is Executive Vice President of FasTech SrL technology company, active in the elastomers and plastics industries. He is a member of the board of China Catalyst Ltd in Hong Kong and a member of the board of the Consultants Professional Association (APCO).

REVIEWERS AND ADVISORY BOARD MEMBERS

Gennady E. Zaikov, DSc

Gennady E. Zaikov, DSc, is Head of the Polymer Division at the N. M. Emanuel Institute of Biochemical Physics, Russian Academy of Sciences, Moscow, Russia, and Professor at Moscow State Academy of Fine Chemical Technology, Russia, as well as Professor at Kazan National Research Technological University, Kazan, Russia. He is also a prolific author, researcher, and lecturer. He has received several awards for his work, including the the Russian Federation Scholarship for Outstanding Scientists. He has been a member of many professional organizations and on the editorial boards of many international science journals.

A. K. Haghi, PhD

A. K. Haghi, PhD, holds a BSc in urban and environmental engineering from University of North Carolina (USA); a MSc in mechanical engineering from North Carolina A&T State University (USA); a DEA in applied mechanics, acoustics and materials from Université de Technologie de Compiègne (France); and a PhD in engineering sciences from Université de Franche-Comté (France). He is the author and editor of 65 books as well as 1000 published papers in various journals and conference proceedings. Dr. Haghi has received several grants, consulted for a number of major corporations, and is a frequent speaker to national and international audiences. Since 1983, he served as a professor at several universities. He is currently Editor-in-Chief of the *International Journal of Chemoinformatics and Chemical Engineering* and *Polymers Research Journal* and on the editorial boards of many international journals. He is a member of the Canadian Research and Development Center of Sciences and Cultures (CRDCSC), Montreal, Quebec, Canada.

CONTENTS

LIST OF CONTRIBUTORS

G. Adamek
Poznan University of Technology, Institute of Materials Science and Engineering, M. Sklodowska-Curie 5 Sq., 60-965 Poznan, Poland.

R. Ansari
University of Guilan, Rasht, Iran.

Heinrich Badenhorst
SARChI Chair in Carbon Materials and Technology, Department of Chemical Engineering, University of Pretoria, Lynwood Road, Pretoria, Gauteng, 0002, South Africa.
P.O. Box 66464, Highveld Ext. 7, Centurion, Gauteng, 0169, South Africa.
Tel.: +27 12 420 4173; Fax: +27 12 420 2516
Email: heinrich.badenhorst@up.ac.za

A. Yu. Bedanokov
D. I. Mendeleev Russian University for Chemical Technology, Moscow, Russia.

B. A. Bhanvase
Chemical Engineering Department, Vishwakarma Institute of Technology, 666, Upper Indiranagar, Bibwewadi, Pune-411037, Maharashtra, India.

M. I. Buzin
Institution of Russian Academy of Sciences AN. Nesmeyanov Institute of Organoelement Compounds, RAS, Vavilova Str. 28, 119991 Moscow, Russian Federation.

E. Ya. Davidov
N. M. Emanuel Institute of Biochemical Physics of Russian Academy of Sciences, 4 Kosygin str., 119334 Moscow, Russia.
E-mail: Chembio@sky.chph.ras.ru

S. A. Derkach
SE "Mechnicov Institute of Microbiology and Immunology of Academy of Medical Sciences of Ukraine".

I. L. Dubnikova
N. N. Semenov Institute of Chemical Physics of Russian Academy of Sciences, 4 Kosygin str., 119991 Moscow, Russia.

Alexei V. Dubrovsky
Institute of Theoretical and Experimental Biophysics RAS, Institutskaya Street 3, 142290 Pushchino, Moscow region, Russia.

R. A. Dvorikova
Institution of Russian Academy of Sciences A. Nesmeyanov Institute of Organoelement Cjmpounds, RAS, Vavilov St. 28, 119991 Moscow, Russian Federation.

M. R. El-Aassar
Polymer materials research Department, Institute of Advanced Technology and New Material, City of Scientific Research and Technology Applications, New Borg El-Arab City 21934, Alexandria, Egypt.

M. Esmaeili
University of Guilan, Rasht, Iran.

Walter Focke
SARChI Chair in Carbon Materials and Technology, Department of Chemical Engineering, University of Pretoria, Lynwood Road, Pretoria, Gauteng, 0002, South Africa.
P.O. Box 66464, Highveld Ext. 7, Centurion, Gauteng, 0169, South Africa.
Tel.: +27 12 420 4173, Fax: +27 12 420 2516
Email: heinrich.badenhorst@up.ac.za

A. K. Haghi
University of Guilan, Rasht, Iran.

E. A. Hassan
Department of Chemistry, Faculty of Science, Al-Azhar University, Cairo/Egypt.

H. Hlídková
Institute of Macromolecular Chemistry Academy of Sciences of the Czech Republic, v.v.i.
Heyrovský Sq. 2, 162 06 Prague 6, Czech Republic.

D. Horák
Institute of Macromolecular Chemistry Academy of Sciences of the Czech Republic, v.v.i. Heyrovský Sq. 2, 162 06 Prague 6, Czech Republic.
E-mail: horak@imc.cas.cz

N. V. Ilyashenko
Department of Biology, Tver State Medical Academy, Sovetskaya Str., 4, 170100 Tver, Russia.

N. N. Ivanova
SE "Institute of Dermatology and Venerology of Academy of Medical Sciences of Ukraine" str. Chernishevskaya 7/9, 61057 Kharkov, Ukraine.
Fax: +38-057- 706-32-03
E-mail: jet-74@mail.ru

D.V.S. Jain
Panjab University, Chandigarh, India

J. Jakubowicz
Poznan University of Technology, Institute of Materials Science and Engineering, M. Sklodowska-Curie 5 Sq., 60-965 Poznan, Poland.

S. E. Karekar
Chemical Engineering Department, Vishwakarma Institute of Technology, 666, Upper Indiranagar, Bibwewadi, Pune-411037, Maharashtra, India.

L. I. Kazakova
Institute of Theoretical and Experimental Biophysics Russian Academy of Science, 142290 Pushchino, Moscow region 142290, Russia.

A. R. Khokhlov
Institution of Russian Academy of Sciences A. Nesmeyanov Institute of Organoelement Compounds, RAS, Vavilova Str. 28, 119991 Moscow, Russian Federation.

Z. S. Klemenkova
Institution of Russian Academy of Sciences A. Nesmeyanov Institute of Organoelement Compounds, RAS, Vavilova Str. 28, 119991 Moscow, Russian Federation.

Yu. V. Korshak
D. Mendeleev University for Chemical Technology of Russia, Miusskaya Pl. 9, 125047 Moscow, Russian Federation.

A. Kostopoulou
Institute of Electronic Structure & Laser (IESL) Foundation for Research & Technology - Hellas (FORTH) P.O. Box 1385, Vassilika Vouton 71110 Heraklion, Crete.

E. V. Kotsar
SE "Institute of Dermatology and Venerology of Academy of Medical Sciences of Ukraine" str. Chernishevskaya 7/9, 61057 Kharkov, Ukraine.
Fax +38-057- 706-32-03
E-mail: jet-74@mail.ru

G. V. Kozlov
Institute of Applied Mechanics of Russian Academy of Sciences, Leninskii pr., 32 A, Moscow-119991, Russian Federation.

R. Kozlowski
Institute of Natural Fibres, 71b Wojska Polskiego str., 60-630 Poznan, Poland.

L. A. Kurbatova
Department of Biology, Tver State Medical Academy, Sovetskaya Str., 4, 170100 Tver, RUSSIA.

A. Lappas
Institute of Electronic Structure & Laser (IESL) Foundation for Research & Technology - Hellas (FORTH) P.O. Box 1385, Vassilika Vouton 711 10 Heraklion, Crete.

S. M. Lomakin
NM Emanuel Institute of Biochemical Physics of Russian Academy of Sciences, 4 Kosygin str., 119334 Moscow, Russia.
E-mail: Chembio@sky.chph.ras.ru

G. I. Mavrov
SE "Institute of Dermatology and Venerology of Academy of Medical Sciences of Ukraine" str. Chernishevskaya 7/9, 61057 Kharkov, Ukraine.
E-mail: jet-74@mail.ru, fax +38-057- 706-32-03

V. K. Meena
Central Scientific Instruments Organisation (CSIR-CSIO), Chandigarh, India

A. K. Mikitaev
Kabardino – Balkarian State University, Nalchik, Russia.

M. A. Mikitaev
L. Ya. Karpov Research Institute, Moscow, Russia.

A. I. Minett
ARC Centre of Excellence for Electromaterials, Intelligent Polymer Research Institute, University of Wollongong, Northfields Avenue, Wollongong, NSW 2522, Australia.

M. S. Mohy Eldin
Polymer materials research Department, Institute of Advanced Technology and New Material, City of Scientific Research and Technology Applications, New Borg El-Arab City 21934, Alexandria, Egypt.

M. S. Mohy Eldin
Polymer Materials Research Department, Advanced Technology and New Materials Research Institute, City for Scientific Research and Technological Applications, New Boarg Elarab City, 21934, Alexandria, Egypt.

V. Mottaghitalab
University of Guilan, Rasht, Iran.

B. U. Nair
Chemical Laboratory, Central Leather Research Institute, Council of Scientific and Industrial Research, Adyar, Chennai, 600 020 India.

M. Nidhin
Chemical Laboratory, Central Leather Research Institute, Council of Scientific and Industrial Research, Adyar, Chennai, 600 020 India.

L. N. Nikitin
Institution of Russian Academy of Sciences A. Nesmeyanov Institute of Organoelement Compounds, RAS, Vavilova Str. 28, 119991 Moscow, Russian Federation.

A. M. Omer
Polymer Materials Research Department, Advanced Technology and New Materials Research Institute, City for Scientific Research and Technological Applications, New Boarg Elarab City, 21934, Alexandria, Egypt.

P. M. Pakhomov
Tver State University, Zheliabova Str., 33, 170100 Tver, Russia
E mail: nadya_bioecology@mail.ru

N. V. Pavlova
Department of Biology, Tver State Medical Academy, Sovetskaya Str., 4, 170100 Tver, Russia.

M. B. Petrova
Department of Biology, Tver State Medical Academy, Sovetskaya Str., 4, 170100 Tver, Russia.

A. D. Rakhimkulov
N. M. Emanuel Institute of Biochemical Physics of Russian Academy of Sciences, 4 Kosygin str., 119334 Moscow, Russia.
E-mail: Chembio@sky.chph.ras.ru

Brian Rand
SARChI Chair in Carbon Materials and Technology, Department of Chemical Engineering, University of Pretoria, Lynwood Road, Pretoria, Gauteng, 0002, South Africa.
P.O. Box 66464, Highveld Ext. 7, Centurion, Gauteng, 0169, South Africa.
Tel.: +27 12 420 4173; Fax: +27 12 420 2516
Email: heinrich.badenhorst@up.ac.za

S. D. Razumovskii
N. M. Emanuel Institute of Biochemical Physics, Russian Academy of Sciences, 4, Kosygin str., Moscow 119334, Russia.
E-mail: chembio@sky.chph.ras.ru

A. L. Rusanov
Institution of Russian Academy of Sciences A. Nesmeyanov Institute of Organoelement Cjmpounds, RAS, Vavilov St. 28, 119991 Moscow, Russian Federation.

Lyudmila I. Shabarchina
Institute of Theoretical and Experimental Biophysics RAS, Institutskaya Street 3, 142290 Pushchino, Moscow region, Russia.
E-mail: shabarchina@rambler.ru

Ekaterina A. Saburova,
Institute of Theoretical and Experimental Biophysics RAS, Institutskaya Street 3, 142290 Pushchino, Moscow region, Russia.

V. A. Shanditsev
Institution of Russian Academy of Sciences A. Nesmeyanov Institute of Organoelement Cjmpounds, RAS, Vavilov Str. 28, 119991 Moscow, Russian Federation.

A. N. Shchegolikhin
NM Emanuel Institute of Biochemical Physics of Russian Academy of Sciences, 4 Kosygin str.,
119334 Moscow, Russia.
E-mail: Chembio@sky.chph.ras.ru

M. L. Singla
Central Scientific Instruments Organisation (CSIR-CSIO), Chandigarh, India.

Suman Singh
Central Scientific Instruments Organisation (CSIR-CSIO), Chandigarh, India

E. A. Soliman
Polymer Materials Research Department, Advanced Technology and New Materials Research Institute, City
for Scientific Research and Technological Applications, New Boarg Elarab City, 21934, Alexandria, Egypt.

S. H. Sonawane
Department of Chemical Engineering National Institute of Technology, Warangal, 506004 AP India.

G. M. Spinks
ARC Centre of Excellence for Electromaterials, Intelligent Polymer Research Institute, University of
Wollongong, Northfields Avenue, Wollongong, NSW 2522, Australia.

K. J. Sreeram
Chemical Laboratory, Central Leather Research Institute, Council of Scientific and Industrial Research,
Adyar, Chennai, 600 020 India.

G. B. Sukhorukov
Institute of Theoretical and Experimental Biophysics Russian Academy of Science,
142290 Pushchino, Moscow region 142290, Russia.
School of Engineering & Materials Science, Queen Mary University of London, London, UK.

Sergey A. Tikhonenko
Institute of Theoretical and Experimental Biophysics RAS, Institutskaya Street 3, 142290 Pushchino,
Moscow region, Russia.
E-mail: tikhonenkosa@gmail.com

G. G. Wallace
ARC Centre of Excellence for Electromaterials, Intelligent Polymer Research Institute, University of
Wollongong, Northfields Avenue, Wollongong, NSW 2522, Australia

B. Xi
ARC Centre of Excellence for Electromaterials, Intelligent Polymer Research Institute, University of
Wollongong, Northfields Avenue, Wollongong, NSW 2522, Australia.

Yu. G. Yanovsky
Institute of Applied Mechanics of Russian Academy of Sciences, Leninskii pr., 32 A, Moscow-119991,
Russian Federation.

G. E. Zaikov
N. M. Emanuel Institute of Biochemical Physics of Russian Academy of Sciences, 4 Kosygin str.,
119334 Moscow, Russia Federation.
E-mail: Chembio@sky.chph.ras.ru

LIST OF ABBREVIATIONS

ABC	Atomistic-based continuum
BD	Brownian dynamics
CNT	Carbon nanotube
CV	Cyclic voltammetry
CXR	Cyclohexane regain
DA	Dirt adherence
DFT	Density functional theory
DMA	Dynamic mechanical analysis
DPD	Dissipative particle dynamics
DSC	Differential scanning calorimetery
EP	Electrostatic properties
EPR	Electron paramagnetic resonance
FEM	Finite element method
FP	Ferrocene-containing polymers
FTIR	Fourier transform infrared spectroscopy
FTT	Fire testing technology
HVSEM	High vacuum scanning electron microscopy
LB	Lattice Boltzmann
LbL	Layer-by-Layer
LVSEM	Low vacuum scanning electron microscopy
MC	Monte Carlo
MD	Molecular dynamics
MIC	Minimum inhibitory concentration
MM	Molecular mechanics
MWCNT	Multi-walled carbon nanotube
PEBBLE	Photonic explorers for bioanalyse with biologically localized embedding system
PMC	Polyelectrolyte microcapsules
PXRD	Powder X-ray diffraction
RVE	Representative volume element
SEM	Scanning electron microscopy
SUSHI	Simulation utilities for soft and hard interfaces
TDGL	Time-dependent Ginzburg–Landau method
TEM	Transmission electron microscopy
TGA	Thermogravimetric analysis
XRD	X-Ray powder diffraction
ZFC	Zero-field cooled

PREFACE

This book provides a broad overview on current studies in the engineering of polymers and chemicals complexity of various origins, on scales ranging from single molecules and nano-phenomena to macroscopic chemicals. The book consists of 15 chapters that survey the current progress in particular research fields. The chapters, prepared by leading international experts, yield together a fascinating picture of a rapidly developing research discipline that brings chemical engineering to new frontiers.

The aim of Chapter 1 is to demonstrate conditions under which communicating pores are formed enabling high permeability of PHEMA scaffolds that is crucial for future cell seeding. Thermal properties of the polypropylene/multi-walled carbon nanotube composites are studied in Chapter 2. Chapter 3 explains that the polymeric nanocomposites could be considered as polymers filled with nanoparticles that interact with the polymeric matrix on the molecular level, contrary to the macrointeraction in composite materials. Mentioned nanointeraction results in high adhesion hardness of the polymeric matrix to the nanoparticles. New magnetic nanomaterials have been synthesized from ferrocene-containing polyphenylenes in Chapter 4. In Chapter 5 the stability of polymeric products to oxidizing and hydrolytic destructions is investigated in detail.

The aim of Chapter 6 is to demonstrate a particular example of a sensor system, which combines catalytic activity for urea and, at the same time, enabling monitoring enzymatic reaction by optical recording. The proposed sensor system is based on multilayer polyelectrolyte microcapsules containing urease and a pH-sensitive fluorescent dye, which translates the enzymatic reaction into a fluorescently registered signal. A study on activity of liposomal antimicrobic preparations is reported in Chapter 7. Polyelectrolyte ensym-bearing microdiagnosticum as a new step in clinical-biochemistry analysis is presented in Chapter 8 of this book. A detailed review of mathematical modeling and experimental case studies on nanofibers and CNTs is presented in Chapter 9. In Chapter 10, the orientation controlled immobilization strategy for β-galactosidase on alginate beads is investigated. Chapter 11 reports a study on the preparation, characterization, and evaluation of water-swellable hydrogel via grafting cross-linked polyacrylamide chains onto gelatin backbone by free radical polymerization. The aim of this chapter is to increase the water holding capacity of gelatin to wide its applications as soil conditioners. Chapter 12 presents an experimental study on the chemical composition and anatomic structure of polluted higher aquatic plants by making use of combined physical methods of characterization band Fourier transform infrared spectroscopy, scanning electron microscopy, and X-ray microanalysis. Several case studies presented in Chapter 13 on complexities in nanomaterials. In Chapter 14 a detailed discussion presented on microstructures of graphite. In the last chapter, theoretical treatment of disperse nanofiller aggregation process in butadiene-styrene rubber matrix within the frameworks of irreversible aggregation models was carried out.

This book provides innovative chapters on the growth of educational, scientific, and industrial research activities among chemists, biologists, and polymer and chemical engineers and provides a medium for mutual communication between international academia and the industry.

— LinShu Liu, PhD, and Antonio Ballada, PhD

CHAPTER 1

HYDROGEL-BASED SUPPORTS: DESIGN AND SYNTHESIS

D. HORÁK and H. HLÍDKOVÁ

CONTENTS

1.1   INTRODUCTION

Superporous poly(2-hydroxyethyl methacrylate) (PHEMA) supports with pore size from tens to hundreds micrometers were prepared by radical polymerization of 2-hydroxyethyl methacrylate (HEMA) with 2 wt% ethylene dimethacrylate (EDMA) with the aim to obtain a support for cell cultivation. Superpores were created by the salt leaching technique using NaCl or $(NH_4)_2SO_4$ as a porogen. Addition of liquid porogen (cyclohexanol/dodecan-1-ol (DOH) (CyOH/DOH) = 9/1 w/w) to the polymerization mixture did not considerably affect formation of meso- and macropores. The prepared scaffolds were characterized by several methods including water and CXR by centrifugation, water regain by suction, scanning electron microscopy (SEM), mercury porosimetry, and dynamic desorption of nitrogen. High vacuum scanning electron microscopy (HVSEM) confirmed permeability of hydrogels to 8 μm microspheres, whereas low vacuum scanning electron microscopy (LVSEM) at cryo-conditions showed the undeformed structure of the frozen hydrogels. Interconnection of pores in the PHEMA scaffolds was proved. Water regain determined by centrifugation method did not include volume of large superpores (imprints of porogen crystals), in contrast to water regain by suction method. The porosities of the constructs ranging from 81 to 91% were proportional to the volume of porogen in the feed.

Polymer supports have received much attention as microenvironment for cell adhesion, proliferation, migration, and differentiation in tissue engineering and regenerative medicine. The three-dimensional scaffold structure provides support for high level of tissue organization and remodelling. Regeneration of different tissues, such as bone [1], cartilage [2], skin [3], nerves [4], or blood vessels [5] is investigated using such constructs. An ideal polymer scaffold should thus mimic the living tissue, that is possess a high water content, with possibility to incorporate bioactive molecules allowing a better control of cell differentiation. At the same time it requires a range of properties including biocompatibility and/or biodegradability, highly porous structure with communicating pores allowing high cell adhesion and tissue in-growth. The material should be sterilizable and also possess good mechanical strength. Both natural and synthetic hydrogels are being developed. The advantage of synthetic polymer matrices consists in their easy proccessability, tunable physical and chemical properties, susceptibility to modifications, and possibly controlled degradation.

Many techniques have been developed to fabricate highly porous constructs for tissue engineering. They include for instance solvent casting [6], gas foaming [7] or/and salt leaching [8], freeze-thaw procedure [9,10], supercritical fluid technology [11] (disks exposed to CO_2 at high pressure), and electrospinning (for nanofiber matrices) [12]. A wide range of polymers was suggested for scaffolds. In addition to natural materials, such as collagen, gelatine, dextran [13], chitosan [14], phosphorylcholine [15], alginic [16], and hyaluronic acids, it includes also synthetic polymers, for example poly(vinyl alcohol) [17], poly(lactic acid) [1, 18], polycaprolactone [19], poly(ethylene glycol) [20], polyacrylamide [21], polyphosphazenes [22], and as well as polyurethane [23].

Among various kinds of materials being used in biomedical and pharmaceutical applications, hydrogels composed of hydrophilic polymers or copolymers find a unique place. They have a highly water-swollen rubbery three-dimensional structure which is similar to natural tissue [24,25]. In this report, PHEMA was selected as a suitable hydrogel intended for cell cultivation. The presence of hydroxyl and carboxyl groups makes this polymer compatible with water, whereas the hydrophobic methyl groups and backbone impart hydrolytic stability to the polymer and support the mechanical strength of the matrix [26]. The PHEMA hydrogels are known for their resistance to high temperatures, acid and alkaline hydrolysis and low reactivity with amines [27]. Previously, porous structure in PHEMA hydrogels was obtained by phase separation using a low molecular weight or polymeric porogen, or by the salt leaching method. The material was used as a mouse embryonic stem cell support [8, 28,29]. The aim of this report is to demonstrate conditions under which communicating pores are formed enabling high permeability of PHEMA scaffolds which is crucial for future cell seeding.

1.2 EXPERIMENTAL DETAILS

1.2.1 REAGENTS

The HEMA (Röhm GmbH, Germany) and EDMA (Ugilor S.A., France), were purified by distillation. The 2,2′-Azobisisobutyronitrile (AIBN, Fluka) was crystallized from ethanol and used as initiator. Sodium chloride G. R. (Lach-Ner, s.r.o. Neratovice, Czech Republic) was classified, particle size 250–500 μm and ammonium sulfate needles (100 × 600 μm, Lachema, Neratovice, Czech Republic) were utilized as porogens. Cyclohexanol (CyOH, Lachema, Neratovice, Czech Republic) was distilled, DOH and all other solvents and reagents were obtained from Aldrich and used without purification. Ammonolyzed PGMA microspheres (2 μm) were obtained by the previously described procedure [30]. Sulfonated polystyrene (PSt) microspheres (8 μm) Ostion LG KS 0803 were purchased from Spolek pro chemickou a hutní výrobu, Ústí n. L., Czech Republic. Polyaniline hydrochloride microspheres (PANI, 200–400 nm) were prepared according to literature [31].

1.2.2 HYDOGEL PREPARATION

Crosslinked hydrogel constructs were prepared by the bulk radical polymerization of a reaction mixture containing monomer (HEMA), crosslinking agent (EDMA), initiator (AIBN) and NaCl or/and liquid diluent as a porogen (CyOH/DOH = 9/1 w/w). The compositions of polymerization mixtures are summarized in Table 1.

TABLE 1 Preparation of PHEMA hydrogels, conditions, and properties[a]

Run	NaCl (vol.%)	Water regain (ml/g)		CX regain[d] (ml/g)	Cumulative pore volume[f] (ml/g)
1	41.4	0.84[d]	7.52[e]	0.13	0.35

TABLE 1　*(Continued)*

2	40.8	0.88[d]	7.40[e]	0.23	0.47
3	40.0	1.04[d]	5.34[e]	0.56	1.03
4	39.1	0.81[d]	4.05[e]	0.33	1.24
5	37.9	0.78[d]	4.04[e]	0.21	1.60
6	37.0	0.79[d]	3.64[e]	0.34	1.83
7	35.9	0.75[d]	3.32[e]	0.32	1.70
8[b]	37.9	0.89[d]	4.30[e]	0.45	1.65
9	42.3[c]	0.84[d]	2.11[e]	0.08	0.08

[a] Crosslinked with 2 wt% EDMA, 1 wt% AIBN, NaCl in vol% relative to polymerization mixture (HEMA + EDMA + NaCl).
[b] One half of the HEMA/EDMA feed was replaced by CyOH/DOH = 9/1 w/w.
[c] $(NH_4)_2SO_4$.
[d] Centrifugation method.
[e] Suction method.
[f] Mercury porosimetry.

The amount of crosslinker (2 wt%) and AIBN (1 wt%) dissolved in monomers was the same in all experiments, while the amount of porogen in the polymerization batch was varied from 35.9–41.4 vol%. Optionally, needle-like $(NH_4)_2SO_4$ crystals (42.3 vol%) together with saturated $(NH_4)_2SO_4$ solution were used as a porogen instead of NaCl crystals, allowing thus formation of hydrogels with communicating pores (Run 9, Table 1). For the sake of comparison, a copolymer was prepared with a mixture of solid (NaCl) and liquid low molecular weight porogen (diluent), amounting to 50% of the polymerization feed (Run 8). The thickness of the hydrogel was adjusted with a 3 mm thick silicone rubber spacer between the Teflon plates (10 × 10 cm), greased with a silicone oil and covered with cellophane. The reaction mixture was transferred onto a hollow plate and covered with a second plate, clamped and heated at 70°C for 8 hr. After polymerization, the hydrogels obtained with inorganic salt as a porogen were soaked in water and washed until the reaction of chloride or sulfate ions disappeared. The scaffolds prepared in the presence of a liquid diluent, were washed with ethanol/water mixtures (98/2, 70/30, 40/60, and 10/90 v/v) and water to remove the diluent, unreacted monomers and initiator residues. The washing water was then changed every day for 2 weeks.

1.2.3　METHODS

MICROSCOPY

The LVSEM was performed with a microscope Quanta 200 FEG (FEI, Czech Republic). Neat hydrated hydrogels were cut with a razor blade into ~ 5 mm cubes, flash frozen in liquid nitrogen and placed on the sample stage cooled to −10°C. Before microscopic observation, the top of a frozen sample was cut off using a sharp blade. During the observation, the conditions in the microscope (–10°C, 100 Pa) caused slow sublimation of ice from the sample which made it possible to visualize its 3D morphology. All samples were observed with a low vacuum secondary electron detector, using the accelerating voltage 30 kV. Lyophilized PHEMA hydrogels Run 3 and Run 8 filled with microspheres were also examined by LVSEM; however, the microspheres were washed out of the pores during freezing and, consequently, they were scarcely observed on the micrographs. The HVSEM was carried out with an electron microscope Vega TS 51355 (Tescan, Czech Republic). Permeability of the water imbibed hydrogels (Runs 7 and 9) was investigated by the flow of water suspension of the polymer microspheres. Before observation, the wet hydrogel was placed on a wet filtration paper and a droplet of a suspension of 8 μm PSt microspheres in water was placed on the top. The sample was dried at ambient temperature and cut with a sharp blade in the direction of the microsphere flow. Samples showing the top, bottom, and cross sections of flowed through hydrogels were sputtered with a 8 nm layer of platinum using a vacuum sputter coater (Baltec SCD 050), fixed with a conductive paste to a brass support and viewed in a scanning electron microscope in high vacuum (10^{-3} Pa), using the acceleration voltage 30 kV and a secondary electrons detector. This technique made it possible to observe both microspheres and pores of the hydrogel.

SOLVENT REGAIN

The solvent (water or cyclohexane - CX) regain was determined in 1×2 cm sponge pieces of hydrogel kept for 1 week in deionized water, which was exchanged daily. Water regain was measured by two methods: (i) centrifugation [32] (WR_c) and (ii) suction (WR_s). In centrifugation method, solvent swollen samples were placed into glass columns with fritted disc, centrifuged at 980 g for 10 min and immediately weighed (w_w – weight of hydrated sample), then vacuum dried at 80°C for 7 hr and again weighed (w_d - weight of dry sample). In the second method, excessive water was removed from the imbibed hydogel by suction and the hydrogel weighted to determine w_w. Weight of dry sample w_d was determined as above. Water regains WR_c or WR_s (ml/g) were calculated according to the equation:

$$WR_c \left(WR_s \right) = \frac{w_w - w_d}{w_d} \qquad (1)$$

The results are average values of two measurements for each hydrogel. To measure cyclohexane regain (CXR) by centrifugation, equilibrium water-swollen hydrogels were successively washed with ethanol, acetone, and finally cyclohexane. Using the solvent exchange, a thermodynamically good (swelling) solvent in the swollen gel was replaced by a thermodynamically poor solvent (non-solvent). Porosity of the hydrogels (p) was calculated from the water and CXRs (Table 1) and PHEMA density ($\rho = 1.3$ g/ml) according to the equation:

$$p = \frac{R \times 100}{R + \dfrac{1}{\rho}} \ (\%) \tag{2}$$

Where $R = WR_c$, WR_s, or CXR (ml/g).

MERCURY POROSIMETRY

Pore structure of freeze dried PHEMA scaffolds was characterized on a mercury porosimeter Pascal 140 and 440 (Thermo Finigan, Rodano, Italy). It works in two pressure intervals, 0–400 kPa and 1–400 MPa, allowing determination of meso- (2–50 nm), macro- (50–1000 nm) and small superpores (1–116 μm). The pore volume and most frequent pore diameter were calculated under the assumption of a cylindrical pore model by the PASCAL program. It employed Washburn's equation describing capillary flow in porous materials [33]. The volumes of bottle and spherical pores were evaluated as the difference between the end values on the volume/pressure curve. Porosity was calculated according to Equation 2, where cumulative pore volume (meso-, macro- and small superpores) from mercury porosimetry was used for R.

1.3   DISCUSSION AND RESULTS

1.3.1   MORPHOLOGY OF HYDROGELS

The prepared PHEMA constructs had always an opaque appearance indicating a permanent porous structure. Pores are generally divided into micro-, meso-, macropores, and small and large superpores. Morphology of water-swollen PHEMA hydrogels was investigated by LVSEM as shown in Figure 1.

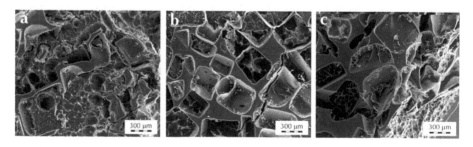

FIGURE 1   The LVSEM micrographs showing frozen cross section of PHEMA hydrogels prepared with (a) 41.4 vol% - Run 1, (b) 40 vol% - Run 3, and (c) 37.9 vol% NaCl (250–500 μm) - Run 5. PHEMA crosslinked with 2 wt% EDMA (relative to monomers).

Large 200–500 μm superpores were developed as imprints of NaCl crystals, which were subsequently washed out from the hydrogel; the interstitial space between them was filled with the polymer. During the observation, ice crystals filling soft polymer net were clearly visible in the centre of the hydrogel Run 1 (Table 1) prepared at the highest content of NaCl (41.4 vol%) in the feed (Figure 1 (a)). The internal surface area was too small to be determined. Figure 1 (b) and (c) show hydrogels from Run 3 and 5 (40 vol% and 37.9 vol% NaCl), respectively, documenting their more compact structure accompanied by thicker walls between large superpores as compared with the hydrogel from Run 1 (Figure 1 (a)). According to LVSEM, *ca.* 8 µm pores, the presence of which was confirmed by mercury porosimetry (volume about 1 ml/g), were observed in the walls between the large superpores. Longitudinal cracks in the material structure (Figure 1 (b)) were obviously caused by sample handling and fast freezing in liquid nitrogen. Nevertheless, the LVSEM micrographs displayed only cross sections of hydrogels and it was not clear whether their pores are interconnected.

Interconnection of pores is of vital importance for cell ingrowths in future applications to tissue regeneration. This feature was tested by the permeability of the whole hydrogels for different kinds of microspheres under two microscopic observations. First, cross sections of the frozen hydrogels filled with microspheres were observed in LVSEM. Second, the water-swollen hydrogels were flowed through by a suspension of microspheres in water and their dried cross sections were then viewed in HVSEM.

The LVSEM showed water-swollen morphology of hydrogel constructs which preserved due to their freezing in liquid nitrogen. The PHEMA Run 3 prepared with neat NaCl (Figure 2 (a)) was compared with Run 8 obtained in the presence of NaCl together with a mixture of CyOH/DOH (Figure 2 (d)). Addition of liquid porogens did not change the morphology; however, it increased the pore volume (from 0.21 to 0.45 ml CX/g, Table 1) and softness of the hydrogel. As a result, it had a tendency to disintegrate during the washing procedure. The LVSEM of both PHEMA hydrogels filled with 2 µm ammonolyzed PGMA and 200–400 nm PANI microspheres is illustrated in Figure 2 (b), e and Figure 2 c, f, respectively. The micrographs showed undistorted morphology of the frozen hydrogels, but just a few microspheres and/or their agglomerates. This was attributed to the fact that most of them were washed out during preparation of the sample for LVSEM.

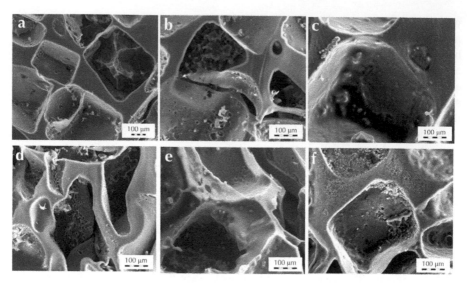

FIGURE 2 The LVSEM micrographs showing frozen cross section of PHEMA constructs; (a–c) - Run 3, (d–f) - Run 8; (a, d) neat and filled with (b, e) 2 μm ammonolyzed PGMA and (c, f) 200–400 nm PANI microspheres.

The morphology of the PHEMA hydrogels flowed through by a suspension of microspheres was observed by HVSEM. Figure 3 shows HVSEM micrographs of cross sections of the top and bottom part of the hydrogels from Runs 3, 5, and 8 flowed through by a suspension of 8 μm sulfonated PSt microspheres in water. While the microspheres flowed through the hydrogel construct from Run 3, they did not penetrate the ones from Run 5 and 8 prepared in the presence of a rather low content of NaCl (37.9 vol%). At the same time, surface and inner structure of the hydrogels slightly differed. Figure 4 shows HVSEM of the longitudinal section of the PHEMA scaffold obtained with cubic NaCl crystals as a porogen (Run 7). While Figure 4 (a) shows the bulk, Figures 4 (b–d) detailed sections. Again, small superpores with an average size about 13 μm were in the walls between the large superpores, forming small channels through which water flowed. To prove or exclude the interconnection of at least some pores, a suspension of 8 μm sulfonated polystyrene (PSt) microspheres in water was poured on the centre of the top side of the gel. While water flowed through the hydrogel bulk, the microspheres were retained on the surface of the hydrogel or penetrated only superficial layers due to the surface cracks (Figure 4 (a, b)). This confirmed that the pores of PHEMA hydrogels obtained with a low content of NaCl porogen (35.9 vol%) did not communicate. In contrast, Figure 5 presents longitudinal section of the PHEMA hydrogel from Run 9 (both bulk and detailed) obtained with needle-like $(NH_4)_2SO_4$ crystals as a porogen. This porogen allowed formation of connected pores which is explained by the needle-like structure of ammonium sulfate crystals that are linked to the gel structure. At the same time, the crystals grew to large structures due to the presence of saturated $(NH_4)_2SO_4$ solution in the feed. As a result, long interconnected large superpores - channels - were formed. This is documented in

Figure 5 (a–d) by the fact that suspension of 8 µm sulfonated PSt microspheres in water deposited in the centre of the top side of the hydrogel flowed through. The captured microspheres are well visible in Figure 5 (b–d). They accumulated at the places of pore narrowing; their majority, however, was found on the bottom part of the hydrogel. In such a way, the flow of water suspension of microspheres in the hydrogel was traced.

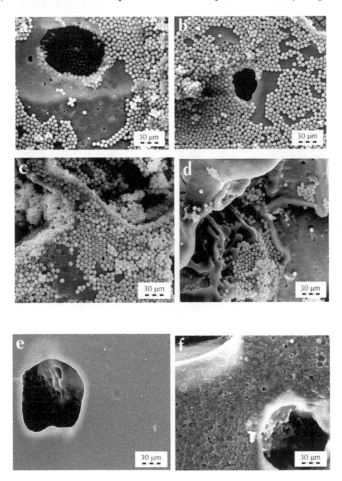

FIGURE 3   The HVSEM micrographs of PHEMA hydrogels Run 3 (a, d), Run 5 (b, e) and Run 8 (c, f) showing top (a–c) and bottom (d–f) of the hydrogels after the flow of a suspension of 8 µm sulfonated PSt microspheres in water.

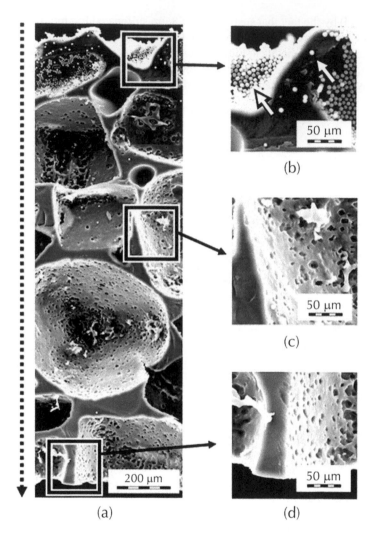

FIGURE 4   Selected HVSEM micrographs showing longitudinal section of PHEMA hydrogel 3 mm thick (Run 7) obtained with NaCl (250–500 µm) as a porogen after passing of a suspension of 8 µm sulfonated PSt microspheres in water (in the direction of the dotted line). (a) The whole cross section through the hydrogel and selected details from (b) top, (c) centre, and (d) bottom. The PSt microspheres are denoted with white arrows.

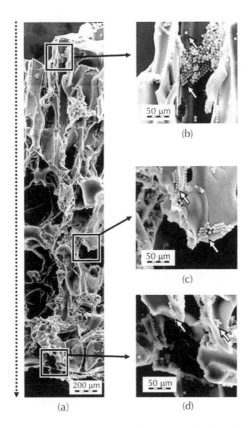

FIGURE 5 Selected HVSEM micrographs showing longitudinal section of PHEMA hydrogel 3 mm thick (Run 9) obtained with $(NH_4)_2SO_4$ (100×600 µm) as a porogen after passing of a suspension of 8 µm sulfonated PSt microspheres in water (in the direction of the dotted line). (a) The whole section through the construct and selected details from (b) top, (c) centre, and (d) bottom. The PSt microspheres are denoted with white arrows.

Mechanical properties of the porous constructs were sensitive to the concentration of porogen in the feed. Hydrogels with lower contents of NaCl and therefore higher proportion of PHEMA had thicker walls between the pores and were more compact allowing increased swelling of polymer chains in water. Two PHEMA hydrogels with the highest contents of NaCl in the feed (41.4 vol% - Run 1 and 40.8 vol% - Run 2) possessing thin polymer walls between large superpores easily disintegrated as well as hydrogel prepared using $(NH_4)_2SO_4$ (42.3 vol% - Run 9).

1.3.2 CHARACTERIZATION OF POROSITY BY SOLVENT REGAIN

The dependences of porosity of PHEMA hydrogels calculated from water or CXR and also from mercury porosimetry on NaCl content in the polymerization feed showed similar behavior (Figure 6). Porosities 81–91 and 49–57% for water regain were obtained by suction and centrifugation, respectively, 14–42% for CXR and 31–70% for mercury porosimetry. The porosity determined by centrifugation of samples soaked with water and cyclohexane (solvents with different affinities to polar methacrylate chain) consists of two contributions: filling of the pores and swelling (solvation) of PHEMA chains. The uptake of cyclohexane, a thermodynamically poor solvent which cannot swell the polymer, is a result of the former contribution only, reflecting thus the pore volume. The water regain from centrifugation was always higher than the CXR demonstrating thus swelling of polymer chains with water (Table 1). Solvent regains were affected by the concentration of NaCl porogen in the polymerization feed. Porosities according to both water and CXRs by centrifugation slightly increased with increasing volume of NaCl porogen in the polymerization feed from 35.9 to 40 vol% and then decreased with a further NaCl increase up to 41.4 vol% (Figure 6). In the latter range of NaCl, the porosity evaluated by mercury porosimetry exhibited an analogous dependence. This decrease in solvent and mercury regains can be explained by thin polymer walls between large superpores inducing collapse of the porous structure. In the concentration range of NaCl in the feed 35.9–40 vol%, mercury porosimetry provided higher porosities than those obtained from regains by centrifugation at 980 g because it obviously did not retain solvents in large superpores. Retained water reflected thus only small superpores, closed pores, and solvation of the polymer in water similarly as observed earlier for macroporous PHEMA scaffolds [34]. Water regain was determined also by the suction method (Table 1) which gave the values several times higher (3.3–7.5 ml/g) than by centrifugation due to filling all the pores in the polymer structure, including large superpores. As expected, porosity by the suction method increased with increasing volume of NaCl porogen in the polymerization feed (Figure 6).

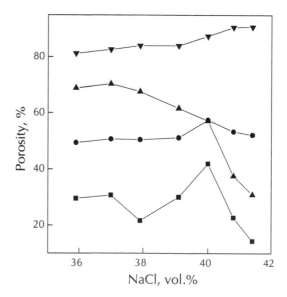

FIGURE 6 Dependence of porosity of PHEMA hydrogels determined from cyclohexane (■) and water regain measured by centrifugation (●) or suction (▼) and mercury porosimetry (▲) on the content of NaCl (250–500 μm) porogen in the polymerization feed.

The hydrogel from Run 8 formed in the presence of NaCl and CyOH/DOH porogen showed higher solvent regains and mercury penetration than the comparable hydrogel from Run 5 obtained with the same content of neat NaCl (Table 1). This can be explained by the higher total amount of porogen in the former hydrogel. In contrast, the hydrogel from Run 9 prepared with needle-like $(NH_4)_2SO_4$ crystals as a porogen had the lowest solvent and mercury regains of all the samples. The exception was water regain by centrifugation which was identical with that of sample Run 1 (Table 1) having a similar content of the NaCl porogen in the feed. This can imply that only large continuous superpores were present in this hydrogel and small superpores, macro- and mesopores were almost absent as evidenced by the low values of solvent and mercury regains.

1.3.3 CHARACTERIZATION OF POROSITY BY MERCURY POROSIMETRY

The advantage of mercury porosimetry is that it provides not only pore volumes, but also pore size distribution not available by other techniques. The method measures samples dried by lyophilization, which does not distort the pore structure. As already mentioned, porosities determined by mercury porosimetry were lower than those obtained from water regain by the suction which included large superpores, and higher than those from water and CXR detected by centrifugation. This was due to better filling of the compact xerogel structures obtained at lower contents of NaCl in the feed with mercury under a high pressure than with water or cyclohexane under atmospheric pressure.

Figure 7 shows the dependence of most frequent mesopore size of PHEMA scaffolds and their pore volumes on the NaCl porogen content in the polymerization feed. Predominantly, 4–5 nm mesopores were detected with their volume increasing from 0.03 to 0.1 ml/g with increasing NaCl content in the polymerization feed. Macropores were absent and very low values of specific surface areas (< 0.1 m^2/g) were found. The presence of CyOH/DOH porogen (Run 8) did not substantially affect the formation of meso- and macropores (volume 0.022 ml/g), because the amount of crosslinker in the polymerization feed was limited to only 2 wt%. The separation of the polymer from the porogen phase could not thus occur and porous structure was not formed. Both in hydrophobic styrene-divinylbenzene [35,36] and polar methacrylate copolymers [37] prepared in the presence of liquid porogens, phase separation, and formation of macroporous structure occurred at crosslinker contents higher than 10 wt%.

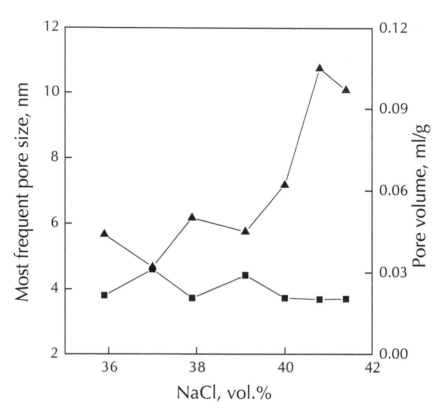

FIGURE 7 Dependence of pore volume (▲) and most frequent mesopore size (■) of PHEMA hydrogels on the content of NaCl (250–500 μm) porogen in the polymerization feed according to mercury porosimetry.

Figure 8 represents the dependence of most frequent small superpore size of PHE-MA constructs and their pore volume on the content of NaCl porogen in the polymerization. Pore size increased up to 28–69 μm with increasing NaCl volume. This was pronounced in the range 40–41.4 vol% NaCl probably due to the aggregation of NaCl crystals in the mixture at their high contents. All the investigated samples contained small superpores, the volume of which was about 20 times higher than that of mesopores. The volume of small superpores continuously decreased from 1.8 to 0.2 ml/g with raising NaCl amount in the feed. This could be explained by collapse of the pore structure and destruction of last two hydrogels with the highest content of NaCl porogen (Runs 1 and 2) under high pressure as mentioned above. The size of small superpores according to mercury porosimetry was by an order of magnitude smaller than the particle size of the used NaCl porogen (250–500 μm) because the method was able to distinguish the superpores only in the size range 1–116 μm (Figure 9). Large superpores (imprints of NaCl crystals) were detected by LVSEM. Figure 9 exemplifies a typical cumulative pore volume and a derivative pore size distribution curve of PHEMA Run 4 with a decisive contribution of small superpores 25 μm in size.

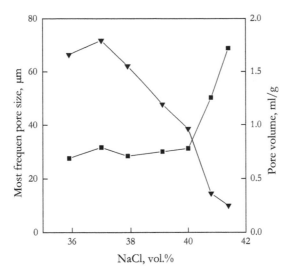

FIGURE 8 Dependence of pore volume (▼) and most frequent small superpore size (■) of PHEMA hydrogels on the content of NaCl (250–500 μm) porogen in the polymerization feed according to mercury porosimetry.

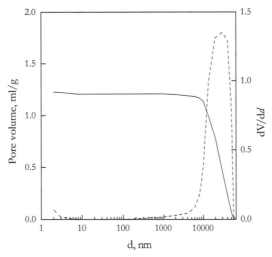

FIGURE 9 Cumulative pore volume (—) and pore size distribution (---) of the Run 4 hydrogel determined by mercury porosimetry in the range 1.88 nm–116 μm.

1.4 CONCLUSION

Superporous PHEMA constructs were prepared by bulk radical copolymerization of HEMA and EDMA in the presence of NaCl or/and liquid diluent (CyOH/DOH) or

$(NH_4)_2SO_4$ crystals. Morphology of the prepared scaffolds was characterized by several methods including scanning electron microscopy both in swollen (LVSEM) and dry (HVSEM) state, solvent (water and cyclohexane) regains high and low pressure mercury porosimetry of lyophilized samples and dynamic desorption of nitrogen. Morphology and porous structure of the hydrogels were preferentially affected by the character and amount of the used porogen – NaCl, CyOH/DOH mixture or $(NH_4)_2SO_4$. After washing out of the salts and solvents from PHEMA, three types of pores were detected by microscopic and mercury porosimetry methods, including large superpores (hundreds of micrometers) as imprints of salt crystals. The hydrogels formed can be divided into two groups, with disconnected and interconnected pores. The latter allowed the passage of suspension of microspheres in water, which was observed only for the samples with ammonium sulfate and the highest content of NaCl crystals used as a porogen in the feed. Interconnected pores are crucial for potential application of the scaffolds as living cell supports. The LVSEM showed the undistorted (frozen) structure of the hydrogels, but only few flowed-through microspheres could be observed as they tended to escape from the pores during sample preparation. The HVSEM seemed to be the best microscopic technique especially for viewing permeability of hydrogels to 8 μm microspheres. Hydrogels were initially flowed through by the particles in their natural wet state, but the specimens were then dried before SEM observation. The microparticles could be traced both on the upper/lower parts of the hydrogels and on the longitudinal sections.

Mercury porosimetry provided detailed description of morphology of PHEMA constructs with pore sizes from units of nanometers to tens of micrometers. The drawback of the method is that the hydrogels are not measured in the swollen, but dry state, as xerogels. But comparison of the data in both wet and dry states showed that lyophilization did not change the pore structure. The mesopores and small superpores detected by mercury porosimetry cannot be formed by the imprinting mechanism. While mesopores present only in very small amounts may be formed by phase separation, small superpores arise by polymer contraction in the walls of large superpores. Small 28–69 μm superpores were mainly present in the porous structure apart from the large superpores (200–500 μm imprints of solid porogen crystals), the volume of which was several times higher than that of other pores, as confirmed by water regain obtained by the suction method.

KEYWORDS

- **Hydrogel**
- **2-Hydroxyethyl methacrylate**
- **Porosity**
- **Scaffold**

ACKNOWLEDGMENT

Financial support of the Grant Agency of the Czech Republic (project No. P304/11/0731) is gratefully acknowledged.

REFERENCES

1. Kofron, M. D., Cooper, J. A., Kumbar, S. G., and Laurencin, C. T. Novel tubular composite matrix for bone repair. *J Biomed Mater Res Part A*, **82**, 415–425 (2007).
2. Moroni, L., Hendriks, J. A. A., Schotel, R., De Wijn, J. R., and Van Blitterswijk, C. A. Design of biphasic polymeric 3-dimensional fiber deposited scaffolds for cartilage tissue engineering applications. *Tissue Eng*, **13**, 361–371 (2007).
3. Dvořánková, B., Holíková, Z., Vacík, J., Königová, R., Kapounková, Z., Michálek, J., Přádný, M., and Smetana, K. Reconstruction of epidermis by grafting of keratinocytes cultured on polymer support - clinical study. *Int J Dermatol*, **42**, 219–223 (2003).
4. Bhang, S. H., Lim, J. S., Choi, C. Y., Kwon, Y. K., and Kim BS. The behavior of neural stem cells on biodegradable synthetic polymers. *J Biomater Sci, Polym Ed*, **18**, 223–239 (2007).
5. Bianchi, F., Vassalle, C., Simonetti, M., Vozzi, G., Domenici, C., and Ahluwalia, A. Endothelial cell function on 2D and 3D micro-fabricated polymer scaffolds: Applications in cardiovascular tissue engineering. *J Biomater Sci, Polym Ed*, **17**, 37–51 (2006).
6. Sander, E. A., Alb, A. M., Nauman, E. A., Reed, W. F., and Dee, K. C. Solvent effects on the microstructure and properties of 75/25 poly(D,L-lactide-co-glycolide) tissue scaffolds. *J Biomed Mater Res, Part A*, **70**, 506–513 (2004).
7. Nam, Y. S., Yoon, J. J., and Park, T. G. A novel fabrication method of macroporous biodegradable polymer scaffolds using gas foaming salt as a porogen additive. *J Biomed Mater Res, Appl Biomater*, **53**, 1–7 (2000).
8. Kroupová, J., Horák, D., Pacherník, J., Dvořák, P., and Šlouf M. Functional polymer hydrogels for embryonic stem cell support. *J Biomed Mater Res, Part B, Appl Biomater*, **76B**, 315–325 (2006).
9. Tighe, B. and Corkhill, P. Hydrogels in biomaterials design: Is there life after polyHEMA? *Macromol Rep*, **A31**, 707–713 (1994).
10. Plieva, F. M., Galaev, I. Y., and Mattiasson, B. Macroporous gels prepared at subzero temperatures as novel materials for chromatography of particulate-containing fluids and cell culture applications. *J Sep Sci*, **30**, 1657–1671 (2007).
11. Mooney, D. J., Baldwin, D. F., Suh, N. P., Vacanti, J. P., and Langer, R. Novel approach to fabricate porous sponges of poly(D,L-lactic-co-glycolic acid) without the use of organic solvents. *Biomaterials*, **17**, 1417–1422 (1996).
12. Chung, H. J. and Park, T. G. Surface engineered and drug releasing pre-fabricated scaffolds for tissue engineering. *Adv Drug Delivery Rev*, **59**, 249–262 (2007).
13. Ferreira, L. S., Gerecht, S., Fuller, J., Shieh, H. F., Vunjak-Novakovic, G., and Langer, R. Bioactive hydrogel scaffolds for controllable vascular differentiation of human embryonic stem cells. *Biomaterials*, **28**, 2706–2717 (2007).
14. Tangsadthakun, C., Kanokpanont, S., Sanchavanakit, N., Pichyangkura, R., Banaprasert, T., Tabata, Y., and Damrongsakkul, S. The influence of molecular weight of chitosan on the physical and biological properties of collagen/chitosan scaffolds. *J Biomater Sci, Polym Ed*, **18**, 147–163 (2007).
15. Wachiralarpphaithoon, C., Iwasaki, Y., and Akiyoshi, K. Enzyme-degradable phosphorylcholine porous hydrogels cross-linked with polyphosphoesters for cell matrice. *Biomaterials*, **28**, 984–993 (2007).
16. **Treml, H., Woelki, S., and Kohler, H. H.** Theory of capillary formation in alginate gels. *Chem Phys*, **293**, 341–353 (2003).
17. Konno, T. and Ishihara, K. Temporal and spatially controllable cell encapsulation using a water-soluble phospholipid polymer with phenylboronic acid moiety. *Biomaterials*, **28**, 1770–1777 (2007).

18. Heckmann, L., Schlenker, H. J., Fiedler, J., Brenner, R., Dauner, M., Bergenthal, G., Mattes, T., Claes, L., and Ignatius, A. Human mesenchymal progenitor cell responses to a novel textured poly(L-lactide) scaffold for ligament tissue engineering. *J Biomed Mater Res Part B, Appl Biomater*, **81**, 82–90 (2007).

19. Darling, A. L. and Sun, W. 3D microtomographic characterization of precision extruded poly-ε-caprolactone scaffolds. *J Biomed Mater Res Part B, Appl Biomater*, **70**, 311–317 (2004).

20. Rhee, W., Rosenblatt, J., Castro, M., Schroeder, J., Rao, P. R., Harner, C. F. H., and Berg, R. A. In vivo stability of poly(ethylene glycol)-collagen composites, in: Poly(Ethylene Glycol) Chemistry and Biological Applications. J. M Harris and S. Zalipsky (Eds.), *ACS Symp Ser*, **680**, 420–440 (1997).

21. Savina, I. N., Galaev, I. Y., and Mattiasson, B. Ion-exchange macroporous hydrophilic gel monolith with grafted polymer brushes. *J Mol Recognit*, **19**, 313–321 (2006).

22. Carampin, P., Conconi, M. T., Lora, S., Menti, A. M., Baiguera, S., Bellini, S., Grandi, C., and Parnigotto, P. P. Electrospun polyphosphazene nanofibers for in vitro rat endothelial cells proliferation. *J Biomed Mater Res Part A*, **80**, 661–668 (2007).

23. Zhang CH, Zhang N, Wen XJ. Synthesis and characterization of biocompatible, degradable, light-curable, polyurethane-based elastic hydrogels. *J Biomed Mater Res, Part A*, **82**, 637–650 (2007).

24. Castner DG, Ratner BD. Biomedical surface science: Foundation to frontiers. *Surf Sci*, **500**, 28–60 (2002).

25. Lee, K. Y. and Mooney, D. J. Hydrogels for tissue engineering. *Chem Rev*, **101**, 1869–1879 (2001).

26. Refojo, M. F. Hydrophobic interactions in poly(2-hydroxyethyl methacrylate) homogeneous hydrogel. *J Polym Sci Part A1, Polym Chem*, **5**, 3103–3108 (1967).

27. Ratner, B. D. and Hoffman, A. S. Hydrogels for Medical and Related Applications. *ACS Symp Ser*, **31**, 1–36 (1976).

28. Horák, D., Dvořák, P., Hampl, A., and Šlouf, M. Poly(2-hydroxyethyl methacrylate-*co*-ethylene dimethacrylate) as a mouse embryonic stem cell support, *J Appl Polym Sci*, **87**, 425–432 (2003).

29. Horák, D., Kroupová, J., Šlouf, M., and Dvořák, P. Poly(2-hydroxyethyl methacrylate)-based slabs as a mouse embryonic stem cell support. *Biomaterials*, **25**, 5249–5260 (2004).

30. Horák, D. and Shapoval, P. Reactive poly(glycidyl methacrylate) microspheres prepared by dispersion polymerization. *J Polym Sci Part A, Polym Chem Ed*, 38, 3855–3863 (2000).

31. Stejskal, J., Kratochvíl, P., Gospodinova, N., Terlemezyan, L., and Mokreva, P. Polyaniline dispersions: Preparation of spherical particles and their light-scattering characterization. *Polymer*, 33, 4857–4858 (1992).

32. Štamberg, J. and Ševčík, S. Chemical transformations of polymers III. Selective hydrolysis of a copolymer of diethylene glycol methacrylate and diethylene glycol dimethacrylate. *Collect Czech Chem Commun*, 31, 1009–2016 (1966).

33. Porosimeter Pascal 140 and Pascal 440, Instruction manual, p. 8.

34. Hradil, J. and Horák, D. Characterization of pore structure of PHEMA-based slabs. *React Funct Polym*, 62, 1–9 (2005).

35. Millar, J. A., Smith, D. G., Marr, W. E., and Kresmann, T. R. E. Solvent modified polymer networks. Part 1. The preparation and characterization of expanded-networks and macroporous styrene-DVB copolymers and their sulfonates. *J Chem Soc*, pp. 218–225 (1963).

36. Kun KA, Kunin R. Macroreticular resins III. Formation of macroreticular styrene-divinylbenzene copolymers. *J Polym Sci, Part A1, Polym Chem*, 6, 2689–2701 (1968).

37. Hradil, J., Křiváková, M., Starý, P., and Čoupek, J. Chromatographic properties of macroporous copolymers of 2-hydroxyethyl methacrylate and ethylene dimethacrylate. *J Chromatogr*, 79, 99–105 (1973).

CHAPTER 2

THERMAL PROPERTIES OF THE POLYPROPYLENE/MULTI-WALLED CARBON NANOTUBE COMPOSITES

G. E. ZAIKOV, A. D. RAKHIMKULOV, S. M. LOMAKIN, I. L. DUBNIKOVA, A. N. SHCHEGOLIKHIN, E. YA. DAVIDOV, and R. KOZLOWSKI

CONTENTS

2.1 INTRODUCTION

Studies of thermal and fire resistant properties of the polypropylene/multi-walled carbon nanotube composites (PP/MWCNT) prepared by means of melt intercalation are discussed. The sets of the data acquired with the aid of non-isothermal thermogravimetric (TG) experiments have been treated by the model kinetic analysis. The thermal-oxidative degradation behavior of PP/MWCNT and stabilizing effect caused by addition of multi-walled carbon nanotube (MWCNT) has been investigated by means of thermogravimetric analysis (TGA) and electron paramagnetic resonance (EPR) spectroscopy.

The results of cone calorimetric tests lead to the conclusion that char formation plays a key role in the mechanism of flame retardation for nanocomposites. This could be explained by the specific antioxidant properties and high thermal conductivity of MWCNT which determine high performance carbonization during thermal degradation process.

The comparative analysis of the flammability characteristics for PP-clay/MWCNT nanocomposites was provided in order to emphasize the specific behavior of the nanocomposites under high temperature tests.

At present time the great attention is given to the study of properties of polymeric nanocomposites produced on the basis of well known thermoplastics (PP, PE, PS, PMMA, polycarbonates, and polyamides) and carbon nanotubes (CN). The CNs are considered to have the wide set of important properties like thermal stability, reduced combustibility, electroconductivity, and so on [1-7]. Thermoplastic polymer nanocomposites are generally produced with the use of melting technique [1-12].

Development of synthetic methods and the thermal characteristics study of PP/MWCNT nanocomposites were taken as an objective in this chapter.

A number of papers pointed at synthesis and research of thermal properties of nanocomposites (atactic polypropylene (aPP)/MWCNT) were reported [10-12]. It is remarkable that PP/MWCNT composites with minor level of nanocarbon content (1–5% by weight) were determined to obtain an increase in thermal and thermal-oxidative stability in the majority of these publications.

Thermal stability of aPP and aPP/MWCNT nanocomposites with the various concentrations of MWCNT was studied in the chapter [10]. It was shown that thermal degradation processes are similar for aPP and aPP/MWCNT nanocomposites and initial degradation temperatures are the same. However, the maximum mass loss rate temperature of PP/MWCNT nanocomposites with 1 and 5% wt of MWCNT raised by 40–70°C as compared with pristine aPP.

Kashiwagi et al. published the results of study of thermal and combustion properties of PP/MWCNT nanocomposites [11,12]. A significant decrease of maximum rate of heat release (RHR) was detected during combustion research with use of cone calorimeter. A formation of char network structure during the combustion process was considered to be the main reason of combustibility decrease. The carbonization influence upon combustibility of polymeric nanocomposites was widely presented in literature [10-13]. Notably, Kashiwagi et al. [11,12] were the first to hypothesize that abnormal dependence of maximum RHR upon MWCNT concentration is closely related with thermal conductivity growth of PP/MWCNT nanocomposites during high temperature pyrolysis and combustion.

2.2 EXPERIMENTAL DETAILS

2.2.1 MATERIALS

Isotactic polypropylene (melting flow index = 0.7 g/10 min) was used as a polymer matrix in this chapter. The MWCNT (purchased from Shenzhen Nanotechnologies Co. Ltd.) were used as a carbon-containing nanofillers. This product contains low amount of amorphous carbon (less than 0.3 wt%) and could be produced with different size characteristics - different length and different diameter and therefore different diameter to length ratio. Size characteristics for three MWCNT used in this chapter are given in Table 1. Sizes and structure of initial MWCNT were additionally estimated by scanning electron microscopy (SEM) (Figure 1).

TABLE 1   Properties of MWCNT

Designation	D, nm	L, µm	Density, g/cm³	Specific surface area, m²/g
MWCNT (K1)	<10	5–15	2	40–300
MWCNT (K2)	40–60	1–2	2	40–300
MWCNT (K3)	40–60	5–15	2	40–300

2.2.2 NANOCOMPOSITE PROCESSING

Compositions were prepared by blending CN with melted polymer in a laboratory mixer Brabender at 190°C. TOPANOL® (1,1,3-tris(2-methyl-4-hydroxy-5-t-butylphenyl)butane) and dilaurylthiodipropionate (DLTP) were added in the amount of 0.3 wt% and 0.5 wt% as antioxidants to prevent thermal-oxidative degradation during polymer processing.

A number of different covalent and non-covalent nanotube modifications (organofillization) were reported to be used to achieve greater structure similarity and therefore greater nanotube distribution in a polymer matrix [14-23]. In order to functionalize MWCNT we used preliminary ozone treatment of MWCNT followed by ammonolysis of epoxy groups on the MWCNT surface. The selective ozonization of MWCNT was carried out with ozone-oxygen mixture (ozone concentration was 2.3×10^{-4} mol/L) in a bubble reactor. Then the ammonolysis of oxidized MWCNT has been carried out by *tert*-butylamine in the ultrasonic bath (35 kHz) at 50°C for 120 min with following evaporation of *tert*-butylamine excess. Infrared (IR) transmission spectra of tablet specimens of MWCNTs in KBr matrix was analyzed by using Perkin-Elmer

1725X Fourier transform infrared spectroscopy (FTIR) spectrometer and the presence of the alkylamine groups at the MWCNT surface was confirmed by the appearance of the characteristic band ~ 1210 cm^{-1} corresponding to the valency vibration of the bond $-C-N\leftarrow$.

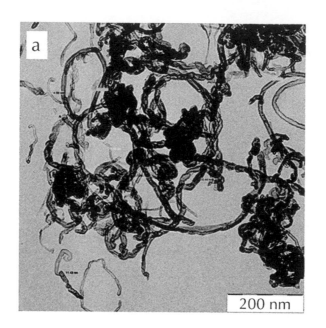

200 nm

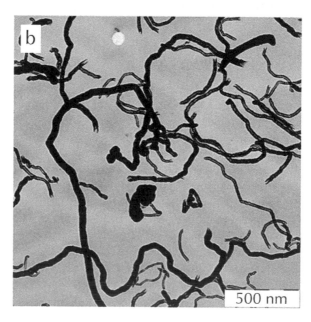

500 nm

FIGURE 1 *(Continued)*

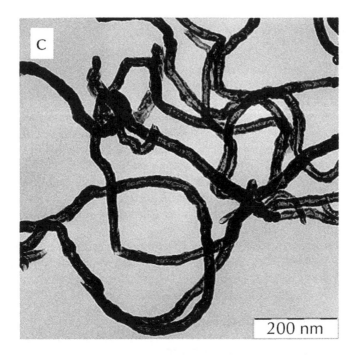

FIGURE 1   The SEM images of original MWCNTs: (a) MWCNT(1), (b) MWCNT(2), and (c) MWCNT(3).

2.2.3 INVESTIGATION TECHNIQUES

SCANNING ELECTRON MICROSCOPY (SEM)

The degree of MWCNT distribution in polymer matrix was analyzed with scanning electron microscope JSM-35. Low temperature chips derived from film type samples were used for this analysis.

TRANSMISSION ELECTRON MICROSCOPY (TEM)

The degree of nanotube dispersion in polymer matrix was studied with transmission electron microscopy (LEO912 AB OMEGA, Germany). Microscopic sections with 70–100 nm width prepared with ultramicrotome "Reichert–Jung Ultracut" with diamond cutter at –80°C. Microscopic analysis was made with accelerating potential of about 100 kV without chemical sample staining.

THERMOGRAVIMETRIC ANALYSIS

A NETZSCH TG 209 F1 Iris thermomicrobalance has been employed for TGA measurements in oxidizing (oxygen) atmosphere. The measurements were carried out at a heating rate of 20 K/min.

Combustibility characteristics (cone calorimeter) were performed according to the standard procedures ASME E1354/ISO 5660 using a DUAL CONE 2000 cone calorimeter (fire testing technology (FTT)). An external radiant heat flux of 35 kW/m² was applied. All of the samples having a standard surface area of 70 × 70 mm and identical masses of 13.0 ± 0.2 g were measured in the horizontal position and wrapped with thin aluminum foil except for the irradiated sample surface.

Heat capacity and heat conductivity were determined with the use of NETZSCH 457 MicroFlash.

The electron paramagnetic resonance spectroscopy measurements were performed in air with the PP/MWCNT (10 wt%) samples using a Mini-EPR SPIn Co. Ltd spectrometer with 100 kHz field modulation. The g factor and EPR intensity (X-band) were measured with respect to a standard calibrating sample of Mn^{2+} and ultramarine.

2.3   DISCUSSION AND RESULTS

2.3.1   NANOCOMPOSITE STRUCTURE

Dispersion analysis of MWNT in nanocomposites. PP/MWNT nanocomposites with original and modified MWNT were produced. Filler concentration varied from 1 to 7 wt% weight percent (0.5–3.5 volume percent correspondingly). Distribution pattern for composites with modified and nonmodified nanotubes was studied with TEM methods (Figure 2). According to TEM images the addition of 1% by weight leads to sufficiently uniform distribution. However, agglomeration of nanotubes was detected for more concentrated nanocomposites, especially for PP/MWNT with MWNT average diameter less than 10 nm (K1).

The TEM images for PP nanocomposites with 5 wt% modified and nonmodified MWNT (2.5% by volume) are shown on Figure 2. According to Figure 2, it could be stated that modified nanotubes (K2 and K3) used during melding process are present as individual particles in nanocomposite in most cases. The number and size of agglomerates is reduced due to increased organophility and improved thermodynamic compatibility with nonpolar polymer.

However, preliminary modification does not lead to uniform filler distribution for nanocomposites containing thin nanotubes (K1). This could be explained by the fact that interaction energy of CNT is more dependent on nanotube diameter than of its length. Molecular dynamic computation given in chapter [24] showed that blending polymers with nanotubes becomes more thermodynamically favorable with increase of nanotubes diameter, owing to the fact that cohesion energy is decreased between the nanotubes and remains almost the same between nanotubes and polymer.

Thus, mixing the thinnest nanotubes (K1) with PP leads to inevitable nanotube agglomeration in nanocomposite sample volume. Nanotube surface modification used in this chapter did not result in complete overcome of nanotube agglomeration tendency for K1 nanotubes [25]. Therefore, the PP/MWNT(K3) nanocomposites presented the main subjects of inquiry in the present study.

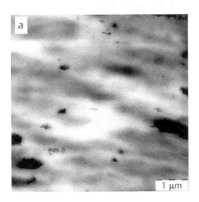

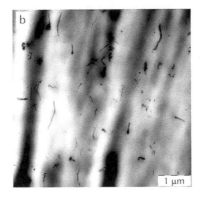

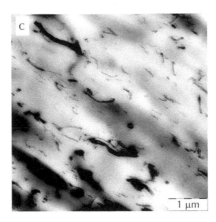

FIGURE 2   The TEM images of PP/MWCNT nanocomposites showing dispersion of MWCNT in a polymer matrix: (a) PP/MWCNT(1), (b) PP/MWCNT(2), and (c) PP/MWCNT(3).

2.4 THERMAL-OXIDATIVE DEGRADATION OF PP/MWCNT NANOCOMPOSITES

The diverse behavior of PP and PP/MWCNT nanocomposites with 1 wt%, 3 wt%, and 5 wt% of MWCNT(3). Figure 3 shows that the influence of MWCNTs on the thermal oxidation process resulted in higher thermal-oxidative stability of PP/MWCNT nano-composites. It is possible to see a regular increase in the temperature values of the maximum mass loss rates (up to 60°C) for the PP/MWCNT as compared to pristine PP (Figure 3).

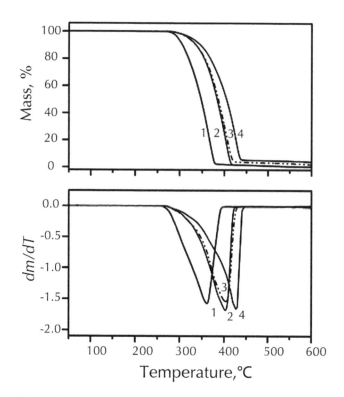

FIGURE 3 The TG and DTG curves for PP (1) and PP/MWCNT(3) composites with 1 wt% (2), 3 wt% (3), and 5 wt% (4) filler loadings.

Detailed analysis of TGA graphs (Figure 3) allows claiming that thermal stability increase is achieved even by addition of 1 wt% of MWCNT to PP, while further addition does not lead to such fundamental growth. In addition, Figure 4 shows the comparative results for onset degradation temperatures (T_{on}) and the maximum mass loss temperatures (T_{max}) of PP/MWCNT nanocomposites with the different types and concentrations of MWCNT. One can see nonlinear relation of (T_{on}) and (T_{max}) versus MWCNT concentration in the PP compositions (Figure 4).

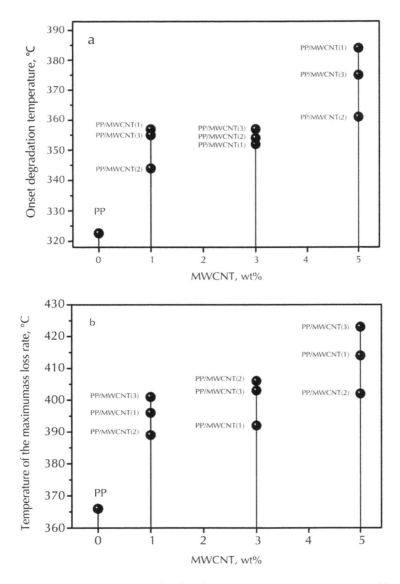

FIGURE 4 Comparative diagrams showing the onset degradation temperatures (a) and the maximum mass loss temperatures (b) for PP and PP/MWCNT nanocomposites with the different types and concentrations of MWCNT.

At the present time the nature of thermal stability effect caused by MWCNT addition to polymers is an object of comprehensive study. Most likely, MWCNTs could be considered as high temperature stabilizers (antioxidants) in reactions of thermaloxidative degradation by analogy with fullerenes [26]. Stabilizing effect caused by addition of MWCNT was previously detected for PP/MWCNT nanocomposites [8]: the temperature of the maximum mass loss rates of PP/MWCNT (9 wt%) was increased by 50°C as compared with pristine PP.

Results achieved in this study confirm the previous findings of the inhibiting effect of MWNCT upon the PP/MWCNT nanocomposite thermal-oxidative degradation. Obviously, the complex nature of this effect is closely related to radical acceptor properties of MWCNT resulting in chain termination reactions of alkyl/alkoxyl radical, which lead to cross-linking and carbonization process in PP/MWCNT nanocomposites. Carbonization phenomenon was reported previously in papers aimed at PP/MWCNT heat resistance and flame retardancy study [11,12].

2.4.1   KINETIC ANALYSIS OF THERMAL DEGRADATION OF PP/MWNT

The kinetic studies of material degradation have long history, and there exists a long list of data analysis techniques employed for the purpose. Often, TGA is the method of choice for acquiring experimental data for subsequent kinetic calculations, and namely this technique was employed here. It is commonly accepted that the degradation of materials follows the base Equation (1) [27]

$$dc/dt = -F(t, T, c, p) \tag{1}$$

where: t—time, T—temperature, c_o—initial concentration of the reactant, and p—concentration of the final product. The right hand part of the equation F(t, T, c, p) can be represented by the two separable functions, k(T) and f(c, p):

$$F(t, T, c, p) = k [T(t) \cdot f(c, p)] \tag{2}$$

Arrhenius equation (4) will be assumed to be valid for the following:

$$k(T) = A \cdot exp(-E/RT) \tag{3}$$

Therefore,

$$dc/dt = -A \cdot exp(-E/RT) \cdot f(c, p) \tag{4}$$

All feasible reactions can be subdivided onto classic homogeneous reactions and typical solid state reactions, which are listed in Table 2. The analytical output must provide good fit to measurements with different temperature profiles by means of a common kinetic model.

TABLE 2 Considered reaction models $dc/dt = -A \cdot exp(-E/RT)f(c, p)$

Reaction models	f(c, p)
First order (F_1)	c
Second order (F_2)	c^2
n-order (F_n)	c^n
Two-dimensional phase boundary (R_2)	$2 \cdot c^{1/2}$
Three-dimensional phase boundary (R_3)	$3 \cdot c^{2/3}$
One-dimensional diffusion (D_1)	$0.5/(1-c)$
Two-dimensional diffusion (D_2)	$-1/\ln(c)$
Three-dimensional diffusion, Jander's type (D_3)	$1.5c^{1/3}(c^{-1/3}-1)$
Three-dimensional diffusion, Ginstling–Brounstein (D_4)	$1.5/(c^{-1/3}-1)$
One-dimensional diffusion (Fick law) (D_{1F})	-
Three-dimensional diffusion (Fick law) (D_{3F})	-
Prout-Tompkins equation (B_1)	$c \cdot p$
Expanded Prout–Tompkins equation (B_{na})	$c^n \cdot p^a$
First order reaction with autocatalysis by X (C_{1-X})	$c \cdot (1+K_{cat} X)$
n-order reaction with autocatalysis by X (C_{n-X})	$c^n \cdot (1+K_{cat} X)$
Two-dimensional nucleation, Avrami–Erofeev equation (A_2)	$2 \cdot c \cdot (-\ln(c))^{1/2}$
Three-dimensional nucleation, Avrami–Erofeev equation (A_3)	$3 \cdot c \cdot (-\ln(c))^{2/3}$
n-dimensional nucleation, Avrami–Erofeev equation (A_n)	$n \cdot c \, (-\ln(c))^{(n-1)/n}$

Thermogravimetric analysis of PP and PP nanocomposite degradation was carried out in dynamic conditions at the rates of 2.5 K/min, 5 K/min, and 10 K/min on air.

Model-independent estimation of activation energy using Friedman approach [28] was taken to get preliminary model analysis for thermal degradation and selection of initial conditions. According to this evaluation, a two-step process ($A \rightarrow X_1 \rightarrow B \rightarrow X_2 \rightarrow C$) was chosen for PP degradation. Taking into account the carbonization stage the more complex three-step process ($A \rightarrow X_1 \rightarrow B \rightarrow X_2 \rightarrow C \rightarrow X_3 \rightarrow D$) was selected for PP/MWNT degradation [27,29].

According to the results of nonlinear regression and taking the set of reaction models into consideration we computed the values of active kinetic parameters, which represent the best approximation of experimental TGA graphs (Figure 5, Table 3).

TABLE 3   Kinetic parameters for thermal degradation of (a) PP (Fn → Fn) and (b) PP/MWNT nanocomposite (Fn →D1→ Fn). The TGA analysis was performed in air flow with the use of multiple nonlinear regression analysis for model processes.

(a)

Reaction model	Kinetic parameters	Values	Correlation coefficient
Fn→Fn	$\log A_1$, s^{-1}	9.53	
	E_1, kJ/mol	110.25	
	n_1	1.89	
			0.9996
	$\log A_2$, s^{-1}	15.25	
	E_2, kJ/mol	150.65	
	n_2	1.50	

(b)

Reaction model	Kinetic parameters	Values	Correlation coefficient
	$\log A_1$, s^{-1}	6.3	
	E_1, kJ/mol	105.1	
	n_1	0.91	
	$\log A_2$, s^{-1}	7.4	
Fn→D1→Fn	E_2, kJ/mol	120.4	0.9996
	$\log A_3$, s^{-1}	16.7	
	E_3, kJ/mol	229.5	
	n_3	0.5	

Two-step PP thermal-oxidative degradation in dynamic heating conditions was confirmed by obtained data [30]. At the first stage the values of activation energy and pre-exponential factor are 110.25 kJ/mol and $10^{9.5}$ s^{-1} correspondingly, while the reaction order is close to 2 (1.89). The values of activation energy and pre-exponential factor are larger on the second stage (E_2 = 150.65 kJ/mol, A_2 = $10^{15.3}$ s^{-1}) with effective reaction order of n_2 = 1.50.

The preferred model for PP/MWNT thermal-oxidative degradation and with respect to statistical analysis of kinetic parameters is composed of three consecutive reactions $F_n \rightarrow D_1 \rightarrow F_n$, where D_1—one-dimensional diffusion and F_n—n-order reaction

(Figure 4(b), Table 3(b)). In this case the first step activation energy is equal to 105.1 kJ/mol, reaction order is close to 1 ($n_1 = 0.91$). On the second step, which is described as one-dimensional diffusion, the value of activation energy is equal 120.4 kJ/mol, while the value is almost twice large for the third step ($E_3 = 229.5$ kJ/mol) with effective reaction order of $n_3 = 0.5$ (Table 3(b)).

The comparison of thermal oxidative degradation parameters for PP/MWNT with layered silicate PP/montmorillonite (MMT) showed that the values of activation energy of the second and the third stages are higher for PP/MWNT:

$E_2 = 120.4$ kJ/mol and $E_3 = 229.5$ kJ/mol for PP/MWNT;

$E_2 = 100.0$ kJ/mol and $E_3 = 199.8$ kJ/mol for PP/MPP/MMT correspondingly [30].

This data may testify to more intensive carbonization in case of PP/MWNT than in case of PP/MMT, which finally leads to decrease in RHR value.

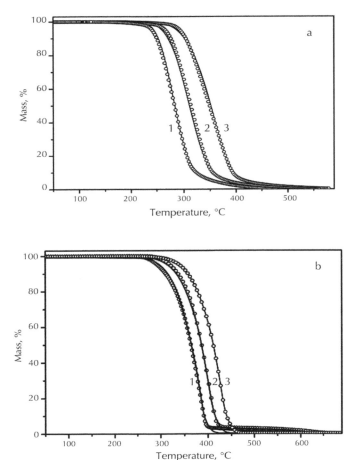

FIGURE 5 Nonlinear kinetic modeling of (a) PP and (b) PP/MWCNT(3) thermal-oxidative degradation in air. Comparison between experimental TG data (dots) and the model results (firm lines) at several heating rates: (1) –2.5, (2) –5, and (3) –10 K/min.

2.5   COMBUSTIBILITY OF PP/MWCNT NANOCOMPOSITES

Figure 6 depicts the plots of the RHR, as basic flammability characteristic, versus time for PP, as well as for the PP/MWCNT nanocomposites.

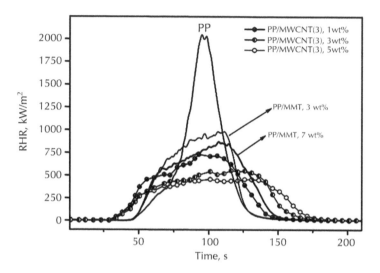

FIGURE 6 The RHR versus time for PP, PP/MWCNT and PP/MMT (Cloisite 20A) nanocomposites obtained by cone calorimeter at the incident heat flux of 35 kW m^{-2}.

From Figure 6, it could be seen that the maximum RHR for pristine PP is 2076 kW/m^2, whereas that for the PP/MWCNT (%) nanocomposites (1 wt%), PP/MCWNT (3 wt%) and the PP/MCWNT (5 wt%) RHR values are 729 kW/m^2, 552.8 kW/m^2, and 455.8 kW/m^2, respectively; thus, the peak RHR decreases by 65%, 73%, and 78%.

The observed flame retardancy effect is associated with solid-phase carbonization reactions, by analogy with layered silicates [31,32]. In early paper [30] we have found that additions of 3 wt% and 7 wt% of layered silicate (Cloisite 20A) to the PP compositions PP lead to RHR decrease by 51% and 57% as compared with pristine PP (Figure 5).

We believe that a higher carbonization effectiveness of MWCNTs depends on their heat conductivity. It is well known that, PP has a low thermal conductivity at standard conditions, and characterized by a minor increase with temperature up to melting point (~0.2 W/m K). On the other hand the heat conductivity of individual MWCNT is extremely high and equals to 3000 W/m K [33,34]. During the high temperature pyrolysis of PP/MWCNT composition at temperatures above 300–400°C the heat conductivity can rise up to 20 W/m K [34] due to actual increase of MWCNT concentration in composition caused by volatilization of polypropylene degradation products (Figure 7). The induced heating of PP/MWCNT intensifies a steady carbonization and charring of the samples and leads to decrease of RHR peak value (Figure 6).

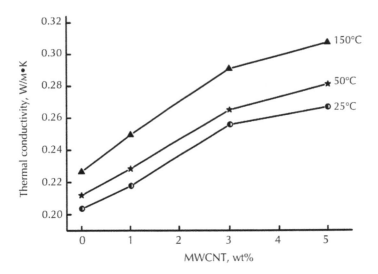

FIGURE 7 Temperature dependence of the thermal conductivity of PP/MWCNT (3) nanocomposite with different loadings of MWCNT.

Figures 8 and 9 show graphs for the specific extinction area and effective heat of combustion, correspondingly, for PP and PP/MWCNT(3) nanocomposites. Calculated values of effective heat of combustion for PP and PP/MWNT demonstrate invariant shift of this parameter for these nanocomposites.

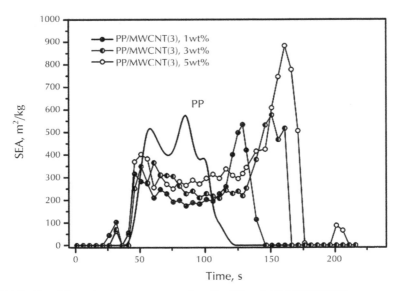

FIGURE 8 Specific extinction area versus time for PP and PP/MWCNT(3) nanocomposites obtained by cone calorimeter at the incident heat flux of 35 kW m^{-2}.

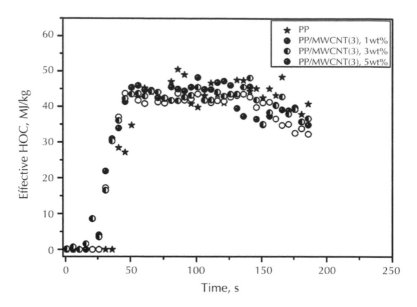

FIGURE 9 Effective heat of combustion versus time for PP and PP/MWCNT(3) nanocomposites obtained by cone calorimeter at the incident heat flux of 35 kW m^{-2}.

In the present study EPR research were performed to follow formation of stable radicals, responsible for carbonization process, upon isothermal heating of PP/MC-WNT (10 wt%) in air at 350°C.

Figure 10(a) shows EPR spectrum of the stable paramagnetic centers formed in the samples of PP/MCWNT (10 wt%) heating in air at 350°C. When a heated in air PP/MCWNT specimen was placed into an EPR sample tube, a narrow singlet signal with a line width of $\Delta H_{1/2} = 0{,}69$ mT and a g value of 2.003 was detected due to the stable radicals generation, analogous to those previously registered during polymers carbonization process [35]. No EPR signal similar to that of PP/MCWNT samples were observed in the samples of pristine PP and MCWNT samples heated at 350°C in air. It should be noted that although iron impurity from MWNCT has been mentioned in other studies on pyrolysis of polymer nanocomposites as the radical traps [36, 11], the EPR analyses in the current study showed the presence of paramagnetic centers relating to carbonaceous stable radicals only.

As it is seen from Figure 10(b), the formation of stabilized radicals occurs with pronounced induction period which is related to antioxidant properties of MWNCT. Such a type of kinetic dependence is coincided with an oxygen uptake kinetics observed during inhibited polyolefines thermal oxidation. Moreover, no EPR signals were observed in the samples of the PP/MCWNT samples heated at 350°C in inert Ar.

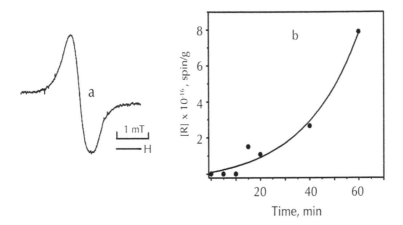

FIGURE 10   The EPR spectrum of (a) the stable paramagnetic centers formed in the samples of PP/MCWNT (10 wt%) heating in air at 350°C and (b) kinetic dependence of stable radicals generation from PP/MCWNT (10% wt) under isothermal heating at 350°C in air.

Thus, multi-walled carbon nanotubes are considered to be more effective filling agents than layered silicates in the terms of improvement of thermal properties and flame retardancy of PP matrix. This could be explained by the specific antioxidant properties and high thermal conductivity of MWCNT, which determine the carbonization reactions during thermal-oxidative degradation process.

KEYWORDS

- **Multi-walled carbon nanotubes (MWCNT)**
- **Polypropylene**
- **Scanning electron microscopy (SEM)**
- **Thermogravimetric analysis (TG)**
- **Transmission electron microscopy (TEM)**

REFERENCES

1.   Shaffer, M. S. P and Windle, A. H. *Adv Mater*, **11**, 937 (1999).
2.   Qian, D., Dickey, E. C., Andrews, R., and Rantell, T. *Appl Phys Lett*, **76**, 2868 (2000).
3.   Jin, Z., Pramoda, K. P., Xu, G., and Goh, S. H. *Chem Phys Lett*, **337**, 43 (2001).
4.   Thostenson, E. T. and Chou, T. W. *J Phys D: Appl Phys*, **35**, L77 (2002).
5.   Bin, Y., Kitanaka, M., Zhu, D., and Matsuo, M. *Macromolecules*, **36**, 6213 (2003).
6.   Potschke, P., Dudkin, S. M., and Alig, I. *Polymer*, **44**, 5023 (2003).
7.   Safadi, B., Andrews, R., and Grulke, E. A. *J Appl Polym Sci*, **84**, 2660 (2002).
8.   Watts, P.C.P., Fearon, P. K., Hsu, W. K., Billingham, N. C., Kroto, H. W., and Walton, D. R. M. *J. Mater. Chem.*, **13**, 491 (2003).

9. Watts, P. C. P., Hsu, W. K., Randall, D. P., Kroto, H. W. and Walton, D. R. M. *Phys. Chem. Chem. Phys.*, **4**, 5655 (2002).
10. Yang, J., Lin, Y., and Wang, J. *J Appl Polym Sci*, **98**(3), 1087 (2005)
11. Kashiwagi, T., Grulke, E., Hilding, J., Groth, K., Harris, R. H., Butler, K., Shields, J. R., Kharchenko, S., and Douglas, J. *Polymer*, **45**, 4227–4239 (2004).
12. Kashiwagi, T., Grulke, E., Hilding, J., Harris, Jr R. H., Awad, W. H., and Dougls, J. *Macromol Rapid Commun*, **23**, 761–765 (2002).
13. Lomakin, S. M., Novokshonova, L. A., Brevnov, P. N., and Shchegolikhin, A. N. *Journal of Materials Science*, **43**(4), 1340 (2008).
14. Chen, J., Hamon, M. A., Hu, H., Chen, Y., Rao, A. M., Eklund, P. C., and Haddon, R. C. *Science*, **282**, 95 (1998).
15. Stevens, J. L., Huang, A. Y., Peng, H., Chiang, I. W., Khabashesku, V. N., and Margrave, J. L. *Nano Lett.*, **3**, 331 (2003).
16. Eitan, A., Jiang, K., Dukes, D., Andrews, R., and Schadler, L. S. *Chem. Mater.*, **15**, 3198 (2003).
17. Hu, H., Ni, Y., Montana, V., Haddon, R. C., and Parpura, V. *Nano Lett.*, **4**, 507 (2004).
18. Kong, H., Gao, C., and Yan, D. *J. Am. Chem. Soc.*, **126**, 412 (2004).
19. Holzinger, M., Vostrowsky, O., Hirsch, A., Hennrich, F., Kappes, M., Weiss, R., and Jellen, F. *Angew. Chem. Int. Edn*, **40**, 4002 (2001).
20. Holzinger, M., Abraham, J., Whelan, P., Graupner, R., Ley, L., Hennrich, F., Kappes, M. and Hirsch, A. *J. Am. Chem. Soc.*, **125**, 8566 (2003).
21. Yao, Z., Braidy, N., Botton, G. A., and Adronov, A. *J. Am.Chem. Soc.* **125**, 16015 (2003).
22. Ying, Y., Saini, R. K., Liang, F., Sadana, A. K., and Billups, W. E. *Org. Lett.*, **5**, 1471 (2003).
23. Alvaro, M., Atienzar, P., de la Cruz, P., Delgado, J. L., Garcia, H., and Langa, F. *J. Phys. Chem. B*, **108** 12691 (2004).
24. Nyden, M. R. and Stoliarov, S. I. *Polymer*,**49**, 635–641 (2008).
25. Rakhimkulov, A. D., Lomakin, S. M., Alexeyeva, O. V., Dubnikova, I. L., Schegolikhin, A. N., and Zaikov, G. E. In Proceedings of IBCP International Conference, **56** (2006).
26. Krusic, J., Wasserman, E., Keizer, P. N., Morton, J. R., and Preston, K. F. *Science*, **254**, 1183 (1991).
27. Opfermann, J. *J Thermal Anal Cal.*, **60**(3), 641 (2000).
28. Friedman, H. L. *J Polym. Sci.*, **C6**, 175 (1965).
29. Opfermann, J. and Kaisersberger, E. *Thermochim Acta.*, **11**(1), 167 (1992).
30. Lomakin, S. M., Dubnikova, I. L., Berezina, S. M., and Zaikov, G. E. *Polymer Science*, **48**, Ser. A, **1**, 72 (2006).
31. Gilman, G. W., Jackson, C. L., Morgan, A. B., Harris, R. H., Manias, E., Giannelis, E. P., Wuthenow, M., Hilton, D., Phillips, S. *Chem. Mater.*, **12**, 1866 (2000).
32. Kashiwagi, T., Harris, R. H. Jr., Zhang, X., Briber, R. M., Cipriano, B. H., Raghavan, S. R., Awad, W. H., and Shields, J. R. *Polymer*, **45**, 881 (2004).
33. Kim, P., Shi, L., Majumdar, A., and McEuen, P. L. *Phys Rev Lett*, **87**, 215502 (2001).
34. Yi, W., Lu, L., Zhang, D. L., Pan, Z. W., and Xie, S. S. *Phys Rev B*, **59**, R9015–R9018 (1999).
35. Echevskii, G. V., Kalinina, N. G., Anufrienko, V. F., and Poluboyarov, V. A. *React. Kinet. Catal. Lett.*, **33**(8), 305 (1987).
36. Zhu, J., Uhl, F., Morgan, A. B., and Wilkie, C. A. *Chem Mater*, **13**, 4649–54 (2001).

CHAPTER 3

RECENT ADVANCES IN POLYMERIC NANOCOMPOSITES: STRUCTURE, MANUFACTURE, AND PROPERTIES

A. K. MIKITAEV, A. YU. BEDANOKOV, and M. A. MIKITAEV

CONTENTS

3.1   INTRODUCTION

The polymeric nanocomposites are the polymers filled with nanoparticles which interact with the polymeric matrix on the molecular level in contrary to the macro interaction in composite materials. Mentioned nano interaction results in high adhesion hardness of the polymeric matrix to the nanoparticles [1, 52].

Usual nanoparticle is less than 100 nanometers in any dimension, 1 nanometer being the billionth part of a meter [1,2].

The analysis of the reported studies tells that the investigations in the field of the polymeric nanocomposite materials are very promising.

The first notion of the polymeric nanocomposites was given in patent in 1950 [3]. Blumstain pointed in 1961 [4] that polymeric clay-based nanocomposites had increased thermal stability. It was demonstrated using the data of the thermogravimetric analysis that the polymethylmetacrylate intercalated into the Na+ - methylmetacrylate possessed the temperature of destruction 40–50°C higher than the initial sample.

This branch of the polymeric chemistry did not attract much attention until 1990 when the group of scientists from the Toyota Concern working on the polyamide-based nanocomposites [5-9] found two times increase in the elasticity modulus using only 4.7 wt% of the inorganic compound and 100°C increase in the temperature of destruction, both discoveries widely extending the area of application of the polyamide. The polymeric nanocomposites based on the layered silicates began being intensively studied in state, academic, and industrial laboratories all over the world only after that.

3.2   STRUCTURE OF THE LAYERED SILICATES

The study of the polymeric nanocomposites on the basis of the modified layered silicates (broadly distributed and well-known as various clays) is of much interest. The natural layered inorganic structures used in producing the polymeric nanocomposites are the montmorillonite [10-12], hectorite [13], vermiculite [14], kaolin, saponine [15], and others. The sizes of inorganic layers are about 220 and 1 nanometers in length and width, respectively [16,17].

The perspective ones are the bentonite breeds of clays which include at least 70% of the minerals from the montmorillonite group.

Montmorillonite $(Na,K,Ca)(Al,Fe,Mg)[(Si,Al)_4O_{10}](OH)_2 = nH_2O$, named after the province Montmorillion in France, is the high dispersed layered aluminous silicate of white or gray color in which appears the excess negative charge due to the non-stoichiometric replacements of the cations of the crystal lattice, charge being balanced by the exchange cations from the interlayer space. The main feature of the montmorillonite is its ability to adsorb ions, generally cations, and to exchange them. It produces plastic masses with water and may enlarge itself 10 times. Montmorillonite enters the bentonite clays (the term "bentonite" is given after the place Benton in USA).

The inorganic layers of clays arrange the complexes with the gaps called layers or galleries. The isomorphic replacement within the layers (such as Mg_2^+ replacing Al_3^+ in octahedral structure or Al_3^+ replacing Si_4^+ in tetrahedral one) generates the negative charges which electrostatically are compensated by the cations of the alkali or alkali earth metals located in the galleries (Figure 1) [18]. The bentonite is very hydrophilic because of this. The water penetrates the interlayer space of the montmorillonite, hydrates its surface, and exchanges cations what results in the swelling of the bentonite. The further dilution of the bentonite in water results in the viscous suspension with bold tixotropic properties. The more pronounced cation exchange and adsorption properties are observed in the bentonites montmorillonite of which contains predominantly exchange cations of sodium.

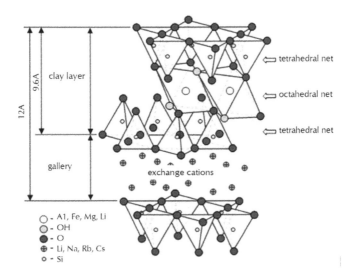

FIGURE 1   The structure of the layered silicate

3.3   MODIFICATION OF THE LAYERED SILICATES

Layered silicates possess quite interesting properties—sharp drop of hardness at wetting, swelling at watering, dilution at dynamical influences, and shrinking at drying.

The hydrophility of aluminous silicates is the reason of their incompatibility with the organic polymeric matrix and is the first hurdle need to be overridden at producing the polymeric nanocomposites.

One way to solve this problem is to modify the clay by the organic substance. The modified clay (organoclay) has at least two advantages: (1) it can be well dispersed in polymeric matrix [19] and (2) it interacts with the polymeric chain [13].

The modification of the aluminous silicates can be done with the replacement of the inorganic cations inside the galleries by the organic ones. The replacing by the cationic surface-active agents like bulk ammonium and phosphonium ions increases the room between the layers, decreases the surface energy of clay and makes the surface of the clay hydrophobic. The clays modified such a way are more compatible with the polymers and form the layered polymeric nanocomposites [52]. One can use the non-ionic modifiers besides the organic ones which link themselves to the clay surface through the hydrogen bond. Organoclays produced with help of non-ionic modifiers in some cases become more chemically stable than the organoclays produced with help of cationic modifiers (Figure 2.I) [20].

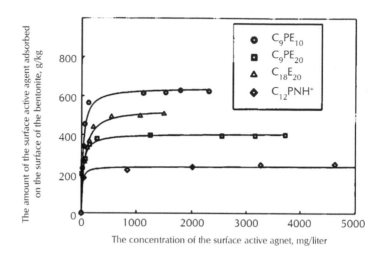

FIGURE 2.I   The adsorption of different modifiers on the clay surface

The least degree of desorption is observed for non-ionic interaction between the clay surface and organic modifier (Figure 2.II). The hydrogen bonds between the ethylenoxide grouping and the surface of the clay apparently make these organoclays more chemically stable than organoclays produced with non-ionic mechanism.

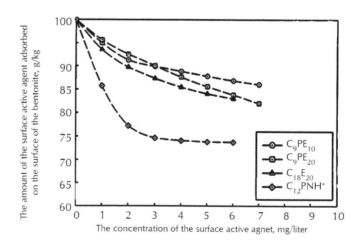

FIGURE 2.II.   The desorption of different modifiers from the clay surface: C9PE10 – $C_9H_{19}C_6H_4(CH_2 \ CH_2O)_{10}OH$, C_9PE_{20} – $C_9H_{19}C_6H_4(CH_2 \ CH_2O)_{20}OH, C_{18}E_{20}$ – $C_{18}H_{37}(CH_2 \ CH_2O)_{20}OH$, $C_{12}PNH+$ – $C_{12}H_{25}C_6H_4NH^+Cl^-$.

3.4   STRUCTURE OF THE POLYMERIC NANOCOMPOSITES ON THE BASIS OF THE MONTMORILLONITE

The study of the distribution of the organoclay in the polymeric matrix is of great importance because the properties of composites obtained are in the direct relation from the degree of the distribution.

According Giannelis [21], the process of the formation of the nanocomposite goes in several intermediate stages (Figure 3). The formation of the tactoid happens on the first stage – the polymer surrounds the agglomerations of the organoclay. The polymer penetrates the interlayer space of the organoclay on the second stage. Here the gap between the layers may reach 2–3 nanometers [22]. The further separation of the layers, third stage, results in partial dissolution and disorientation of the layers. Exfoliation is observed when polymer shifts the clay layers on more than 8–10 nanometers.

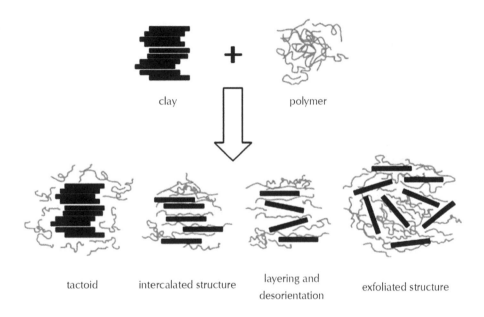

clay   polymer

tactoid   intercalated structure   layering and desorientation   exfoliated structure

FIGURE 3   The schematic formation of the polymeric nanocomposite [24]

All mentioned structures may be present in real polymeric nanocomposites in dependence from the degree of distribution of the organoclay in the polymeric matrix. Exfoliated structure is the result of the extreme distribution of the organoclay. The excess of the organoclay or bad dispersing may born the agglomerates of the organoclays in the polymeric matrix what finds experimental confirmation in the X-ray analysis [11, 12,21,23].

In the following subsections we describe a number of specific methods used at studying the structure of the polymeric nanocomposites.

3.4.1   DETERMINATION OF THE INTERLAYER SPACE

The X-ray determination of the interlayer distance in the initial and modified layered silicates as well as in final polymeric nanocomposite is one of the main methods of studying the structure of the nanocomposite on the basis of the layered silicate. The peak in the small-angle diapason ($2\theta = 6$–$8°C$) is characteristic for pure clays and responds to the order of the structure of the silicate. This peak drifts to the smaller values of the angle 2θ in organo-modified clays. If clay particles are uniformly distributed in the bulk of the polymeric matrix then this peak disappears, what witnesses on the disordering in the structure of the layered silicate. If the amount of the clay exceeds the certain limit of its distribution in the polymeric matrix, then the peak reappears again. This regularity was demonstrated on the instance of the polybutylenterephtalate (Figure 4) [11].

The knowledge of the angle 2θ helps to define the size of the pack of the aluminous silicate consisting of the clay layer and interlayer space. The size of such pack increases in a row from initial silicate to polymeric nanocomposite according to the increase in the interlayer space. The average size of that pack for montmorillonite is 1.2–1.5 nanometers but for organo-modified one varies in the range of 1.8–3.5 nanometers.

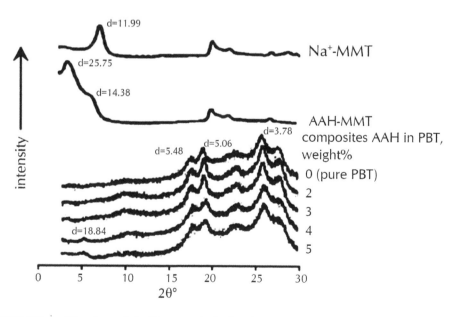

FIGURE 4   The data of the X-ray analysis for clay, organoclay and nanocomposite PBT/organoclay

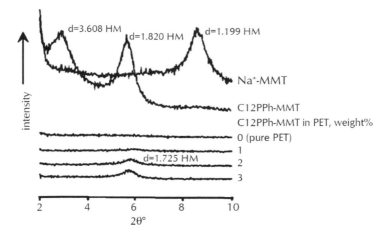

FIGURE 5 The data of the X-ray analysis for clay, organoclay and nanocomposite PET/ organoclay

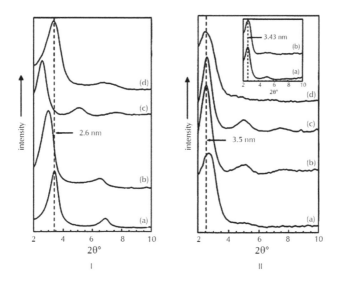

FIGURE 6 The data of the X-ray analysis for: I. (a) dimetyldioctadecylammonium (DMDODA) – hectorite,(b) – 50% polystyrene (PS)/50% DMDODA – hectorite,(c) – 75% polyethylmetacrylate (PEM)/25% DMDODA,(d) – 50% PS/50 % DMDODA – hectorite after 24 hr of etching in cyclohexane. II. mix of PS, PEM and ofganoclay: (a) – 23.8% PS/71.2% PEM/5% DMDODA – hectorite, (b) – 21.2% PS/63.8% PEM/15% DMDODA – hectorite, (c) – 18.2% PS/54.8% PEM/27% DMDODA – hectorite, (d) – 21.2% PS/63.8% PEM/15% DMDODA – hectorite after 24 hr of etching in cyclohexane.

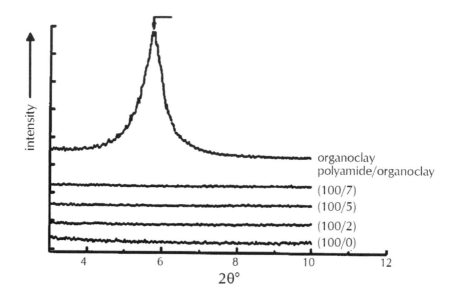

FIGURE 7 The data of the X-ray analysis for organoclay and nanocomposite polyamide acid/organoclay.

Summing up we conclude that comparing the data of the X-ray analysis for the organoclay and nanocomposite allows for the determination of the optimal clay amount need be added to the composite. The data from the scanning tunneling microscopes (STM) and transmission electron microscopes (TEM) [27,28] can be used as well.

3.4.2 THE DEGREE OF THE DISTRIBUTION OF THE CLAY PARTICLES IN THE POLYMERIC MATRIX

The two structures, namely the intercalated and exfoliated ones, could be distinguished with the respect to the degree of the distribution of the clay particles, Figure 8. One should note that clay layers are quite flexible though they are shown straight in the figure. The formation of the intercalated or exfoliated structures depends on many factors, for example the method of the production of the nanocomposite or the nature of the clay etc [29].

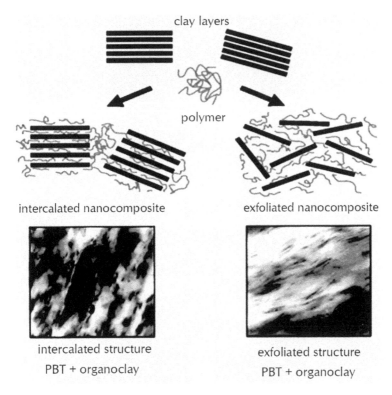

FIGURE 8   The formation of the intercalated and exfoliated structures of the nanocomposite

The TEM images of the surface of the nanocomposites can help to find out the degree of the distribution of the nano-sized clay particles, see plots (a) to (d) in the Figure 9.

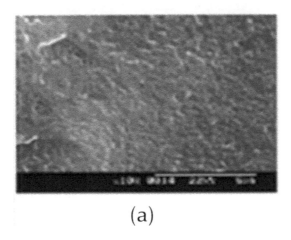

(a)

FIGURE 9   *(Continued)*

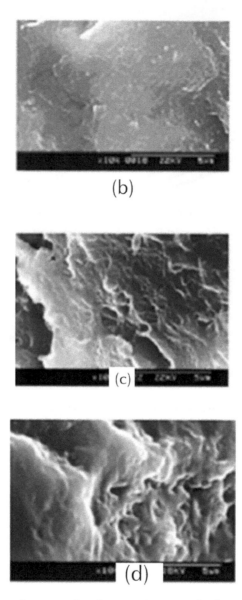

FIGURE 9   The images from scanning electron microscope for the nanocomposite surfaces: (a) pure PBT,(b) 3 wt% of organoclay in PBT, (c) 4 wt% of organoclay in PBT, and (d) 5 wt% of organoclay in PBT.

The smooth surface tells about the uniform distribution of the organoclay particles. The surface of the nanocomposite becomes deformed with the increasing amount of the organoclay, see plots (a) to (d) in the Figure 10. Probably, this is due to the influence of the clay agglomerates [30,31].

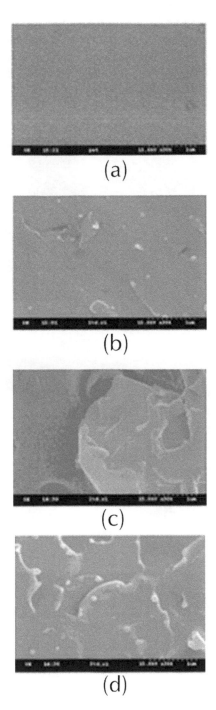

FIGURE 10   The images from scanning electron microscope for the nanocomposite surfaces: (a) pure PET, (b) 3 wt% of organoclay in PET, (c) 4 wt% of organoclay in PET, and (d) 5 wt% of organoclay in PET.

Also one can use the STM images to judge on the degree of the distribution of the organoclay in the nanocomposite, Figures 11 and 12. If the content of the organoclay is 2–3 wt% then the clay layers are separated by the polymeric layer of 4–10 nanometers width, Figure 11. If the content of the organoclay reaches 4–5 wt% then the majority of the clay becomes well distributed however the agglomerates of 4–8 nanometers may appear.

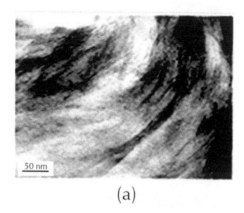

(a)

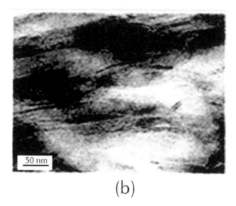

(b)

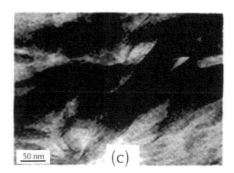

(c)

FIGURE 11   *(Continued)*

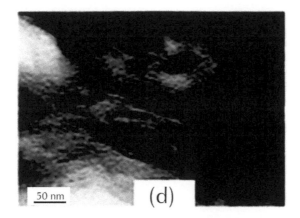

FIGURE 11 The images from tunneling electron microscope for the nanocomposite surfaces: (a) 2 wt% of organoclay in PBT, (b) 3 wt% of organoclay in PBT, (c) 4 wt% of organoclay in PBT, (d) 5 wt% of organoclay in PBT.

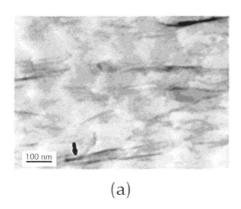

(a)

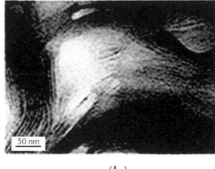

(b)

FIGURE 12 *(Continued)*

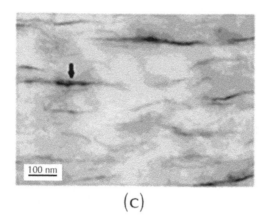

(c)

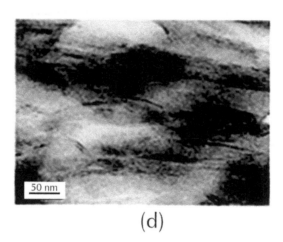

(d)

FIGURE 12   The images from tunneling electron microscope for the nanocomposite surfaces: (a) 1 wt% of organoclay in PET, (b) 2 wt% of organoclay in PET, (c) 3 wt% of organoclay in PET, and (d) 4 wt% of organoclay in PET.

So, the involvement of the X-ray analysis and the use of the microscopy data tell that the nanocomposite consists of the exfoliated clay at the low content (below 3 wt%) of the organoclay.

3.5   PRODUCTION OF THE POLYMERIC NANOCOMPOSITES ON THE BASIS OF THE ALUMINOUS SILICATES

The different groups of authors [32-35] offer following methods for obtaining nano-composites on the basis of the organoclays: (1) in the process of the synthesis of the polymer [33,36,37], (2) in the melt [38,39], (3) in the solution [40-46], and (4) in the sol-gel process [47-50].

The most popular ones are the methods of producing in melt and during the process of the synthesis of the polymer.

The producing of the polymeric nanocomposite in situ is the intercalation of the monomer into the clay layers. The monomer migrates through the organoclay galleries and the polymerization happens inside the layers [19,51] (Figure 13).

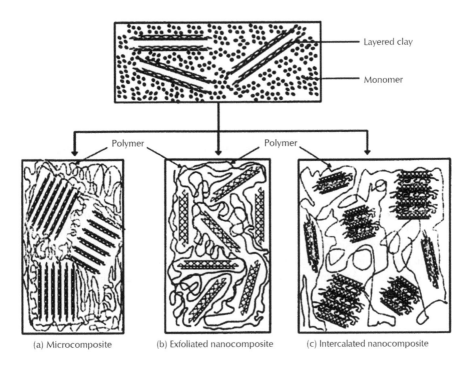

FIGURE 13   The production of the nanocomposite *in situ*: (a) microcomposite, (b) exfoliated nanocomposite, (c) intercalated nanocomposite [51].

The polymerization may be initiated by the heat, irradiation, or other source. Obviously, the best results on the degree of the distribution of the clay particles in the polymeric matrix must emerge if using given method. This is associated with the fact that the separation of the clay layers happens in the very process of the inclusion of the monomer in the interlayer space. In other words, the force responsible for the separation of the clay layers is the growth of the polymeric chain whereas the main factor for reaching the necessary degree of the clay distribution in solution or melt is just satisfactory mixing. The most favorable condition for synthesizing the nanocomposites is the vacuuming or the flow of the inert gas. Besides, one has to use the fast speeds of mixing for satisfactory dispersing of the organoclay in the polymeric matrix.

The method of obtaining the polymeric nanocomposites in melt (or the method of extrusion) is the mixing of the polymer melted with the organoclay. The polymeric chains lose the considerable amount of the conformational entropy during the intercalation. The probable motive force for this process is the important contribution of the enthalpy of the interaction between the polymer and organoclay at mixing. One should add that polymeric nanocomposite on the basis of the organoclays could be successively produced by the extrusion [22]. The advantage of the extrusion method is the absence of any solvents what excludes the unnecessary leaks. Moreover the speed of the process is several times more and the technical side is simpler. The extrusion method is the best one in the industrial scales of production of the polymeric nano-composites what acquires the lesser source expenses and easier technological scheme.

If one produces the polymer-silicate nanocomposite in solution then the organo-silicate swells in the polar solvent such as toluene or N,N-dimethylformamide. Then the added is the solution of the polymer which enters the interlayer space of the silicate. The removing of the solvent by means of evaporation in vacuum happens after that. The main advantage of the given method is that "polymer-layered silicate" might be produced from the polymer of low polarity or even non-polar one. However this method is not widely used in industry because of much solvent consumption [52].

The sol-gel technologies find application at producing nanocomposites on the basis of the various ceramics and polymers. The initial compounds in these technologies are the alcoholates of specific elements and organic oligomers.

The alcoholates are firstly hydrolyzed and obtained hydroxides being polycondensated then. The ceramics from the inorganic 3D net is formed as a result. Also the method of synthesis exists in which the polymerization and the formation of the inorganic glass happen simultaneously. The application of the nanocomposites on the basis of ceramics and polymers as special hard defensive coverage and like optic fibers [53] is possible.

3.6 PROPERTIES OF THE POLYMERIC NANOCOMPOSITES

Many investigations in physics, chemistry, and biology have shown that the jump from macro objects to the particles of 1–10 nanometers results in the qualitative transformations in both separate phases and systems from them [54].

One can improve the thermal stability and mechanical properties of the polymers by inserting the organoclay particles into the polymeric matrix. It can be done by means of joining the complexes of properties of both the organic and inorganic substances that is combining the light weight, flexibility, and plasticity of former and durability, heat stability, and chemical resistance of latter.

Nanocomposites demonstrate essential change in properties if compared to the non-filled polymers. So, if one introduces modified layered silicates in the range of 2–10 wt% into the polymeric matrix then he observes the change in mechanical (tensile, compression, bending, and overall strength), barrier (penetrability and stability to the solvent impact), optical, and other properties. The increased heat and flame resistance even at low filler content is among the interesting properties too. The formation of the thermal isolation and negligible penetrability of the charred polymer to the flame provide for the advantages of using these materials.

The organoclay as a nanoaddition to the polymers may change the temperature of the destruction, refractoriness, rigidity, and rupture strength. The nanocomposites also possess the increased rigidity modulus, decreased coefficient of the heat expansion, low gas-penetrability, and increased stability to the solvent impact and offer broad range of the barrier properties [54]. In Table 1, we gather the characteristics of the nylon-6 and its derivative containing 4.7 wt% of the organo-modified montmorillonite.

TABLE 1   The properties of the nylon-6 and composite based on it [54]

	Rigidity modulus, GPa	Tensile strength, MPa	Temperature of the deformation, °C	Impact viscosity, kJ/m²	Water consumption, wt%	Coefficient of the thermal expansion (x, y)
Nylon-6	1.11	68.6	65	6.21	0.87	13×10^{-5}
Nanocomposite	1.87	97.2	152	6.06	0.51	6.3×10^{-5}

It is important that the temperature of the deformation of the nanocomposite increases on 87°C.

The thermal properties of the polymeric nanocomposites with the varying organoclay content are collected in Table 2.

TABLE 2   The main properties of the polymeric nanocomposites

Property	Composition								
	Polybutylene terephtalate + *AAX*-montmorillonite					Polyethylene terephtalate + C_{12}PPh-montmorillonite			
	Organoclay content, %								
	0	2	3	4	5	0	1	2	3
Viscosity, dliter/g	0.84	1.16	0.77	0.88	0.86	1.02	1.26	0.98	1.23
T_g, °C	27	33	34	33	33	---	---	---	---
T_m, °C	222	230	230	229	231	245	247	245	246
T_d, °C	371	390	388	390	389	370	375	384	386
W_{tR}^{600c}, %	1	6	7	7	9	1	8	15	21
Strength limit, MPa	41	50	60	53	49	46	58	68	71

TABLE 2   *(Continued)*

Rigidity modulus, GPa	1.37	1.66	1.76	1.80	1.86	2.21	2.88	3.31	4.10
Relative enlargement, %	5	7	6	7	7	3	3	3	3

The inclusion of the organoclay into the polybutyleneterephtalate leads to the increase in the glass transition temperature (Tg) from 27 to 33°C if the amount of the clay raises from 0 to 2 wt%. That temperature does not change with the further increase of the organoclay content. The increase in Tg may be the result of two reasons [56-59]. The first is the dispersion of the small amount of the organoclay in the free volume of the polymer, and the second is the limiting of the mobility of the segments of the polymeric chain due to its interlocking between the layers of the organoclay.

The same as the Tg, the melting temperature Tm increases from the 222 to 230°C if the organoclay content raises from 0 to 2 wt% and stays constant up to 5 wt%, see Table 2. This increase might be the consequence of both complex multilayer structure of the nanocomposite and interaction between the organoclay and polymeric chain [60,61]. Similar regularities have been observed in other polymeric nanocomposites also.

The thermal stability of the nanocomposites polybutylene terephtalate, briefly PBT, (or polyethylele terephtalate)/organoclay determined by the thermogravimetric analysis is presented in Table 3 and in Figusrs 14 and 15 [11,12].

TABLE 3   The thermal properties of the fibers from PET with varying organoclay content

Organoclay content, wt%	η_{inh}[a]	T_m (°C)	DH_m[b] (J/g)	T_D[c] (°C)	Wt_R^{600}[d] (%)
0 (pure PET)	1.02	245	32	370	1
1	1.26	247	32	375	8
2	0.98	245	33	384	15
3	1.23	246	32	386	21

[a]—viscosities were measured at 30°C using 0.1 gram of polymer on 100 milliliters of solution in mix phenol/tetrachlorineethane (50/50),
[b]—change in enthalpy of melting,
[c]—initial temperature of decomposition,
[d]—weight percentage of the coke remnant at 600°C.

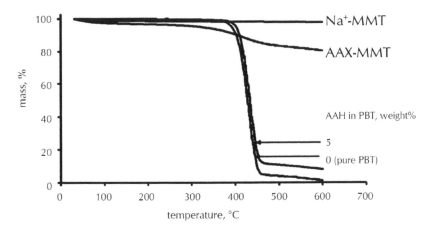

FIGURE 14 The thermogravimetrical curves for the montmorillonites, PBT, and nanocomposites PBT/organoclay

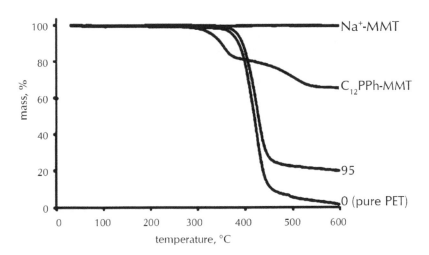

FIGURE 15 The thermogravimetrical curves for the montmorillonites, PET, and nanocomposites PE/organoclay

The temperature of the destruction, TD, increases with the organoclay content up to 350°C in case of the composite PBT/organoclay. The thermogravimetrical curves for pure and composite PBTs have similar shapes below 350°C. The values of temperature TD depends on the amount of organoclay above 350°C. The organoclay added becomes a barrier for volatile products being formed during the destruction [61,62]. Such example of the improvement of the thermal stability was studied in papers [63, 64]. The mass of the remnant at 600°C increases with organoclay content.

Following obtained data authors draw the conclusion that the optimal results for thermal properties are being obtained if 2 wt% of the organoclay is added [11,12,19].

The great number of studies on the polymeric composite organoclay-based materials show [11–13,19] that the inclusion of the inorganic component into the organic polymer improves the thermal stability of the latter, see Tables 3 and 4.

TABLE 4   The basic properties of the nanocomposite based on PBT with varying organoclay content

Organoclay content, weight %	I.V.[a]	T_g	T_m (°C)	T_D^{i} [b] (°C)	Wt_R^{600} [c] (%)
0 (pure PBT)	0.84	27	222	371	1
2	1.16	33	230	390	6
3	0.77	34	230	388	7
4	0.88	33	229	390	7
5	0.86	33	231	389	9

[a]—viscosities were measured at 30°C using 0.1 gram of polymer on 100 milliliters of solution in mix phenol/tetrachlorineethane (50/50),

[b]—initial temperature of the weight loss,

[c]—weight percentage of the coke remnant at 600°C.

The values of the melting temperature increase from 222 to 230°C if the amount of the organoclay added reaches 2 wt% and then stay constant. This effect can be explained by both thermal isolation of the clay and interaction between the polymeric chain and organoclay [43,64]. The increase in the glass transition temperature also occurs what can be a consequence of several reasons [55,56,58]. One of the main among them is the limited motion of the segments of the polymeric chain in the galleries within the organoclay.

If the organoclay content in the polymeric matrix of the PBT reaches 2 wt% then both the temperature of the destruction increases and the amount of the coke remnant increase at 660°C and then both stay practically unchanged with the further increase of the organoclay content up to 5 wt%. The loss of the weight due to the destruction of the polymer in pure PBT and its composites looks familiar in all cases below 350°C. The amount of the organoclay added becomes important above that temperature because the very clays possess good thermal stability and make thermal protection by their layers and form a barrier preventing the volatile products of the decomposition to fly off [43,60]. Such instance of the improvement in thermal properties was observed in many polymeric composites [64-68]. The weight of the coke remnant increases with the rising organoclay content up to 2 wt% and stays constant after that. The increase of the remnant may be linked with the high thermal stability of the organoclay itself. Also it is worth noticing that the polymeric chain closed in interlayer space of the organoclay has fewer degrees of oscillatory motion at heating due to the limited interlayer space and the formation of the abundant intermolecular bonds between the polymeric chain and the clay surface. And the best result is obtained at 2–3 wt% content of the organoclay added to the polymer.

If one considers the influence of the organoclay added to the polyethelene tere-phtalate, briefly PET, [69] then the temperature of the destruction increases on 16°C at optimal amount of organoclay of 3 wt%. The coke remnant at 600°C again increases with the rising organoclay content, see Table 3.

Regarding the change in the temperature of the destruction in the cases of PET and PBT versus the organoclay content one can note that both trends look similar. However the coke remnant considerably increases would the tripheyldodecylphosphonium cation be present within the clay. The melting temperature does not increase in case of the organoclay added into the PET in contrary to the case of PBT. Apparently this may be explained by the more crystallinity of the PBT and the growth of the degree of crystallinity with the organoclay content.

It becomes obvious after analyzing the above results that the introduction of the organoclay into the polymer increases the thermal stability of the latter according the (1) thermal isolating effect from the clay layers and (2) barrier effect in relation to the volatile products of destruction.

The studying of the mechanical properties of the nanocomposites, see Table 2, have shown that the limit of the tensile strength increases with the organoclay added up to 3 wt% for the majority of the composites. Further addition of the organoclay, up to 5 wt%, results in the decreasing limit of the tensile strength. We explain this by the fact that agglomerates appear in the nanocomposite when the organoclay content exceeds the 3 wt% values [61,70,71]. The proof for the formation of the agglomerates have been obtained from the X-ray study and using the data from electron microscopes.

Nevertheless the rigidity modulus increases with the amount of the organoclay added into the polymeric matrix, the resistance of the clay itself being the explanation for that. The oriented polymeric chains in the clay layers also participate in the increase of the rigidity modulus [72]. The percentage of enlargement at breaking became 6–7 wt% for all mixes.

Using data of the Table 2, we explain the improvement in the mechanical properties of the nanocomposites with added organoclay up to 3 wt% by the good degree of distribution of the organoclay within the polymeric matrix. The degree of the improvement also depends on the interaction between the polymeric chain and clay layers.

The study of the influence of the degree of the extract of fibers on the mechanical properties has shown that the limit of strength and the rigidity modulus both increase in PBT whereas they decrease in nanocomposites (Table 5). This can be explained by the breaking of the bonds between the organoclay and PBT at greater degree of extract. Such phenomena have been observed in numerous polymeric composites [73-75].

TABLE 5   The ability to stretch of the nanocomposites PBT/organoclay at varying degrees of extract

Organoclay content, wt%	Limit of strength, MPa			Rigidity modulus, GPa		
	DR = 1	DR = 3	DR = 6	DR = 1	DR = 3	DR = 6
0 (pure PBT)	41	50	52	1.37	1.49	1.52
3	60	35	29	1.76	1.46	1.39

The first notions on the lowered flammability of the polymeric nanocomposites on the organoclay basis appeared in 1976 in the patent on the composite based on the nylon-6 [5]. The serious papers in the field were absent till the 1995 [76].

The use of the calorimeter is very effective for studying the refractoriness of the polymers. It can help at measuring the heat release, the carbon monoxide depletion, and others. The speed of the heat release is one of the most important parameters defining the refractoriness [77]. The data on the flame resistance in various polymer/organoclay systems such as layered nanocomposite nylon-6/organoclay, intercalated nanocomposites polystyrene (or polypropylene)/organoclay were given in chapter [78] in where the lowered flammability was reported, see Table 6. And the lowered flammability have been observed in systems with low organoclay content, namely in range from 2 to 5 wt%.

TABLE 6   Calorimetric data

Sample	rem-nant (%)±0.5	Peak of the HRR (D%) (kW/m²)	Middle of the HRR (D%) (kW/m²)	Average value H_c (MJ/kg)	Average value SEA (m²/kg)	Average CO left (kg/kg)
Nylon-6	1	1010	603	27	197	0.01
Nylon-6/organoclay, 2%, delaminated	3	686 (32%)	390 (35%)	27	271	0.01
Nylon-6/organoclay, 5%, delaminated	6	378 (63%)	304 (50%)	27	296	0.02
Polystyrene	0	1120	703	29	1460	0.09
PS/organoclay, 3%, bad mixing	3	1080	715	29	1840	0.09

TABLE 6   *(Continued)*

PS/organoclay, 3%, intercalated/delaminated	4	567 (48%)	444 (38%)	27	1730	0.08
PS w/DBDPO/Sb$_2$O$_3$, 30%	3	491 (56%)	318 (54%)	11	2580	0.14
Polypropylene	0	1525	536	39	704	0.02
PP/organoclay, 2%, intercalated	5	450 (70%)	322 (40%)	44	1028	0.02

Hc— heat of combustion,

SEA—specific extinguishing area,

DBDPO—deca-bromine diphenyl oxide,

HRR—speed of the heat release

The curve of the heat release for the polypropylene and the nanocomposite on its basis (organoclay content varying from 2 to 4 wt%) is given in the Figure 16 from which one can see that the speed of the heat release for the nanocomposite enriched with the 4 wt% organoclay (the interlayer distance 3.5 nanometers) is 75% less than for pure polypropylene.

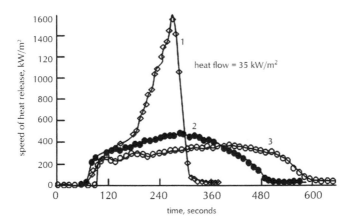

FIGURE 16   The speed of the heat release for: 1—pure polypropylene, 2—nanocomposite with 2 wt% of organoclay, and 3—nanocomposite with 4 wt% of organoclay.

The comparison of the experimental data for the nanocomposites on the basis of the nylon-6, polypropylene, and polystyrene gathered in Table 7 show that the heat of combustion, the smoke release and the amount of the carbon monoxide are almost constant at varying organoclay content. So we conclude that the source for the increased refractoriness of these materials is the stability of the solid phase and not the influence of the vapor phase. The data for the polystyrene with the 30% of the decabromine diphenyl oxide and Sb_2O_3 are given in Table 6 as the proof of the influence of the vapor phase of bromine. The incomplete combustion of the polymeric material in the latter case results in low value of the heat of the combustion and high quantity of the carbon monoxide released [79].

One should note that the mechanism for the increased fire resistance of the polymeric nanocomposites on the basis of the organoclays is not, in fact, clear at all. The formation of the barrier from the clay layers during the combustion at their collapse is supposed to be the main mechanism. That barrier slows down the combustion [80]. In this chapter, we study the influence of the nanocomposite structure on the refractoriness. The layered structure of the nanocomposite expresses higher refractoriness comparing to that in intercalated nanocomposite, see Figure 17.

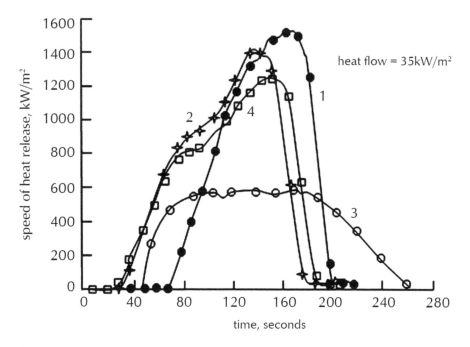

FIGURE 17   The speed of the heat release for:1—pure polystyrene (PS), 2 – PS mixed with 3 wt% of Na+ MMT, 3 – intercalated/delaminated PS (3 wt% 2C18-MMT) extruded at 170°C,4 – intercalated PS (3 weight% C14-FH) extruded at 170°C.

The data on the polymeric polystyrene-based nanocomposites presented in Figure 17 are for (1) initial ammoniumfluorine hectorite and (2) quaternary ammonium montmorillonite. The intercalated nanocomposite was produced in first case whereas the layered intercalated nanocomposite was produced in the second one. But because the chemical nature and the morphology of the organoclay used was quite different and it is very difficult to draw a unique conclusion about the flame resistance in polymeric nanocomposites produced. Nonetheless, one should point out that good results of the same quality were obtained for both layered and intercalated structures when studying the aliphatic groupings of the polyimide nanocomposites based on these clays. The better refractoriness is observed in case of polystyrene embedded in layered nanocomposite while intercalated polystyrene-based nanocomposite (with MMT) also exhibits increased refractoriness.

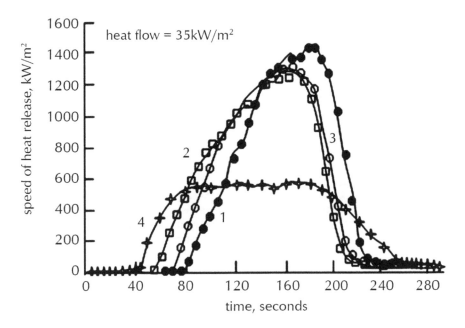

FIGURE 18   The speed of the heat release for:1—pure polysterene (PS), 2—polystyrene with Na-MMT, 3—intercalated PS with organomontmorillonite obtained in extruder at 185°C, 4—intercalated/layered PS with organoclay obtained in extruder at 170°C in nitrogen atmosphere or in vacuum.

As one can see the from the Figure 18, the speed of the heat release for the nano-composite produced in nitrogen atmosphere at 170°C is much lower than for other samples. Probably, the reason for the low refractoriness of the nanocomposite produced in extruder without the vacuuming at 180°C is the influence of the high temperature and of the oxygen from the air what can lead to the destruction of the polymer in such conditions of the synthesis.

It is impossible to give an exact answer on the question about how the refractoriness of organoclay – based nanocomposites increases basing on only the upper experimental data but the obvious fact is that the increased thermal stability and re-fractoriness are due to the presence of the clays existing in the polymeric matrix as nanoparticles and playing the role of the heat isolators and elements preventing the flammable products of the decomposition to fly off.

There are still many problems unresolved in the field but indisputably polymeric nanocomposites will take the leading position in the chemistry of the advanced materials with high heat and flame resistance. Such materials can be used either as itself or in combination with other agents reducing the flammability of the substances.

The processes of the combustion are studied for the number of a polymeric nanocomposites based on the layered silicates such as nylon-6.6 with 5 wt% of Cloisite 15A – montmorillonite being modified with the dimethyldialkylammonium (alkyls studied C18, C16, C14), maleinated polypropylene and polyethylene, both (1.5%) with 10 wt% Cloisite 15A. The general trend is two times reduction of the speed of the heat release. The decrease in the period of the flame induction is reported for all nanocomposites in comparison with the initial polymers [54].

The influence of the nanocomposite structure on its flammability is reflected in the Table 7. One can see that the least flammability is observed in delaminated nanocomposite based on the polystyrene whereas the flammability of the intercalated composite is much higher [54].

TABLE 7   Flammability of several polymers and composites

Sample	Coke remnant, wt%	Max speed of heat release, kW/m²	Average value of heat release, kW/m²	Average heat of combustion, MJ/kg	Specific smoke release, m²/kg	CO release, kg/kg
Nylon-6	1	1010	603	27	197	0.01
Nylon-6 + 2% of silicate (delaminated)	3	686	390	27	271	0.01
Nylon-6 + 5% of silicate (delaminated)	6	378	304	27	296	0.02
Nylon-12	0	1710	846	40	387	0.02
Nylon-12 + 2% of silicate (delaminated)	2	1060	719	40	435	0.02
Polystyrene	0	1562	803	29	1460	0.09
PS + 3% of silicate Na-MMT	3	1404	765	29	1840	0.09
PS + 3% of silicate C14-FH (intercalated)	4	1186	705	28	1790	0.09

TABLE 7   *(Continued)*

PS + 3% of silicate 2C18-MMT (delaminated)	4	567	444	28	1730	0.08
Polypropylene	0	1525	536	39	704	0.02
PP + 2% of silicate (intercalated)	3	450	322	40	1028	0.02

The optical properties of the nanocomposites are of much interest too. The same materials could be either transparent or opaque depending on certain conditions. For example in Figure 19 we see transparency, plot (a), and turbidity, plot (c), of the material in dependence of the frequency of the current applied.

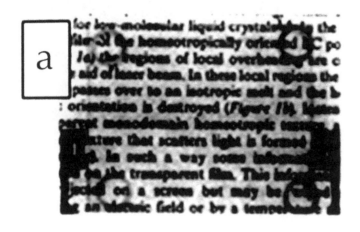

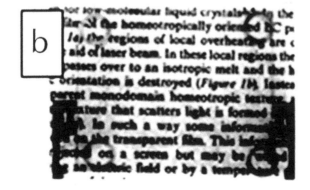

FIGURE 19   *(Continued)*

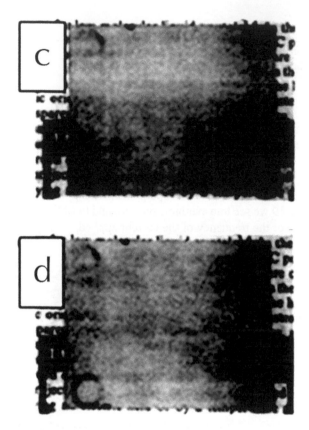

FIGURE 19 The optical properties of the clay-based nanocomposites in dependence of the applied electric current: (a) low frequency, switched on, (b) low frequency, switched off, (c) high frequency, switched on, and (d) high frequency, switched off.

The effect in Figure 19 is reversible and can be innumerately repeated. The transparent and opaque states exhibit the memory effect after the applied current switched off, plots (b) and (d) in Figure 19. The study of the intercalated nanocomposites based on the smectite clays reveals that the optical and elecro-optical properties depend on the degree of intercalation [81].

3.7 CONCLUSION

The quantity of the literature in the field of the nanocomposite polymeric materials has grown multiple times in recent years. The possibility to use almost all polymeric and polycondensated materials as a matrix is shown. The nanocomposites from various organoclays and polymers have been synthesized. Here is just a small part of the compounds for being the matrix referenced in literature: polyacrylate [83], polyamides [82,84,85], polybenzoxazine [86], polybutylene terephtalate [11,82,87], polyimides [88], polycarbonate [89], polymethylmetacrylate [90], polypropylene [91,92], polystyrene [90], polysulphones [93], polyurethane [94], polybuthylene terephtalate and polyethylene terephtalate [10,65,68,79,95,99-107], polyethylene [96], and epoxies [97].

The organo-modified montmorillonite is of the special interest because it can be an element of the nanotechnology and it can also be a carrier of the nanostructure and of asymmetry of length and width in layered structures. The organic modification is being usually performed using the ion-inducing surface-active agents. The non-ionic hydrophobisation of the surface of the layered structures have been reported either. The general knowledge about the methods of the study is being formed and the under-standing of the structure of the nanocomposite polymeric materials is becoming clear. Also scientists come closer to the realizing of the relations between the deformational and strength properties and the specifics of the nanocomposite structure. The growth of researches and their direction into the nanoarea forecasts the fast broadening of the industrial involvement to the novel and attractive branch of the materials science.

KEYWORDS

- **Organoclay**
- **Organo-modified montmorillonite**
- **Polymeric nanocomposites**
- **Polymerization**
- **Sol-gel technologies**

REFERENCES

1   Romanovsky, B. V. and Makshina E. V. Sorosovskii obrazovatelniu zhurnal [in Russian], **8(2)** 50–55 (2004).
2   Golovin, Yu. I. Priroda [in Russian], 1, (2004).
3   Carter, L. W., Hendrics, J. G., and Bolley, D. S. United States Patent №2,531,396, (1950).
4   Blumstain, A. Bull Chem Soc, pp. 899–905 (1961).
5   Fujiwara, S. and Sakamoto, T. Japanese Application № 109,998 (1976).
6   Usuki, A., Kojima, Y., Kawasumi, M., Okada, A., Fukushima, Y., Kurauchi, T., and Kami-gatio, O. Journal of Appl polym science, **55**, 119 (1995).
7   Usuki, A., Koiwai, A., Kojima, Y., Kawasumi, M., Okada, A., Kurauchi, T., and Kami-gaito, O. Journal of Appl polym science, **55**, 119 (1995).
8   Okada, A. and Usuki, A. Mater Sci Engng., **3**, 109 (1995).
9   Okada, A., Fukushima, Y., Kawasumi, M., Inagaki, S., Usuki, A., Sugiyama, S., Kurauchi, T., and Kamigaito, O. United States Patent №4,739,007 (1988).
10 Mikitaev, M. A., Lednev, O. B., Kaladjian, A. A., Beshtoev, B. Z., Bedanokov, A. Yu., and Mikitaev, A. K. Second International Conference, Nalchik (2005).
11 Chang, J. H., An, Y. U., Kim, S. J., and Im, S. Polymer, **44**, 5655–5661 (2003).
12 Mikitaev, A. K., Bedanokov, A. Y., Lednev, O. B., Mikitaev, M. A. Polymer/silicate nano-composites based on organomodified clays/Polymers, Polymer Blends, Polymer Com-posites and Filled Polymers. *Synthesis, Properties, Application.* Nova Science Publishers, New York (2006).

13   Delozier, D. M., Orwoll, R. A., Cahoon, J. F., Johnston, N. J., Smith, J. G., and Connell, J. W. *Polymer*, **43**, 813–822 (2002).

14   Kelly, P., Akelah, A., and Moet, A. J. *Mater. Sci.*, **29**, 2274–2280 (1994).

15   Chang, J. H., An, Y. U., Cho, D., and Giannelis, E. P. Polymer, **44**, 3715–3720 (2003).

16   Yano, K., Usuki, A., and Okada, A. *J Polym Sci, Part A: Polym Chem.*, **35**, 2289 (1997).

17   Garcia-Martinez, J. M., Laguna, O., Areso, S., and Collar, E. P. *J Polym Sci, Part B: Polym Phys.*, **38**, 1564 (2000).

18   Giannelis, E. P., Krishnamoorti, R., and Manias, E. *Advances in Polymer Science*, Springer-Verlag Berlin Heidelberg, **138** (1999).

19   Delozier, D. M., Orwoll, R. A., Cahoon, J. F., Ladislaw, J. S., Smith, J.G., and Connell, J. W. *Polymer*, **44**, 2231–2241 (2003).

20   Shen, Y. H. *Chemosphere*, **44**, 989–995 (2001).

21   Giannelis, E. P. *Adv. Mater.*, **8**, 29–35 (1996).

22   Dennis, H. R., Hunter, D. L., Chang, D., Kim, S., White, J. L., Cho, J. W., and Paul, D. R. *Polymer*, **42**, 9513–9522 (2001).

23   Kornmann, X., Lindberg, H., and Berglund, L. A. Polymer, 42, 1303–1310 (2001).

24   Fornes, T. D. and Paul, D. R. Formation and properties of nylon 6 nanocomposites. São Carlos, *Polímeros*, **13(4)** (Oct/Dec, 2003).

25   Voulgaris, D. and Petridis, D. *Polymer*, **43**, 2213–2218 (2002).

26   Tyan, H. L., Liu, Y. C., and Wei, K. H. *Polymer*, **40**, 4877–4886 (1999).

27   Davis, C. H., Mathias, L. J., Gilman, J. W., Schiraldi, D. A, Shields, J. R, Trulove, P., Sutto, T. E., and Delong, H. C. *J Polym Sci, Part B: Polym Phys.*, **40**, 2661 (2002).

28   Morgan, A. B. and Gilman, J. W. *J Appl Polym Sci.*, **87**, 329 (2003).

29   John, N. and Hay and Steve, J. Shaw. Organic-inorganic hybridssthe best of both worlds? *Europhysics News*, **34(3)** (2003).

30   Chang, J. H., An, Y. U., and Sur, G. S. *J Polym Sci Part B: Polym Phys.*, **41**, 94 (2003).

31   Chang, J. H., Park, D. K., and Ihn, K. J. *J Appl Polym Sci.*, **84**, 2294 (2002).

32   Pinnavaia, T. J. *Science*, **220**, 365 (1983).

33   Messersmith, P. B. and Giannelis, E. P. *Chem Mater.*, **5**, 1064 (1993).

34   Vaia, R. A., Ishii, H., and Giannelis, E. P. *Adv Mater.*, **8**, 29 (1996).

35   Gilman, J. W. *Appl Clay Sci.*, **15**, 31 (1999).

36   Fukushima, Y., Okada, A., Kawasumi, M., Kurauchi, T., and Kamigaito, O. *Clay Miner*, **23**, 27 (1988).

37   Akelah, A. and Moet, A. *J Mater Sci.*, **31**, 3589 (1996).

38   Vaia, R. A., Ishii, H., and Giannelis, E. P. *Adv Mater.*, **8**, 29 (1996).

39   Vaia, R. A., Jandt, K. D, Kramer, E. J., and Giannelis, E. P. *Macromolecules*, **28**, 8080 (1995).

40   Greenland, D. G. *J Colloid Sci.*, **18**, 647 (1963).

41   Chang, J. H. and Park, K. M. *Polym Engng Sci.*, **41** 2226 (2001).

42   Greenland, D. G. *J Colloid Sci.*, **18**, 647 (1963).

43   Chang, J. H., Seo, B. S., and Hwang, D. H. *Polymer*, **43**, 2969 (2002).

44   Vaia, R. A., Jandt, K. D., Kramer, E. J., and Giannelis, E. P. *Macromolecules*, **28**, 8080 (1995).

45   Fukushima, Y., Okada, A., Kawasumi, M., Kurauchi, T., and Kamigaito, O. *Clay Miner.*, **23**, 27 (1988).

46   Chvalun, S. N. *Priroda*, [in Russian], **7** (2000).

47   Brinker, C. J. and Scherer, G. W. *Sol-Gel Science*, Boston (1990).

48   Mascia, L. and Tang, T. *Polymer*, *39* 3045 (1998).

49   Tamaki, R. and Chujo, Y. *Chem Mater.*, **11**, 1719 (1999).

50   Serge Bourbigot, E. A. Investigation of Nanodispersion in Polystyrene–Montmorillon-
      ite Nanocomposites by Solid-State NMR. *Journal of Polymer Science: Part B: Polymer
      Physics*, **41**, 3188–3213 (2003).

51   Lednev, O. B., Kaladjian, A. A., Mikitaev, M. A., and Tlenkopatchev, M. A. *New poly-
      butylene terephtalate and organoclay nanocomposite materials*. Abstracts of the Interna-
      tional Conference on Polymer materials, México (2005)

52   Tretiakov, A. O. *Oborudovanie I instrument dlia professionalov* [in Russian], №02(37)
      (2003)

53   Sergeev, G. B. *Ros. Chem. J.* (The journal of the D.I.Mendeleev Russian chemical society)
      [in Russian], **46(5)** (2002).

54   Lomakin, S. M. and Zaikov, G. E. *Visokomol. Soed. B.* [in Russian], **47(1)**, 104–120
      (2005).

55   Xu, H., Kuo, S. W., Lee, J. S., and Chang, F. C. *Macromolecules*, **35**, 8788 (2002).

56   Haddad, T. S. and Lichtenhan, J. D. *Macromolecules*, **29**, 7302 (1996).

57   Mather, P. T., Jeon, H. G., Romo-Uribe, A., Haddad, T. S., and Lichtenhan, J. D. *Macro-
      molecules*, **29**, 7302 (1996).

58   Hsu, S. L. C. and Chang, K. C. *Polymer*, **43**, 4097 (2002).

59   Chang, J. H., Seo, B. S., and Hwang, D. H. *Polymer*, **43**, 2969 (2002).

60   Fornes, T. D., Yoon, P. J., Hunter, D. L., Keskkula, H., and Paul, D. R. *Polymer.*, **43**, 5915
      (2002).

61   Chang, J. H., Seo, B. S., and Hwang, D. H. *Polymer*, **43**, 2969 (2002).

62   Fornes, T. D., Yoon, P. J., Hunter, D. L., Keskkula, H., and Paul, D. R. *Polymer*, **43**, 5915
      (2002).

63   Wen, J. and Wikes, G. L. *Chem Mater.*, **8**, 1667 (1996).

64   Zhu, Z. K., Yang, Y., Yin, J., Wang, X., Ke, Y., and Qi, Z. *J Appl Polym Sci.*, **3**, 2063
      (1999).

65   Mikitaev, M. A., Lednev, O. B., Beshtoev, B. Z., Bedanokov, A. Yu., and Mikitaev, A. K.
      Second International conference *Polymeric composite materials and covers* [in Russian],
      Yaroslavl (May, 2005).

66   Fischer, H. R., Gielgens, L. H., Koster, T. P. M. Acta Polym, 50, 122 (1999).

67   Petrovic, X. S., Javni, L., Waddong, A., and Banhegyi, G. J. *J Appl Polym Sci.*,**76** 133
      (2000).

68   Lednev, O. B., Beshtoev, B. Z., Bedanokov, A.Yu., Alarhanova, Z. Z., and Mikitaev, A. K.
      Second International Conference (Nalchik) [in Russian] (2005).

69   Chang, J. -H., Kim, S. J., Joo, Y. L., and Im., S. *Polymer*, **45**, 919–926 (2004).

70   Lan, T. and Pinnavaia, T. *J. Chem Mater.*, **6**, 2216 (1994).

71   Masenelli-Varlot, K., Reynaud, E., Vigier, G., and Varlet, J. *J Polym Sci Part B: Polym
      Phys.*, **40**, 272 (2002).

72   Yano, K., Usuki, A., and Okada, A. *J Polym Sci Part A: Polym Chem.*, **35**, 2289 (1997).

73   Shia, D., Hui, Y., Burnside, S. D., and Giannelis, E. P. *Polym Engng Sci.*, **27**, 887 (1987).

74   Curtin, W. A. *J Am Ceram Soc.*, **74**, 2837 (1991).

75   Chawla, K. K. *Composite materials science and engineering*. New York: Springer, (1987).

76   Burnside, S. D. and Giannelis, E. P., *Chem. Mater.* 7, 4597 (1995).

77   Babrauskas, V. and Peacock, R. D., *Fire Safety Journal*, **18**, 225 (1992).

78   Gilman, J., Kashiwagi, T., Lomakin, S., Giannelis, E., Manias, E., Lichtenhan, J., and Jones, P., In: *Fire Retardancy of Polymers: the Use of Intumescence*. The Royal Society of Chemistry, Cambridge, pp 203–221 (1998).
79   Mikitaev, A. K., Kaladjian, A. A., Lednev, O. B., and Mikitaev, M. A. *Plastic masses.*, 12, 45–50 [in Russian] (2004).
80   Gilman, J. and Morgan, A. 10 th Annual BCC Conference, May 24–26, (1999).
81   John, N. Hay, S., and Shaw, *J.Organic-inorganic hybrids: the best of both worlds?* Euro physics News **34(3)** (2003).
82   Delozier, D. M., Orwoll, R. A., Cahoon, J. F., Johnston, N. J., Smith, J. G., and Connell, J. W. *Polymer*, **43**, 813–822 (2002).
83   Chen, Z., Huang, C., Liu, S., Zhang, Y., and Gong, K. *J. Apply Polym Sci.*, **75**, 796–801 (2000).
84   Okado, A., Kawasumi, M., Kojima, Y., Kurauchi, T., and Kamigato, O. *Mater Res Soc Symp Proc.*, **171**, 45 (1990).
85   Leszek, A. U. and Lyngaae-Jorgensen, J. Rheologica Acta., 41, 394–407 (2002).
86   Wagener, R. and Reisinger, T. J. G. *Polymer*, **44**, 7513–7518 (2003).
87   Li, X., Kang, T., Cho, W. J., Lee, J. K., and Ha, C. S. Macromol Rapid Commun.
88   Tyan, H. -L., Liu, Y. -C., and Wei, K. -H. *Polymer*, **40**, 4877–4886 (1999).
89   Vaia, R., Huang, X., Lewis, S., and Brittain, W. *Macromolecules*, **33**, 2000–4 (2000).
90   Okamoto, M., Morita, S., Taguchi, H., Kim, Y., Kotaka, T., and Tateyama, H. *Polymer*, **41**, 3887–90 (2000).
91   Chow, W. S., Mohd Ishak, Z. A., Karger-Kocsis, J., Apostolov, A. A., and Ishiaku, U. S. *Polymer*, **44**, 7427–7440 (2003).
92   Antipov, E. M., Guseva, M. A., Gerasin, V. A., Korolev, Yu. M., Rebrov, A. V., Fisher, H. R., and Razumovskaya, I. V. *Visokomol soed. A.*, **45(11)**, 1885–1899 [in Russian] (2003).
93   Sur, G., Sun, H., Lyu, S., and Mark, J. *Polymer*, **42**, 9783–9 (2001).
94   Wang, Z. and Pinnavaia, T. *Chem Mater.*, **10**, 3769–71 (1998).
95   Bedanokov, A. Yu. and Beshtoev, B. Z. Malij polimernij congress (Moscow) [in Russian] (2005).
96   Antipov, E. M., Guseva, M. A., Gerasin, V. A., Korolev, Yu. M., Rebrov, A. V., Fisher, H. R., and Razumovskaya, I. V. *Visokomol. Soed. A.*, **45(11)**, 1874–1884 [in Russian] (2003).
97   Lan, T., Kaviartna, P., and Pinnavaia, T. Proceedings of the ACS PMSE, 71, 527–8 (1994).
98   Kawasumi, et al. Nematic liquid crystal/clay mineral composites. *Science & Engineering C*, **6**, 135–143, (1998).
99   Lednev, O. B., Kaladjian, A. A., and Mikitaev, M. A. Second International Conference (Nalchik) [in Russian] (2005).
100   Mikitaev, A. K., Kaladjian, A. A., Lednev, O. B., Mikitaev, M. A., and Davidov, E. M. *Plastic masses*, **4**, 26–31[in Russian] (2005).
101   Eid, A., Mikitaev, M. A., Bedanokov, A. Y., and Mikitaev, A. K. *Recycled Polyethylene Terephthalate/Organo-Montmorillanite Nanocomposites, Formation and Properties*. The first Afro-Asian Conference on Advanced Materials Science and Technology (AMSAT 06), Egypt (2006).
102   Mikitaev, A. K., Bedanokov, A. Y., Lednev, O. B., and Mikitaev, M. A. Polymer/silicate nanocomposites based on organomodified clays/ Polymers, Polymer Blends, Polymer Composites and Filled Polymers. *Synthesis, Properties, Application*. Nova Science Publishers. New York (2006).

103 Malamatov, A. H., Kozlov, G. V., and Mikitaev, M. A. *Mechanismi uprochnenenia polimernih nanokompositov* [in Russian], (Moscow, RUChT), p. 240 (2006).
104 Eid, A. Doctor Thesis [in Russian], (Moscow, RUChT) p. 121 (2006).
105 Lednev, O. B. Doctor Thesis (Moscow, RUChT) [in Russian], p. 128 (2006).
106 Malamatov, A. H. Professor Thesis [in Russian], (Nalchik, KBSU), p. 296 (2006).
107 Borisov, V. A., Bedanokov, A. Yu., Karmokov, A. M., Mikitaev, A. K., Mikitaev, M. A., and Turaev, E. R. *Plastic masses* [in Russian], **5** (2007).

CHAPTER 4

METAL NANOPARTICLES IN POLYMERS

R. A. DVORIKOVA, YU. V. KORSHAK, L. N. NIKITIN, M. I. BUZIN,
V. A. SHANDITSEV, Z. S. KLEMENKOVA, A. L. RUSANOV,
A. R. KHOKHLOV, A. LAPPAS, and A. KOSTOPOULOU

CONTENTS

4.1 INTRODUCTION

The new magnetic nanomaterials have been synthesized from ferrocene-containing polyphenylenes. Cyclotrimerization of 1,1'-diacetylferrocene by condensation reaction catalyzed by p-toluenesulfonic acid (p-TSA) in the presence of triethyl orthoformate both in solution and supercritical carbon dioxide (SC-CO$_2$) in the temperature range of 70–200°C is described. Highly branched ferrocene-containing polyphenylenes prepared by this procedure were used as precursors for preparing magnetic nanomaterials. This was achieved by thermal treatment of polyphenylenes in the range of 200–750°C. The emerging of crystal magnetite nanoparticles of magnetite with the average size of 6 to 22 nm distributed in polyconjugated carbonized matrix was observed due to crosslinking and thermal degradation of polyphenylene prepolymers. Saturation magnetization of such materials came up to 32 Gs cm^3/g in a filed of 2.5 kOe.

In the last few years, the enhanced interest to nano-sized materials has been developed due to potentially unusual physical properties of those items compared to common substances. The magnetic properties of such substances are of great interest and presently under intense investigations [1-9].

The magnetically active materials are known to use for many purposes including data recording and information storage, for permanent magnets, in magnetic cooling systems, as magnetic sensors, and so on [1,2]. The precise drug delivery by using magnetically active nanoparticles is of current interest in biomedical field [10] as well as new probabilities may be created in the field of MR-imaging, cell sorting and cell separation, in bio-selection processes, for enzyme immobilizing, in immune analysis, in catalytic processes, and so on [11-14].

The stabilization of nano-sized magnetic particles is a key problem and may be solved by their incorporation in oligomer or polymer matrixes—one of the possible ways applicable for preparation of new magnetic materials [1, 2].

Authors developed a new approach for nano-sized composite preparation by thermal treatment of ferrocene-containing polymers (FP) with reactive terminal groups [15]. Thermal treatment of FP prepared from 1,1'-diacetylferrocene in the range of 150–350°C was accompanied by crosslinking and origination of iron-containing nano-sized particles displaying magnetic-ordering properties. The synthesis of highly-branched ferrocene-containing polyphenylenes was performed also by cyclotrimerization of 1,1'-diacetylferrocene in environmentally friendly solvent—liquid and SC-CO$_2$ [16]. This medium had been much used as a "green environment" for conducting of various chemical processes including synthesis and modification reactions of polymers [17-24]. The new approaches for creation of metal-polymeric systems and magneto-active nanocomposites, in particular, are of special interest [25-33].

It describes using of p-TSA or SiCl$_4$/C$_2$H$_5$OH mixture as catalysts for the reaction [16]. The reaction was performed in liquid and SC-CO$_2$ in the presence of triethyl orthoformate at pressure of 20 MPa and temperature of 20°C and 50°C with the polymer yield around 20%. The DTG-analysis revealed 5% weight loss of the polymer at 400°C with the weight of carbonized residue of 80% after heating at 750°C. When heated at 300°C, the process of crosslinking in the samples occurred and crystalline iron-containing nano-sized magnetic particles with the average size of 10–40 nm with saturation magnetization of 13 Gs cm^3/g in a magnetic field of 2.5 kOe were observed in a polymer matrix. The study of cyclotrimerization reaction of 1,1'-diacetylferrocene has been carried out in solution and in SC-CO$_2$ in a wide temperature range for obtaining higher polymer yields. The controlled thermal treatment of prepared ferrocene-containing polyphenylenes has been carried out for attaining improved magnetic properties.

4.2 EXPERIMENTAL DETAILS

The FP were prepared either by conventional solution method [15] or in SC-CO$_2$ (FPSC) using technique described in [16] at 20 MPa and temperatures varied from 70 to 200 °C.

Conventional polymerization of the monomer was carried out in a flat-bottom flask supplied with a thermometer and magnetic stirrer. The 1.0 g (3.7 mmol) of 1,1'-diacetylferrocene, 3 ml (8.0 mmol) of triethyl orthoformate, and 0.10 g p-TSA were placed into the flask and allowed to react at 70°C during 2.5 hr and then left for 40 hr at room temperature. The dark brown precipitate was filtered, washed with water to neutral reaction, then with ethanol and dried in vacuum. The yield was 0.34 g (24% from theory). Increasing of the temperature up to 140°C raised yields to 71%. The polymerization in SC-CO$_2$ was performed in a high pressure reactor with the inner capacity of 10 ml. The reactor was flushed out with CO$_2$ after loading of the reagents and heated up to the required temperature ($\pm$ 0.5°C). The input of CO$_2$ and applying of the required pressure (20 MPa) was completed by hand-operating press ("High Pressure Equipment", USA). The reaction mixture was agitated with magnetic stirrer and the reaction was carried out from 2.5 to 5 hr. After completing of the reaction the autoclave was cooled down and the pressure released. The resulting polymer was washed successively with ethanol and water to pH = 6–7 and then dried. The maximum yield was almost quantitative.

Magnetoactive materials were prepared by heating of the original samples of FP and FPSC in quartz tubes placed into measurement cell of magnetometer in the range of 200–750°C, or by thermal treatment of the samples in glass tubes under argon in the range of 250–500°C. The appearance of the magnetic order during thermal treatment of the polymers was monitored by vibration magnetometer of the Foner type.

The morphology of the nanocomposites was studied by transmission electron microscopy TEM using LEO 912AB OMEGA instrument. The size distribution of nanoparticles was figured out by statistical treatment of 50–100 particle size.

The X-ray powder diffraction was obtained with Rigaku D/MAX-2000H rotating anode diffractometer with CuKα radiation and secondary pyrolytic graphite monochromator. Direct Current (DC) magnetic susceptibility measurements were carried out with an Oxford Instruments MagLab EXA extraction magnetometer. Thermogravimetric (TG) analysis was carried out with "DERIVATOGRAPH-C" (MOM, Hungary) instrument in air and argon by using samples of ~15 mg at a heating rate 5°C/min.

4.3 DISCUSSION AND RESULTS

The cyclotrimerization reaction of 1,1'-diacetylferrocene catalyzed by p-TSA was studied for determining the optimal conditions for higher yields of polyphenylenes as precursors for materials with improved magnetic properties.

The FP were synthesized both under conventional conditions and in SC–CO$_2$, p-TSA was used as a catalyst for cyclotrimerization of 1,1'-diacetylferrocene. The reaction proceeded in accordance with the Scheme 1 [16] which did not consider side reactions that could bring about some defect units of dypnone (-methylchalcone) structures [34].

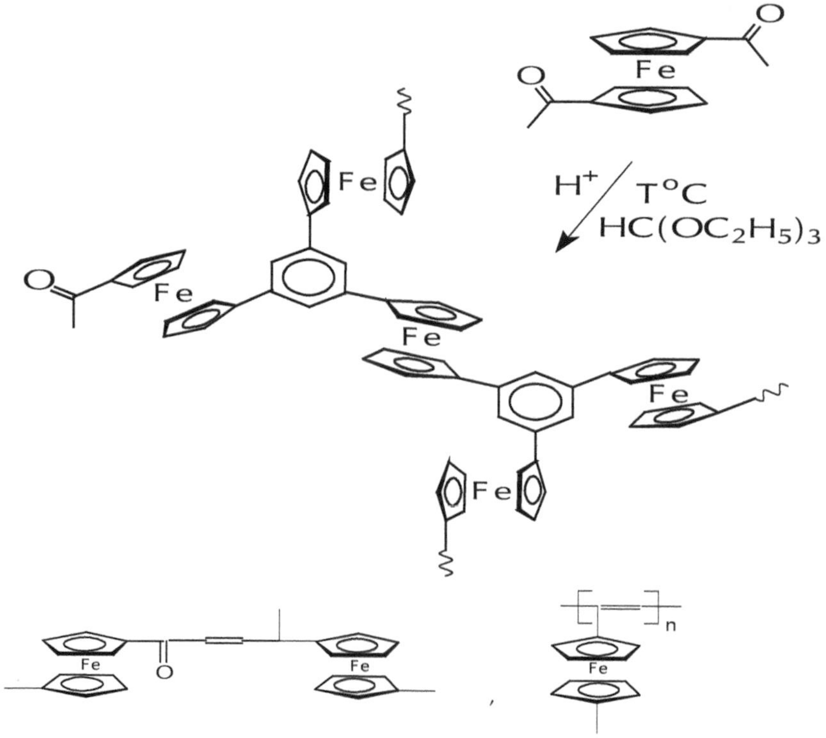

SCHEME 1.

The effect of reaction conditions, such as temperature, duration, catalyst concentration, and post-stirring period at ambient temperature on yield and properties of the final products was studied. The Tables 1 and 2 present some data on the reaction conditions and some properties of the polyphenylenes prepared under conventional conditions (FP) and in SC–CO$_2$. The reaction was carried out in the temperature range of 70–200°C in the presence of triethyl orthoformate, which acts simultaneously as a catalyzing agent and as a solvent. Besides, ditolyl methane was used in some experiments as a solvent. The resulting ferrocene-containing polyphenylenes were obtained as powders of dark-brown color partially soluble in organic solvents such as dioxane, methylene chloride, and benzene.

TABLE 1 Synthesis of FP with p-TSA as a catalyst and specific magnetization values of the

Sample	Temperature, °C	Duration/ Postreaction time at ambient temperature, hrs	Catalyst, %	Yield, %	Elemental analysis, %*				σ_H,max, G·cm³/g (at temperature)
					C	H	Fe	S	
FP-1	70	2.5/40	10	24	66.51	4.86	20.69	0.80	9(500°C)
FP-2	70	5/20	10	25	65.64	4.91	20.02	0.62	10(500°C)
FP-3	100	2	10	16	64.43	4.56	12.83	2.11	21(750°C)
FP-4	100	2	20	24	61.51	4.34	13.10	3.18	21(500°C)
FP-5	100	2/15	10	45	68.81	5.45	18.55	-	12(500°C)
FP-6	100	2/20	10	66	62.85	5.05	19.37	0.65	26(500°C)
FP-7**	110	2/40	10	18	64.22	4.21	14.88	1.65	14(500°C)
FP-8	120	2	10	19	64.95	4.73	12.81	2.11	18(1000°C)
FP-9**	130	2/20	10	19	50.14	4.15	26.67	2.00	21(750°C)
FP-10	140	2/20	20	71	67.32	5.51	20.2	-	32(750°C)

*Calculated for $C_{126}H_{102}Fe_9O_6$: C – 66.88%, H – 4.54%, Fe - 22.21%

TABLE 2 Preparation of FPSC polymers in supercritical CO_2 and specific magnetization values of the samples after heat treatment at 500°C

Sample	Temperature, °C	Duration/Postreaction time at ambient temperature, Hrs	Catalyst, %	Yield, %	Elemental analysis, %				Magnetization $(\sigma_{H'max})$ G·cm³/g,
					C	H	Fe	S	
FPSC-1	70	2.5/20	10	32	67.43	5.17	20.76	0.25	6.0
FPSC-2	140	2/20	10	31	58.98	4.40	16.47	2.73	15.0
FPSC-3	140	2/20	60	98	53.67	4.68	23,36	0	12.5
FPSC-4	160	2/20	10	37	62.09	4.44	16.04	2.01	12.5
FPSC-5	200	2	10	92	63.43	4.95	16.81	0.80	14.5

**Reaction was carried out in ditolylmethane; Molar ratio TEOF/Acetyl group was 1.2

The yields of polymers increased at 100°C from 24 to 66%. Elemental analysis showed good agreement between found and calculated data and, in addition, sulfur was found in the samples in ~2% amount.

The presence of the latter may be due to the chemical interaction between polymer and p-TSA that may also explain the reduced content of iron and carbon in analysis data. The maximum yield of FP was achieved at 100°C and 10% concentration of a catalyst, whereas 20% catalyst concentration provided the highest yields at 140°C.

The increase of the reaction time at 100°C resulted in an enhancing yield from 16 to 66% (Table 1). The same effect was observed when the catalyst concentration was raised up from 10 to 20%.

Preparation of FP in SC-CO$_2$ medium (FPSC) was of special interest. Some data are given in Table 2. Thus, the increase in reaction temperature from 70 to 200°C rose up the polymer yield from 10 to 90%, whereas the increasing the catalyst concentration from 10 to 60% brought about the quantitative yield of polyphenylenes.

The IR-spectra of FP revealed the band at 1600 cм$^{-1}$ typical for stretching modes of CH-groups in 1,3,5-substituted benzene rings (Figure 1(b)), while this band was observed at 1572 cm^{-1} for FPSC (Figure 1(d)). The bands typical for ferrocenyl units were also found IR-spectra of the polymers (stretching modes of CH-groups at 3090 cм$^{-1}$ non-planar bending vibrations of CH-groups in C$_p$ rings at 820–830 cм$^{-1}$, and the bands for doubly degenerated antisymmetric stretching modes Fe-C$_p$ at 484 cм$^{-1}$).

Strong bands at 1275 cm^{-1} are due to symmetric stretching modes of C–C in disubstituted C$_p$ rings and intensive bands at 1670 cm^{-1} and 1700 cm^{-1} may be referred to C=O stretching vibrations in dypnone fragments and terminal acetyl groups. The IR-spectrum of model 1,3,5-triferrocenyl benzene shows the same characteristic bands typical for the polymer samples (Figure 1(a)) except visible broadening of the latter due to higher molecular weight.

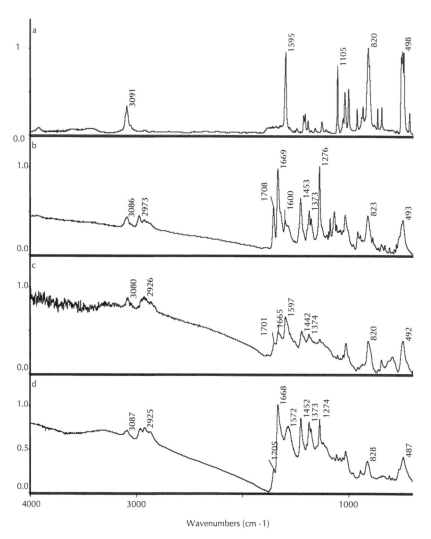

FIGURE 1 The IR-spectra of: (a) 1,3,5-triferrocenylbenzene; (b) FP-10 (table 1), (c) FP-10 after heating at 3500C for 1 hr in measurement cell of magnetometer, and (d) FPSC-5 (table 2).

The amount of the β-methylchalcone fragments and acetyl end groups in polymers was found to reduce after heating samples at 250°C and higher temperatures as evidenced by decreasing intensities of the bands at 1600 cm^{-1} and 1700 cm^{-1}. Simultaneous increase of the regular units in polymer molecules was found due to presence of 1,3,5-substituted benzene rings as evidenced by increasing of the band at 1600 cm^{-1} specific for those groups (Figure 1(c)). At the same time, network formation occurred and arising of the crystal iron-containing particles was observed. The mechanism of the particle formation comprises their spontaneous adjustment at the nano-sized level depending on the polymer structure. Magnetic behavior of the final products is mark-

edly influenced both by temperature regimes during ferrocene-containing polyphenylenes synthesis and subsequent thermal treatment. Thus, formation of magnetic phase
in sample FP-6 (Figure 2) prepared by conventional method at 100°C (FP) begins
after 200°C and the maximum magnetization is equal to 26 Gs cm^3/g. Meanwhile, the
sample FP-10 that was synthesized at 140°C showed the arising of magnetic phase
after heating at 500°C, and at 750°C the magnetization approached to its maximum
value of 32 Gs cm^3/g (Figure 3).

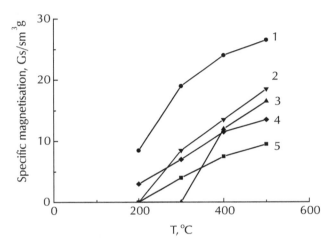

FIGURE 2   Magnetization curves for FPs obtained at different temperature: (1) FP-6, (2) FP-9,
(3) FP-7, (4) FP-8, and (5) FP-1.

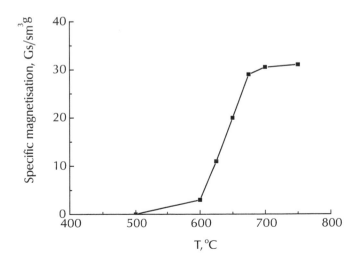

FIGURE 3   Magnetization curve for FP-10.

Figure 4 presents the plot of magnetization values versus heat treatment temperature for ferrocene-containing polyphenylenes synthesized in the range of 70–200°C
in SC-CO$_2$. For such samples formation of magnetic phase stats at 200°C for samples
FPSC-2 and FPSC-4 prepared at 140 and 160°C, respectively. The highest magnetization (~15 Gs cm^3/g) was achieved for samples FPSC-2, FPSC-4, and FPSC-5, which
were synthesized at 140°C, 160°C, and 200°C, respectively (Table 2). However, FPSC-
5 shows magnetic behavior just at 300°C. The lowest magnetization value after heat
treatment was recorded for sample FPSC-1 originally prepared at 70°C. Besides, the
samples synthesized in SC-CO$_2$ medium had the reduced magnetic properties compared to FP that could be explained by their higher thermostability.

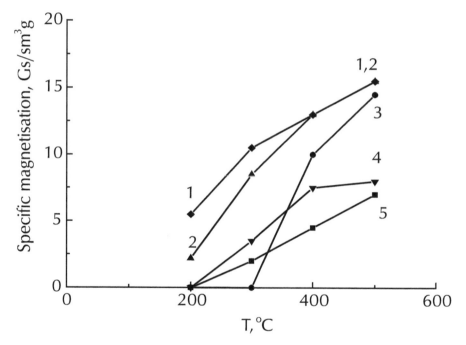

FIGURE 4   Magnetization curves for FPSCs obtained at different temperatures: (1) FPSC-2,
(2) FPSC-4, (3) FPSC-5, (4) FPSC-3, (5) FPSC-1.

The TEM study of thermally treated ferrocene-containing polyphenylenes revealed in all samples presence of iron nanoparticles homogeneously distributed in
polymer matrix with the average size of 6 to 22 nm depending on the synthesis conditions and heating regime. It should be mentioned that by increasing temperature of
thermal treatment from 250 to 500°C one could increase the size of nanoparticles from
8 to 22 nm for the samples obtained in the presence of p-TSA as a catalyst. The TEM
images are shown in Figure 5(a) and 5(b) for samples FPSC-5 and FP-8 subjected to
heating at 250 and 350°C, respectively, in argon for 2 hr. The size of nanoparticles
increased when the temperature of heat treatment rose up. Figure 5(c) shows TEM im-

age revealing nanoparticles with the average size of 22 nm for FP-6 sample after heating at 500°C in argon for 1 hr. In the meantime, the particle size in a sample prepared with 20% of a catalyst and treated at the same temperature of 500°C was half as much and equal to 10.6 nm (Figure 6(a)) and the subsequent heating up at 700°C resulted in reducing to 6.3 nm (Table 3, Figure 6(b)).

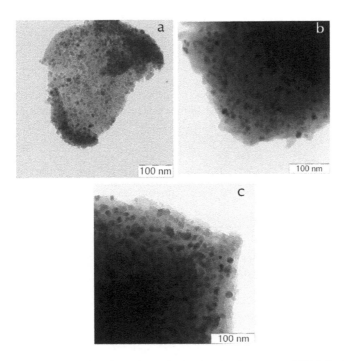

FIGURE 5 Microphotographs of: (a) FPSC-5 after heating in argon at 250°C for 2 hr, average size of nanoparticles—8 nm, (b) FP-8 after heating in argon at 350°C for 2 hr, average size of nanoparticles—13 nm, and (c) FP-6 after heating at 500°C for 1 hr (heated in measurement cell of magnetometer), average size of nanoparticles—22 nm.

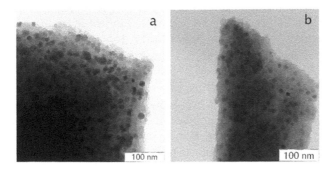

FIGURE 6 *(Continued)*

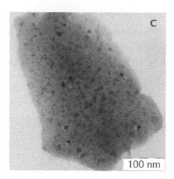

FIGURE 6   Microphotographs for FP-6 (a) and FP-10 (b, c), heated in measurement cell of magnetometer for 1 hr at 500 (a, b) and 700°C (c). The average size of particles—(a) 22, (b) 10.63, and (c) 6.32 nm.

TABLE 3   Size distribution of magnetic nanoparticles as a function of heat treatment temperature

Heat trearment temperature, °C	500	600	650	675	700
Average nanoparticle size, nm	10.63	9.21	6.96	6.66	6.32
Maximum size of nanoparticles, nm	13.63	15.21	25.23	22.07	20.56
Minimum size of nanoparticles, nm	8.55	4.17	2.35	2.09	2.14

Table 3 presents the data on nanoparticle size distribution for FP-7 sample after heat treatment at 500°C, 600°C, 650°C, 675°C, and 700°C. It is evident that increasing of temperature brought about the visible effect on the decreasing of the mean size of nanoparticles. Moreover, the particle size distribution for the sample was rather narrow when it was kept at 500°C compared to higher temperatures.

Thermal and thermo-oxidative stability of synthesized FP was studied by TGA method. It was found that thermal and thermo-oxidative degradation followed the basic features of such processes. Thus, the higher degree of decomposition in air was observed for all investigated polymers compared to that in an inert atmosphere. As an example, Figure 7 shows TGA curves for FP-10 sample. The weight loss of the sample in air was about 70% whereas in inert atmosphere the sample loses essentially less from the initial weight (about 30%). The weight loss, both in air and in argon was practically coinciding and begins at the same temperature (190°C). Then, the thermal decomposition under heating in air was slowed down, and the sample starts to lose weight intensively only in the vicinity of 400°C. Such a behavior can be attributed to a formation of the dense network in the polymeric matrix at the initial stage of the thermo-oxidative degradation that prevents thermal degradation for a while. However, the peroxides formed during the polymer oxidation in air may form redox systems together with ferrous species in ferrocene units to yield additional amounts of free radicals that finally stimulate a deeper degradation of the polymer. On the contrary, in an inert atmosphere the sample slowly loses weight in a wide temperature interval, up to 700°C and then its weight does not change.

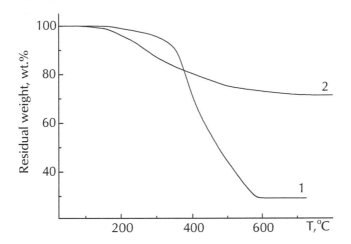

FIGURE 7   The TGA curves for FP-10 in air (1) and in argon (2) at a heating rate 5°C/min.

The most significant observation concerns the comparison of the XRD spectra for the samples prepared by conventional way and in SC-CO$_2$ (FP-6 and FPSK-5) and heated at 250°C, and the models shown in Figure 8(c). We may conclude from Figure 8(a)–8(c) that the only crystalline phase presents in the samples if that of magnetite Fe$_3$O$_4$ and there is an excellent match between the experimental data and Fe$_3$O$_4$ model.

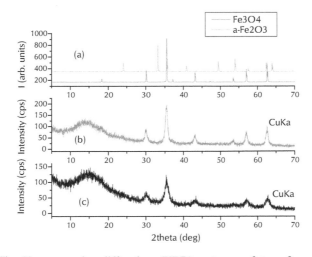

FIGURE 8   The X-ray powder diffraction (XRD) patterns of two ferrocene-containing polyphenylene samples FP-6 (b), and FPSC-5 (c) after heating at 250°C, and the model spectra of two iron-based oxides (a).

    In addition to this, crystalline phase there is a significant volume fraction arising from an amorphous phase, which is identified by the diffuse scattering (broad hump) in the 2Θ range 10–20 degrees. Based on the excess broadening of the Bragg reflections of Fe_3O_4, in sample FPSC-5, we suggest that the magnetite is a little less crystalline than that present in FP-6. Probably this is due to the different conditions of the synthesis for this material, since the sample was prepared in $SC\text{-}CO_2$ medium.

    To this extent the following results, on the magnetic properties of these samples, are mainly dictated by the behavior of the magnetite which is magnetic (ferromagnetic) already at room temperature (Figure 9). Although the measured magnetization, at $T \sim 280$ K, is characteristic of a soft ferromagnetic-like material, with little coercivity (H_c a few Gauss), there are a some qualitative differences (for example steep change of M versus H for FP-6) between the curves of the two materials, probably due to the higher sample crystallinity (FP-6) or/and larger particle size. As far as the magnetic susceptibility is concerned (Figure 10), it find an important reduction of the magnetic moment at relatively low temperature ($T < T_B$) for both samples. Furthermore, FP-6 displays a second, higher "blocking" temperature ($T_{B2} \sim 100$ K) which may be related to the Verwey transition (~118 K) met in bulk magnetite—shifted to relatively lower-T here maybe due to smaller particle-size.

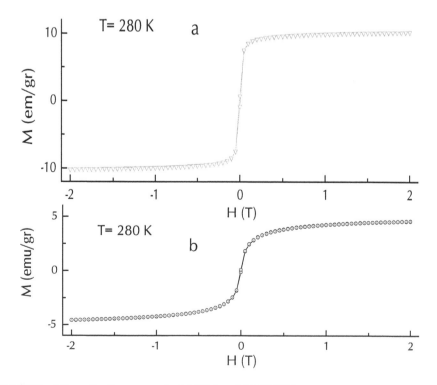

FIGURE 9 Magnetization curves for (a) FP-6 and (b) FPSC-5 at room temperature.

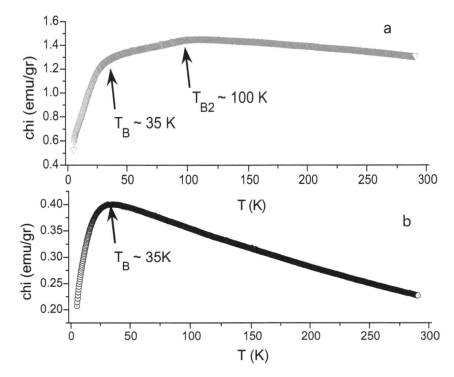

FIGURE 10   Temperature dependence of the zero-field cooled (ZFC) magnetic susceptibility of (a) FP-6 and (b) FPSK-5. The arrows indicate various "blocking" temperatures (T_B) in these samples.

4.4   CONCLUSION

The optimal conditions for synthesis of highly branched ferrocene-containing poly-phenylenes were developed *via* cyclotrimerization of 1,1'-diacetylferrocene by condensation reaction in the range of 70–140°C providing 71% yield of the polymer. The quantitative yields of the polymers were obtained in SC-CO$_2$.

Magnetic nanocomposites with saturation magnetization of 32 Gs cm^3/g in a field of 2.5 kOe were prepared by thermal treatment of such polymers in the range of 250–750°C. The TGA data approved that initial temperature of sample degradation fit in with magnetic phase formation. XRD studies and magnetic measurements showed that crystalline phase of magnetite Fe$_3$O$_4$ presents in the samples with magnetization characteristic of a soft ferromagnetic-like material with small coercivity.

The TEM study of polymer samples approved that the mean size of magnetic nanoparticles in polymer matrixes may be controlled by varying reaction conditions and heat treatment regime of the final polymers.

KEYWORDS

- **Cyclotrimerization**
- **1,1'-Diacetylferrocene**
- **Ferrocene-containing polymers**
- **Magnetoactive materials**
- **Polyphenylenes**

REFERENCES

1. Pomogaylo, A. D., Rosenberg, A. S., and Uflyand, A. S. *Metal Nanoparticals in Polymers*. M., Khimiya, p. 672 (2000).
2. Gubin, S. P., Koksharov, Yu. A., Khomutov, G. B., and Yurkov, G. Yu. *Russ. Chem. Revs.*, **74**(6), 489–520 (2005).
3. Gudoshnikov, S., Liubimov, B., Matveets, L., Ranchinski, M., Usov, N., Gubin, S., Yurkov, G, Snigirev, O., and Volkov, I. *Journal of Magnetism and Magnetic Materials.*,**258–259**, 54–56 (2003).
4. Baker, C., Ismat Shah, S, and Hasanain, S. K. *Journal of Magnetism and Magnetic Materials.*, **280**, №2-3, 412–418 (2004).
5. Kechrakos, D. and Trohidou, K. N. *Applied Surface Science.*, №226, pp. 261–264 (2004).
6. Tackett, R., Sudakar, C. Naik, R., Lawes, G., Rablau, C., and Vaishnava, P. P. *Journal of Magnetism and Magnetic Materials.*, **320**, №21, 2755–2759 (2008).
7. Hayashi, K., Sakamoto, W., and Yogo, T. *Journal of Magnetism and Magnetic Materials.*, **321**, №5, March, 450–457 (2009).
8. Zhang, G., Potzger, K., Zhou, S., Mücklich A, and Yicong Ma, J. *Nuclear Instruments and Methods in Physics Research Section B: Beam Interactions with Materials and Atoms.*, **267**, №8-9, 1596–1599 (2009).
9. Hasegawa, D., Yang, H., Ogawa, T., and Takahashi, M. *Journal of Magnetism and Magnetic Materials.*, **321**, №7, 746–749 (2009).
10. Denkbas, E. B., Kilicay, E., Birlikseven, C., and Ozturk, E. *Reactive and Functional Polymers.*, **50**, 225–232 (2002).
11. Hyeon, T. *Chem. Commun.*, pp. 927–934 (2003).
12. Jun, Y., Choi, J., and Cheon, J. *Chem. Commun.*, pp. 1203–1214 (2007).
13. Kawamura, M., and Sato, K. *Chem. Commun.*,pp. 3404–3405 (2007).
14. Muller, J. L., Klankermayer, J., and Leitner, W. *Chem. Commun.*, pp. 1939–1941 (2007).
15. Dvorikova, R. A., Antipov, B. G., Klemenkova, Z. S., Shanditsev, V. A., Prokof'ev, A. I., Petrovskii, P. V., Rusanov A. L., and Korshak, Yu. V. *Polymer Science, Ser.A.*, **47**, № 11, 1135-1140 (2005).
16. Dvorikova, R. A., Nikitin, L. N., Korshak, Yu. V., Shanditsev, V. A., Rusanov, A. L., Abramchuk, S. S., and Khokhlov, A. R. *Doklady Akademii Nauk.*, **422**, №3, 334–338 (2008).
17. Cooper, A. I. *J. Mater. Chem.* **10**, 207–234 (2000).
18. Nalawade, S. P., Picchioni, F., and Janssen L. P. B. M. *Prog. Polym. Sci.*, **31**, 19–43 (2006).
19. Reverchon, E. and Adami, R. *J. of Supercritical Fluids.* **37**, 1–22 (2006).
20. Zhang, Y. and Erkey, C. *J. of Supercritical Fluids.*, **38**, 252–267 (2006).
21. Erkey, C. *J. of Supercritical Fluids.*, **47**, 517–522 (2009).

22. Yang, J., Hasell, T., Wang, W., and Howdle, S. M. *European Polymer Journal.* **44**, 1331–1336 (2008).
23. Said-Galiyev, Ernest E., Vygodskii, Yakov S., Nikitin, Lev N., Vinokur, Rostislav A., Gallyamov, Marat O., Pototskaya, Inna V., Kireev, Vyacheslav V., Khokhlov, Alexei R., and Schaumburg, Kjeld. *J. of Supercritical Fluids.*, **27**, 121–130 (2003).
24. Nikitin, Lev N., Gallyamov, Marat O., Vinokur, Rostislav A., Nikolaev, Alexander Yu., Said-Galiyev, Ernest E., Khokhlov, Alexei R., Jespersen, Henrik T., and Schaumburg, Kjeld. *J. of Supercritical Fluids.*, **265**, 263–273 (2003).
25. Yuvaraj, H., Woo, M. H., Park, E. J., Jeong, Y. T., and Lim, K. T. *European Polymer Journal.*, **44**, 637–644 (2008).
26. Chen, A. -Z., Kang, Y. -Q., Pu, X. -M., Yin, G. -F., Li, Y., and Hu, J. -Y. Development of Fe_3O_4-poly(l-lactide) magnetic microparticles in supercritical CO_2. *Journal of Colloid and Interface Science.*, **330**, 317–322 (2009).
27. Tsang, S. C., Yu, C. H., Gao, X., and Tam, K. Y. Preparation of nanomagnetic absorbent for partition coefficient measurement. *International Journal of Pharmaceutics.* **327**, 139–144 (2006).
28. Said-Galiyev, E., Nikitin, L., Vinokur, R., Gallyamov, M., Kurykin, M., Petrova, O., Lokshin, B., Volkov, I., Khokhlov, A., and Schaumburg, K. New chelate complexes of copper and iron: synthesis and impregnation into a polymer matrix from solution in supercritical carbon dioxide. *Industrial and Engineering Chem. Research.*, **39**, 4891–4896 (2000).
29. Blackburn, J. M., Long, D. L., Cabanas, A., and Watkins, J. J. Deposition of Conformal Copper and Nickel Films from Supercritical Carbon Dioxide. *Science..* **294**, 141–145 (2001).
30. Vasilkov, A., Naumkin, A., Nikitin, L. Volkov, I., Podshibikhin, V., and Lisichkin, G. Ultrahigh molecular weight polyethylene modified with silver nanoparticles prepared by metal-vapour synthesis. *AIP Conference Proceedings.*, **1042**, 255–257 (2008).
31. Nikitin, L., Vasilkov, A., Vopilov, Yu., Buzin, M., Abramchuk, S., Bouznik, V., and Khokhlov A. Making of metal-polymeric composites. *AIP Conference Proceedings.*, **1042**, 249–251 (2008).
32. Nikitin, L. N., Vasilkov, A. Yu., Naumkin, A. V., Khokhlov, A. R., and Bouznik, V. M. Metal-polymeric composites prepared by supercritical carbon dioxide treatment and metal-vapor synthesis in: Success in Chemistry and Biochemistry: *Mind's Flight in Time and Space*, Volume 4 (A Festschrift in Honor of the 75th Birthday of Professor Gennady E. Zaikov), G. E. Zaikov (Ed.), Nova Science Publishers, Inc. New York, pp. 57–590 (2009).
33. Dvorikova, R. A., Nikitin, L. N., Korshak, Yu. V., Shanditsev, V. A., Rusanov, A. L., Abramchuk, S. S., and Khokhlov, A. R.New magnetic nanomaterials of hyperbranched ferrocenecontaining polyphenylenes prepared in liquid and supercritical carbon dioxide in: Quantitative Foundation of Chemical Reactions, G. E. Zaikov (N.M. Emanuel Institute of Biochemical Physics, Russian Academy of Sciences, Moscow, Russia), Tanislaw Grzegosz (Kaminski Institute of Natural Fibres, Poland), and Lev N. Nikitin (A.N.Nesmeyanov Institute of Organoelement Compounds, Russia), Nova Science Publishers, Inc. New-Y., P. 93–100 (2009).
34. Sasaki, Yu. and Pittman, Ch. U., Jr. Acid-Catalyzed Reaction of Acetylferrocene with Trietyl Orthoformate. *J.Org.Chem.*, **38**(21), 3723–3726 (1973).
35. Seyoum, H. M., Bennet, L. H., and Della Torre, E. /Temporal and temperature variations of dc magnetic aftereffect measurements of Fe_3O_4 powders, *J Appl Phys.*, **5**, 2820–2822 (2003).

STABILITY OF POLYMERIC PRODUCTS TO OXIDIZING AND HYDROLYTIC DESTRUCTIONS

S. D. RAZUMOVSKII and G. E. ZAIKOV

CONTENTS

5.1 INTRODUCTION

The stability of polymeric products to an oxidizing and hydrolysis is one of the important factors describing their weather fastness [1]. However, recently oxidizing and hydrolytic destructions began to apply for surface modification of materials with the purpose of obtaining products with the improved operation properties [2,3]. In this case process of destruction is necessary to stop at the reliable stage so that there was no noticeable change of the complex of positive properties of a polymeric product. All this requires careful learning of kinetic and diffusive regularities having place at degradation, with the purpose of improvement of properties of a surface of a product without deterioration of their bulk properties.

In many cases the surface properties of such products appreciably determine their operation properties, therefore increasing and greater attention is given to surface modification of polymeric products, which alongside with a radiation and chemical grafting can be carried out and with the help of chemical destruction. Certainly, it would be better, if modification by destruction occurred on those molecules of a product, which really are in a surface layer (on a depth 10–15 Å). However actually reaction proceeds not only on a surface, but also in some undersurface bed, which size depends on a relation between speed of diffusion of aggressive substance (oxygen, ozone, water, solutions of salts, basis, and acids) and speed of decomposition of polymeric molecules under action of that substances. Let us consider now separately hydrolytic and oxidizing destruction.

5.2 HYDROLYTIC DEGRADATION

The polymers what have an ability for hydrolytic destruction usually have a heteroatoms in main or in a side chain.

The destruction of polymeric products in the aggressive media proceeds through the following main stages:
- Adsorption of the aggressive media to surface of a polymeric product;
- Diffusion of the aggressive media in volume of a polymeric product;
- Chemical reaction of the aggressive media with chemically nonresistant links of polymer and desorption of destruction products from a surface of product.

Generally, dependence of speed of decay chemically of nonresistant links in polymer from concentration of the aggressive environment can be described by the following equation:

$$w = dc_n/dt = k(c_n^0 - c_n)c_s c_c \qquad (1)$$

where c_n^0-initial concentration chemically of nonresistant links in polymer, c_n - concentration of the breaking links, c_c-concentration of the catalytic agent, and c_s-solvent concentration (for example, H_2O) in polymer.

If the solvent is spent during decay chemically of nonresistant links in polymer, its concentration can be found from the equation:

$$\partial c_s/\partial t = D_s X^2 c_s - k(c_n^0 - c_s) c_s c_c \qquad (2)$$

where D_s – diffusivity coefficient of a solvent, X-Laplace operator. The problem of rate determination of decay chemically of nonresistant links in polymer, thus, is consisted in a joint solution of the Equations (1) and (2). We are interested with a case, when the speed of diffusion of the aggressive media is much lower than speed of chemical reaction. Then, the process of destruction happens in some thin reactionary surface layer or, as it is accepted to speak, on a surface of a polymeric product, which is in external diffusive-kinetic area. For this case the speed of breaking up of hydrolytically nonresistant links is featured by the equation:

$$Dc_n^s/dt = k(c_n^0 - c_n^s)c_c^s c_s^s \qquad (3)$$

where the character s designates values of concentrations in surface layer. It is uneasy to show [4], that is:

$$c_n^s = k_{ef}^s c_c^v c_s^v$$

where the index v corresponds to volume concentration. We shall consider, how the mass of polymer will vary during destruction in this case. Irrespective of a type of decay of polymeric molecules it is possible to write expression for a mass change of formed products:

$$m = k_{ef}^s c_c^v c_s^v tS \qquad (4)$$

where S—area of surface of a polymeric product. If it represents a film, the change of a mass of polymer will be featured by the following Equation [4]:

$$m_n = m_n^0(1 - t/\tau) = m_n^0(1 - t/\tau) = m_n^0(1 - k_{ef}^s c_c^v c_s^v tS/m_n^0) \qquad (5)$$

where, τ—time of complete decomposition of a film, m_n^0—mass of polymer before decomposition.

At carrying out of destruction in external diffusive-kinetic area it is necessary to take into account, that the polymeric molecules in surface layers have, as a rule, properties, which are different from bulk molecules of a sample (molecular weight, degree of crystallinity, orientation etc.). By and large, in micro volumes of polymer with various structures about chemically nonresistant links there can be a various concentration of components of the aggressive media (c_c, c_s), that will lead to various speeds of chemical destruction. All function groups in polymer therefore can conditionally be divided on "accessible" and "unavailable", and the degree of availability is unequal for different components of the media. As is experimentally shown, for polyethylenetherephtalate (PETP) [5,6] decay of a film in an alkaline solution has the following regularities: The reaction has an zero order on polymer; the molecular weight of a polymeric film does not vary down to the termination of the process; In IR-ATR spectrums there are not absorption bands of carboxyl groups forming at decay ether link; the value of an effective rate constant, calculated on the formula (5) varies depending on change of thermodynamic parameters of the media according to the equation:

$$(k_u a_{H2O} K_e/b_0 + k_u{}' a_{H2O})$$

$$k_{ef} = (1 + K_e{}' a_{H2O}/b_0 + K_e{}' a_{H2O}/b_0{}^2)$$

It describs hydrolysis of low-molecular weight compounds in water solutions of alkalis [2,3,5] (a_{H2O} - activity of water, b_0 - basicity of a solution, K_e and $K_e{}'$ - equilibrium constant of the process of the ionized forms formation, k_u and $k_u{}'$ – appropriate rate constants. All these factors allow to count [2-6] that the reactionary zone comes nearer to a monolayer. Microphotos of filaments PETP [2], obtained with the help of a raster microscope confirm this supposition
(Figure 1).

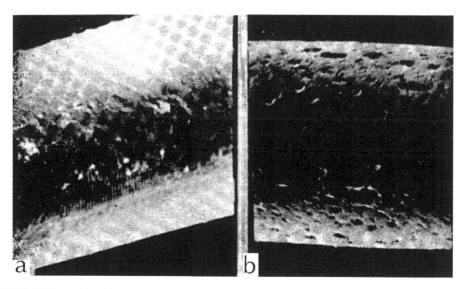

FIGURE 1   Microphotography of PETP-filaments: left – initial, right - after degradation 5 min in 22% KOH solution at 880 C, x3000 [3, 5].

5.3 MODIFICATION OF POLYETHERS BY WATER SOLUTIONS OF ALKALIS WITH THE PURPOSE OF OBTAINING A ROUGH SURFACE

The PE, PETP, polycarbonate (PC) and some other polymers are used as a material for synthetic paper pulp and related products. Recently intensive development of technology for production of rough films is carried out. It is used for production of a long-lived material for cards, draughts, copies, computers, video and sound recording instrumentation, and so on. raw. There are various ways of obtaining of grains of polymeric films. So, for example, they can be received by treatment of films by a corona discharge [7], by orientation [8,9] or chemical pickling of films [10,11], and also as a result of plotting a lacquer coat on its surface [12]. One of the most perspective methods is the method of pickling [13]. For finding optimal modes of pickling of films it is necessary to know the mechanism of this process, first of all reasons of derivation of a rough surface and dependence of the value of a grain on time, temperature, thermodynamic parameters of the media, and from structure of polymer.

5.4 POLYETHYLENTEREPHTALATE FILMS

In case of pickling a thick film PETP (Figure 2) during the process an index of a grain R_a (the average arithmetic deviation of microjaggings from a basic line of a profile) at first transits through a maxima and then remains practically invariable before complete degradation of a film.

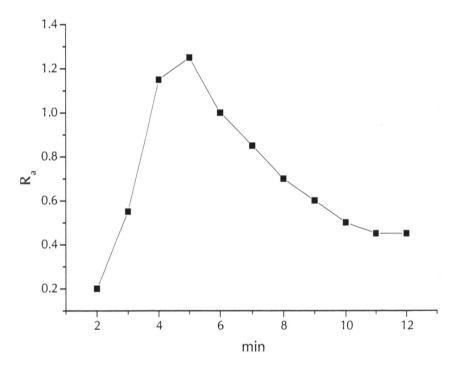

FIGURE 2 The dependence of grain (R_a) on time of pickling for a PETP-film [5]. Film thickness −550 μm, degree of crystallinity − 40%, [KOH] − 4, 9%, 108°C.

Thus the maxima of a grain corresponds to thickness of pickling layer in 25–30 microns. The maximal value of a grain R_a^{max} decreases at increase of degree of crystallinity (Figure 3) and does not depend on degree of orientation.

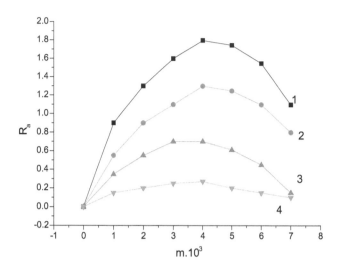

FIGURE 3   The dependence of grain (R_a) on the removed layer mass (m, moles) at different crystallinity %: 5(1); 30(2); 40(3); 50(4) [6].

According to one of existing explanations, the roughness on a surface at pickling will be derived at preferred destroying of an amorphous part, where the miscibility of water, alkalis, salts is rather great, and naked crystal chips form observed jaggings. Apparently, such mechanism of pickling can be carried out, but in a case PETP it is not observed, because the degree of a roughness of amorphous films exceeds a degree of a roughness of crystalline films. Besides it has appeared, that at pickling by the solutions of alkalis of films PETP, obtained in various conditions, but having close degree of crystallinity, one received various values R_a [3]. In this case origin of roughness most likely is stipulated by opening of micro-defects during pickling, which were formed in time of processing a polymer film from a melt [14]. According to modern imaginations, the surface layers of films have more friable structure, smaller density and greater amount of defects [2,3]. With increase of degree of crystallinity the common long of an amorphous phase decreases and happens "curing" of defects in it. All this results are obtaining less rough surfaces at pickling films of high degree of crystallinity. The greatest number of micro-defects is thus on depth 25–30 microns, as stipulates a maximal roughness at pickling the layer of such size.

The value of a roughness is defined only by width of the pickling layer and does not depends neither from temperature, nor from concentration of alkali [2,3,5,6]. Therefore, on thin films PETP (5–20 microns) one was failed to receive enough rough surface, as its, apparently, do not contain enough number of micro-defects. At the formulation of economically expedient and optimal technological modes it is necessary to take into account influence on speed of pickling of alkali concentration and temperature. According to the kinetic data, the change of a mass of a film can be written as [2]:

$$m = m_0(1 - k_{ef}tS)$$

where that m_0—mass of an initial film and S—surface of that film.

Outgoing from the known mechanism of destruction PETP [2-6] we shall receive expression for an effective rate constant:

$$k_{ef} = k_u{''}a_{H2O}/ 1 +K_e b_0$$

because $k_u{''} = Ae^{-E/RT}$, so

$$k_{ef} = Ae^{-E/RT}a_{H2O}/ 1 +K_e b_0$$

Therefore,

$$m = m_0(1 -Ae^{-E/RT}a_{H2O}tS / 1 +K_e b_0)$$

On a site up to maxima of a roughness of the curves of pickling of films:

$$\Delta m = \alpha R_a$$

where and α–ratio coefficient depending on a nature of a material.

From here,

$$R_a = Ae^{-E/RT}a_{H2O}tS / \alpha (1 +K_e b_0)$$

This equation shows dependence of a roughness from temperature, time, and thermodynamic parameters of the media. Inserting substituted values of parameters A, E, and K_e for PETP [2-6], we shall receive:

$$R_a = 6.4\times10^6 exp (-16\ 500/RT)\ a_{H2O}tS/ \alpha (1 + 4\times10^2 b_0)$$

Optimal by technological parameters of obtaining of a roughness on films PETP the following are: degree of crystallinity >10%, concentration KOH 46–49%, temperature 105–110°C [5, 6]. The graphics illustration of the optimal mode choice of obtaining a maximal roughness is given in a Figure 4.

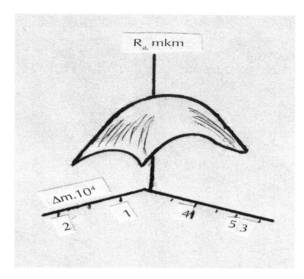

FIGURE 4   Graphic diagram for the election of optimal regime of grain PETP-film preparations (108°C) [2,6].

5.5   POLYCARBONATES FILM

At pickling polycarbonates films (PC) by alkaline solutions, as well as in a case PETP, on their surface will be derivate a rough relief [15, 16]. A source of origin of a roughness is the micro heterogeneities of submicroscopic sizes in structure of PC. Thus of a film of PC obtained by a methods of solution pouring (PCP) and extrusion (PCE) essentially differ. A PCP-film have more loose structure [16] and find out after pickling more advanced micro-relief, than PCE-film obtained by extrusion.

The attempt to change a size of micro-defects by swelling before pickling was made in [15]. Authors treated films by the swelling agent (water solutions of acetone) [17].

Most likely the following mechanism of effect of water solutions of acetone on structure of films of PC takes place [15]. In the beginning there is an intensive swelling (loosing of structure), what increase segmental mobility of macromolecules and therefore, creates conditions for their subsequent ordering in an even amorphous state. The subsequent decrease of a sample weight is bound with partial forcing out of swelling agent. At this time still there is not a crystallization, which testifies about pre-crystal ordering of polymeric chains in structure of amorphous PC, crystallization starts further.

Speed loosing of PC structure at swelling and speed of ordering of packing of chains depend on plastificator composition (in this case from a components ratio in mixture acetone - water). In pure acetone all three sequential stages transit very fast. At addition of water of speed of these processes are retarded, and each of stages becomes separated from others a major time interval.

According to modern imaginations [18], the structure of amorphous polymers is heterogeneous (in sense of presence in it the density fluctuations).

It is possible to suppose [15] that this heterogeneity amplifies in time of treatment films of PC by the plasticizing additives; due to before-crystal improvement of packing of polymeric chains the sizes of low dense areas increase. For this reason it is possible to expect the most developed micro-relief of a surface at pickling for films, swelled in the greatest measure. So, it also is used in practice. From the data Figure 5, it is visible, that with other things being equal major value of a roughness one manages to receive at pickling films beforehand treated by a plasticizing solution.

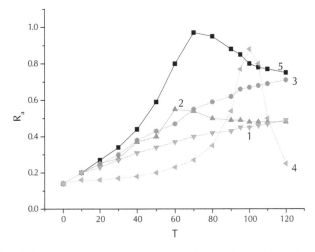

FIGURE 5 The difference in a roughness R_a at the pickling by 50% NaOH solutions at 50°C PC-films, treated by 80% acetone solutions [20]. Time of a treatment by acetone (T, min): 1-untreated 2–15; 3–30; 4–60; 5–120.

By the rather convenient taste for definition of optimal duration of processing of a film by solutions can be the obtaining of the time profiles of specific elongation at tearing up of polymer samples: the moment of appearance of a maxima on these curves corresponds to the moment of maximal swelling of a film [15]; the maximal swelling of a film provides obtaining the maximal value of a roughness after their pickling. In a number of cases of the task of practice require obtaining of enough major swelling on oriented polymeric films. The method of chemical pickling of a surface is now one of the most suitable for creation of a roughness and saving of positive qualities of a product, including oriented structures [19,20]. Outgoing from imaginations about the mechanism of derivation of a roughness on films of PC [15], and also about structural changes in polymer at elongation [21,22] it is possible to suppose [20] that as with propagation of orientation the defects of the molecular and super-molecular order in bulk of polymer "are healed" and decrease their effective size and amount. It means that the possibilities of influence of these micro-defects of polymer in a dimple deepness of a roughness after pickling oriented films are reduced. It is necessary for getting the reliable roughness of films to organize the directional action promoting development of disorder in structure of polymer, the simplest one of which is the process of reorientation of oriented films [20].

The regularities of PCE-films, having initial anisotropy, roughness formation, and changing were researched, especially a change of a roughness at pickling films oriented in a transverse direction up to various coefficients K of film elongation [23].

As shown in a Figure 6, on a surface modified thus of films it is possible under certain conditions to create a rough relief with $R_a = 0.5–1.0$ microns.

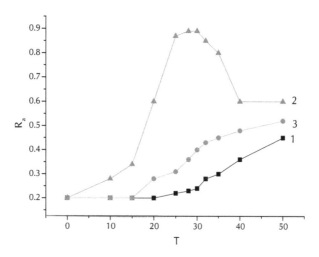

FIGURE 6   Changing of a roughness PC-films oriented across extrusion direction at the pickling. (Degree of elongation: 1 – 1, 5; 2 – 2; 3 – 2, 5; T – time of pickling, min.) [20].

Optimal value of the roughness was obtained here at K = 2.0. Friability and increase the structures defects was at K = 2.0 can be connected with intensive reorientation in structure of polymer [24]. The critical value K, at what there will be an intensive reorientation, in various films should depend on initial structure and anisotropy of a film [20].

At further elongation of a film up to values K noticeably major critical, again there is a compaction of structure of polymer and "healing" of defects. It carries on to depressing a level of a created roughness up to the value, characteristic for films oriented along a direction of tentative expansion during extrusion.

5.6   MODIFICATION OF A SURFACE OF A FILAMENT OF CELLULOSEDIACETATE BY WATER SOLUTIONS OF ALKALIS WITH THE PURPOSE OF DEPRESSING ELECTROSTATIC PROPERTIES (EP) AND DIRT ADHERENCE (DA)

The disadvantages synthetic filaments, including polyesters, are their hardness, strong EP and DA. The formation of charges of a static electricity in time of treatment creates technological and operation difficulties. So, for example, the charges on a surface of polymer promote attraction of a dust and other foreign impurities to end products, and the dirt adherence of clothes made of an electric fabric, can exceed at 300–500 times dirt adherence of clothes from a cotton fabric. The negative consequences of influence of a static electricity on an organism of the man are known also [25]. The neutralizing of charges of a static electricity people try to realize both physical, and chemical methods [2,3]. The introduction in polymers of antistatic agents reduces a static electricity, however, antistatic agents, as a rule, irritates a skin and mucous membrane of an eye and removed in time of using from polymer at washing.

The detailed study of one of chemical ways of neutralizing of a static electricity was carried out on an example of a filament, widely used in an industry, cellulose di acetate (DDA) [26,27]. The difficulty of chemical neutralizing of charges consists in modification a filament at saving the complex of its positive properties. It is well known, that the main amount of charges concentrated on a surface of polymers (though there is their some allocation on thickness). Therefore, main task consists in modification of the surface layer, change it for less electric. If in such surface layer of a DDA-filament to replace strongly electric acetate of groups by hydrolysis on hydrophilic hydroxyl groups, this layer converted to cellulose; it EP on some orders is lower. For such modification acceptable the water solutions of alkalis in concentrations what do not cause destruction of glycoside bonds in a basic chain of macromolecules.

On the Figure 7 is represented the typical S-shape curve of hydrolysis of a DDA-filaments.

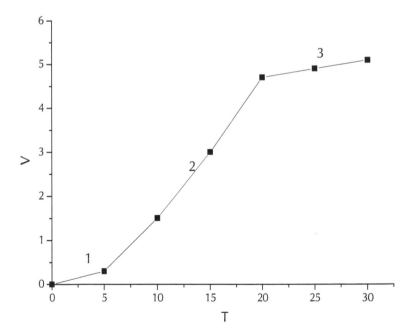

FIGURE 7 The kinetics curve of alkali demand in process of DDA-filaments degradation (T – min, 70°C, NaOH – 0,15% water solution, 1, 2, 3 – selected areas of curve) [27].

The concentration of the broken ether links in DDA c_n can be found from the equation:

$$\partial c_n/\partial t = k(c_n{}^0 - c_n)c_{NaOH},$$

where k—specific reaction rate of hydrolysis of ether links, $c_n{}^0$—initial concentration of ether links in polymer, c_{NaOH}—concentration of alkali in polymer, what it is possible to calculate from the equation:

$$\partial c_{NaOH}/\partial t = D_{NaOH}X^2 c_{NaOH} - k(c_n{}^0 - c_s)\, c_{NaOH,}$$

here D_{NaOH}—effective diffusion coefficient of alkali in volume of polymer.

This set of equations has no analytical solution, but can be solved with the help of numeric calculations. Also, these equations can be solved if to enter various assumptions and suppositions simplifying the reaction schema. From end site 3 of kinetic curves on a Figure 7 are possible to calculate an effective rate constant of hydrolysis of ether links k_{ef}, supposing, that on this site the filament was completely saturated with alkali and the reaction happens in kinetic area. The change of value k_{ef} as a function of temperature and concentrations of alkali is featured by the equation:

$$k_{ef} = 10^{10} \exp(-18000/RT)\, c_{NaOH}$$

In the whole task is reduced to finding conditions, at which the rate of hydrolysis would be much more than speed of diffusion, either by the another words to finding conditions of the most effective destruction process in the narrowest reactionary zone. Such conditions can be reached by two paths: 1) usage of alkali with cations of large volume, for example $(C_4H_9)_3NOH$, that is decrease D_{OH^-} at saving practically of invariable speed of hydrolysis; 2). usage of concentrated solutions of strong alkalis, what increase the rate of hydrolysis at constant speed of diffusion D_{OH^-} (as it is visible on electronic micro-photos of transversal sections of a DDA–filaments the width of regenerated cellulose layer at a conversion ~3% makes 0.5–0.6 micron (Figure 8).

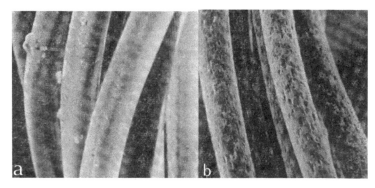

FIGURE 8   Electron microscope photography of DDA-filaments: initial (left) and modified (right) [5,27].

Figure 9 shows data for the dependence of electrical resistance of a surface of a fabric DDA from a degree of transformation of acetate groups (C) after treatment by alkalis with a various size of cations.

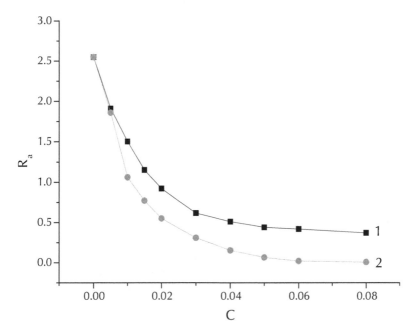

FIGURE 9   The dependence of surface DDA electrical resistance R_a on C for hydrolysis by (1) NaOH and (2) (*tret*-C_4H_9) NOH [28].

Table 1 contains the data indicating influence of a degree of transformation of acetate groups on the mechanic characteristics and electrical resistances of a DDA-filaments. From this data it is visible, that during the process of destruction the electrical resistance of a filament is considerably reduced at a rather small deterioration of the mechanic characteristics [29]. On the basis of the described above kinetic regularities it is possible to choose the optimal regime of modification a DDA-filaments by its hydrolysis.

TABLE 1   Influence of a degree of acetic group transformations in a DDA-filaments on the tensile characteristics and electrical resistance [29]

Degree of transformations	Tearing strength, g/mm²		Elongation, %		electrical resistance, q.10⁻¹¹, ohm
	warp	woof	warp	woof	
0	109.3	115.8	22.7	25.5	24
0.01	108.2	109.0	22.0	22.5	14
0.03	105.9	100.5	21.3	20.3	4
0.08	102.7	91.3	20.5	18.4	3

5.7   RESEARCH OF INFLUENCE OF ALKALINE TREATMENT ON PROPERTIES PETP FILAMENTS

The destruction of filaments PETP, as well as films, proceeds from a surface [4, 28-31]. At destruction of molecules in alkali the decay goes under the law of a case [2,3] with subsequent depolymerisation. It means, that on a surface of a filament always there will be a stationary value stationary concentration ester bonds c_n^0, and the losses of a mass of polymer will be proportional to an amount broken ester bonds:

$$-dm/dt = k_{ef}'S \tag{6}$$

where $k_{ef}' = k_{ef}c_n^0$, a S—surface of a filament.

For a filament $S = 2\pi rl$ and $m = \pi r^2 l\rho$, where r - average radius, l – length of filament, ρ – density PETP. Change of average radius of a filament about time it is possible to find from the following equation:

$$r = r^0 - 2k_{ef}'t/\rho \tag{7}$$

where r^0–initial average radius of a filament.

Integrating the equation (6) and taking into account the equation (7), we shall receive:

$$m_t = m_0[1 - 2k_{ef}'t/\rho r^0(1 - k_{ef}'t/\rho r^0)] \tag{8}$$

In this equation $k_{ef}'/\rho r^0$—inversely value of time of degradation process end. For the small conversions equation (8) can be simplified:

$$m_t = m_0(1 - 2k_{ef}'t/\rho r^0) \tag{9}$$

We shall consider, how varies k_{ef} if we rises alkali concentration in a solution. According to the mechanism of hydrolysis ester bonds of PETP [2,3] for a small degree of ionizations:

$$k_{ef} = k_u'' b_0 a_{H2O}/K_e \qquad (10)$$

Having substituted values of a number of parameters [5, 29] in the equation (10), we shall receive dependence of an effective constant of speed of hydrolysis of ether links on temperature and thermodynamic parameters of the media:

$$k_{ef} = 1.6 \times 10^2 \exp(-1650/RT) b_0 a_{H2O} \qquad (11)$$

At substitution of expression (11) in the equation (9), we shall receive common dependence circumscribing change of a mass of a filament PETP AS a function of temperature, time and thermodynamic parameters of the media:

$$M = m_0[1 - 1,6 \times 10^2 \exp(-16\,500\,RT) b_0 a_{H2O}/\rho r_0] \qquad (12)$$

TABLE 2 Influence of processing PTEP fabrics by 16% water solution NaOH on its properties[2, 9]*.

Samples	Tearing stress, kG		Tearing elongation, %		Electric resistance, om 10^{-12}	KubelkaMuny Index after		Weight loss
	warp	woof	Warp	woof		Soiled	Washing	
Initial fabric	52	32	28	33	5.5	1.00	0.06	0.0
Fabric treated at: 100	48	28	28	34	6	0.65	0.080	6
110	46	20	25	30	5.2	0.59	0.05	14.43
120	38	9	22	24	6	0.78	0.07	3.1
130° C	19		17	15	5	-	-	61.4

* The experiments were carried out in equipment for dyeing under the pressure, time of treatment was 60 min.

At alkaline treatment of PETP electrical resistance of filaments does not change [29], because as a result of hydrolysis on a surface of filaments we have all same PETP (Table 2). Stress load and stress elongation during alkaline treatment of a filament are reduced (Table 2), however alkali treatment considerably reduces dry dirt adherence of a fabric (Kubelka–Munk index decreases with 1.00 up to 0.6) It is ought to mark that the alkaline treatment results in considerable improvement of a fabric dye color deepness [29]. Using the data of kinetic researches, it is possible to find optimal conditions of alkaline processing of a PETP fabric [2,5. 29]. That the destruction happened at the surface layer, specific tensile σ keeps almost constant value:

$$\sigma = P/\pi r^2 \qquad (13)$$

where P—stress load.

Having aggregated the Equations (7), (11), and (13), we shall receive expression circumscribing change P as a function of time, temperature, and thermodynamic properties of solutions of different alkalis:

$$P = \pi\sigma[r_0 - 1,6 \times 10^2 \exp(-16500/RT)b_0 a_{H2O} t/\rho]^2$$

5.8 OXIDIZING DESTRUCTION

Alongside with hydrolytic destruction for modification of polymers the wide application was received methods of oxidizing destruction [32, 33].

Homogeneous oxidizing of PE dissolved in toluene, at the presence of the catalytic agent - cobalt acetate [34] and in the uncatalyzed mode [35] was described as a method of introduction of polar groups in PE for rise of its adhesive power. There are data on processing PE on mixing rolls at 160-200 Ë on air [36]. As a result of such processing was formed functionalized PE with contents of oxygen 0.1–0.5% by weight, it was used for furnish of textile materials due to ability to derivate spontaneous emulsions [37].

The processes of oxidizing destruction of polypropylene are described [38, 39]. At deep oxidizing degradation atactic [40] and isotactic [41] polypropylenes were obtained wax similar products. As the chain reactions of an oxidizing of macromolecules by oxygen proceed rather slowly, for intensification of processes in industry engineers use treatment of polymer materials by particles of high energy [42,43], by a flame [44,45], by electric discharge [44,46,47], by atoms of oxygen [48], by ozone [49]. The action of particles of high energy at the presence of oxygen is accompanied by destruction of a polymer chains, in absence of oxygen predominate the processes of crosslinking [50,51]. At treatment of cellulose or its derivatives in an inert atmosphere there is an intensive destruction [51,52], and at a dose in ~100 Mrad the products become completely soluble in water. The radial method of cellulose derivatives treatment is applied to obtaining of graft copolymers of cellulose and it ethers [53,54].

Treatment of polymer surface by the open flame is also very effective as a method of oxidizing modification. On Figure 10 the schema of flame surface treatment of a polymeric film is shown.

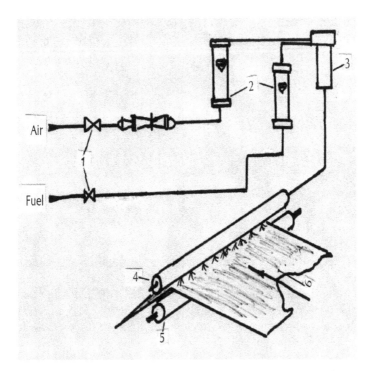

FIGURE 10 Scheme of flame surface treatment of a polymeric films. 1—regulation taps; 2—flow meter; 3—mixer; 4—burner; 5—supporting valve.

For obtaining reproducible results it is necessary to support constant speed of driving of a film (mostly 30–80 m/min) and stationary ratio of mixture air - combustible. Usually recommend to use natural gas or propane. Air takes in major surplus for creation of the oxidizing environment (25:1–32:1). At action of a flame and hot gases on a surface of polymer in surface layer a chain oxidizing proceed, what in a first approximation can be presented by the schema [55]:

$$\sim\sim\sim CH_2\text{-}CH_2\sim\sim + r^* \rightarrow \sim\sim\sim CH^*\text{-}CH_2\sim\sim + rH \qquad (I)$$

$$\sim\sim\sim CH^*\text{-}CH_2\sim\sim + O_2 \rightarrow \sim\sim\sim CH\text{-}CH_2\sim\sim \qquad (II)$$
$$|$$
$$O\text{-}O^*$$
$$\rightarrow \sim\sim\sim CH\text{-}CH_2\sim\sim$$
$$|$$
$$O\text{-}OH$$
$$\rightarrow \sim\sim\sim C\text{-}CH_2\sim\sim + {}^*OH$$

$$O\text{-}O^*$$
$$|$$
$$\sim\sim CH\text{-}CH_2\sim\sim\sim \rightarrow$$

$$\begin{matrix} || \\ O \end{matrix}$$

$$\rightarrow \sim\sim CH{=}CH\sim\sim\sim + HO_2{}^* \qquad\qquad \text{(III)}$$
$$\sim\sim\sim CH + HO\text{-}CH_2\sim\sim\sim$$
$$\begin{matrix} || \\ O \end{matrix}$$
$$\rightarrow \sim\sim\sim C\text{-}OH + {}^*CH_2\sim\sim\sim$$
$$\begin{matrix} || \\ O \end{matrix}$$

$$\sim\sim\sim CH\text{-}CH_2\sim\sim\sim \rightarrow \sim\sim\sim CH\text{-}CH_2\sim\sim\sim + {}^*OH$$
$$\quad\quad | \qquad\qquad\qquad\qquad\quad |$$
$$\quad OOH \qquad\qquad\qquad\quad O^*$$

$$\sim\sim\sim CH\text{-}CH_2\sim\sim\sim + rH \rightarrow \sim\sim\sim CH_2\text{-}CH_2\sim\sim\sim + r^* \qquad \text{(IV)}$$
$$\quad | \qquad\qquad\qquad\qquad\qquad\qquad |$$
$$\quad O^* \qquad\qquad\qquad\qquad\qquad\quad OH$$

$$2RO_2{}^* \rightarrow R{=}O + ROH + O_2 \qquad\qquad\qquad\qquad \text{(V)}$$
$$RO_2{}^* + r^* \rightarrow ROOr \qquad\qquad\qquad\qquad\qquad\quad \text{(VI)}$$
$$RO_2{}^* + {}^*OH \rightarrow ROH + O_2$$

It shows the main paths of polar function groups (carboxyl, hydroxyl, and carbonyl) formation, what stipulate change properties of a surface. The composite dependence with an extreme character between intensity of processing and change of properties of a surface is observed. As an example on Figure 11 are given the test data of adhesion of ink to a surface of PE, treated by a flame of the propane at it different dose (the air delivery was constant) [56].

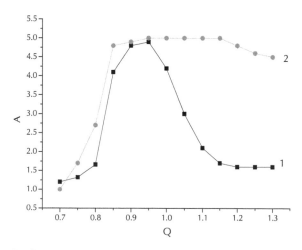

FIGURE 11   Ink adhesion (A) to PE surface treated by flame (1) and electric discharge (2). A – relative units; Q (for 1) – propane demand, m3/hour, Q (for 2) – energy of discharge, Wt.300.

It is seen that adhesion increases from 1 (very poor) up to 5 relative units (good), value 3 corresponds to a valid level. The similar outcomes manage to be received, skipping a film of polymer between two metal electrodes, to what the alternating current by power 8–20 kV is brought. The schema of installation for treatment by electric discharge is given at a fig.12.

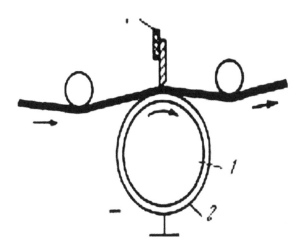

FIGURE 12   The scheme of a system for treatment of film surface by barrier discharge1 – metal valve; 2 – dielectric cover.

The film of polymer or filament is slipped through a zone of barrier discharge, where is exposed to complex action of charged particles, atoms of oxygen, ozone, of nitric oxides [57, 58]. That temperature of processing is usual not strongly differs from room, chain processes proceed (run) in inappreciable degree and function groups, which are accumulated during treatment, formed mainly because of isomerization and decomposition of peroxide radicals (reaction III of the scheme). The dynamic of properties change at treatment by electric discharge is characterized by wider plateau on maxima at the field (Figure 11, curve 2) in comparison with flame treatment, that is important technological advantage. The similar operation is rendered also with many other oxidizing agents. So, for example, the considerable oxidizing of a surface of PE-one of most chemically inert –element is achieved by oxidizing of a hot surface by concentrated hydrogen peroxide, 8% oleum, 9.5 % chlorosulphonic, by fluosulphonic, chromic or nitric acids, solution of potassium bichromate in sulfuric acid, by acid water solution of potassium permanganate and gaseous ozone [50]. Despite of major variety of chemical reactions proceeding at action of mentioned oxidizing agents on a surface of a polymer material, the final outcomes have similar character more often. It is possible to guess, that all effects on polymer are reduced to formation in it peroxide radicals, what further transformations depend on a nature of polymer, temperature of processing and to a lesser degree from a nature of a taken oxidizing agent. In surface layer molecules are partial degraded and it lead to accumulation of oxygen-containing function groups [59-63].

Sometimes surface polymer macromolecules have reactive double bonds. Ozone reacts with such bonds easily, producing ozonides, acids, and aldehydes [64]

$$\sim\sim CH=CH\sim\sim + O_3 \longrightarrow \sim\sim CH\text{--}CH\sim\sim \rightarrow \sim\sim CH=O + {}^-OO\text{--}{}^+CH\sim\sim$$
$$\qquad\qquad\qquad\qquad\qquad\quad |\quad\ |$$
$$\qquad\qquad\qquad\qquad\quad O\text{-}O\text{-}O$$

$$\qquad\qquad\qquad\qquad\qquad\qquad O$$
$$\rightarrow \qquad \sim\sim CH\ \ CH\sim\sim$$

$$\qquad\qquad\qquad O \text{--} O$$
$$\sim\sim CH=O + {}^-OO\text{--}{}^+CH\sim\sim \rightarrow \ \rightarrow \sim\sim CH=O + HO\text{-}CH\sim\sim$$
$$\qquad\qquad\qquad\qquad\qquad\qquad\qquad\qquad ||$$
$$\qquad\qquad\qquad\qquad\qquad\qquad\qquad\qquad O$$

$$\rightarrow \sim\sim CH_2{}^* + CO_2 + O=CH\sim\sim$$

The decomposition of ozonides up to acids and aldehydes goes in main on the ionic mechanism and is accompanied by destruction. However in parallel, the small amounts of free radicals are formed and it can initiate polymerization processes. Due to this before and ozonized rubbers it is easy modify by an grafting [65,66].

Ozone react with C—H bonds more difficult than with double bonds producing alcohols, ketones, acids, peroxides [60,61,64].

$$\qquad\qquad\qquad\qquad\qquad OH \qquad\qquad\qquad\qquad\qquad\qquad\qquad O$$
$$|\qquad\qquad\qquad\qquad\qquad\ \ |\qquad\qquad\qquad\qquad\qquad\qquad\qquad ||$$
$$\qquad\qquad\qquad\rightarrow \sim\sim CH\text{-}CH_2\sim\sim + O_2 \qquad \rightarrow\ \sim\sim COH + {}^*CH_2\sim\sim$$
$$\sim\sim CH_2\text{-}CH_2\sim\sim + O_3 \rightarrow \qquad O\text{-}O^* \qquad\qquad\qquad OOH$$
$$\qquad\qquad\qquad\qquad\qquad\qquad |\qquad\qquad\qquad\qquad\qquad\qquad\qquad |$$
$$\qquad\qquad\qquad\rightarrow \sim\sim CH_2\text{-}CH_2\sim\sim + {}^*OH \quad \rightarrow \ \rightarrow \sim\sim CH\text{-}CH_2\sim\sim$$
$$\qquad\qquad\qquad\qquad\qquad\qquad\qquad\qquad\qquad \rightarrow \sim\sim C\text{-}CH_2\sim\sim + {}^*OH$$
$$\qquad\qquad\qquad\qquad\qquad\qquad\qquad\qquad\qquad\qquad ||$$
$$\qquad\qquad\qquad\qquad\qquad\qquad\qquad\qquad\qquad\qquad O$$

The contents of function groups usually makes ~2 equivalents in calculation on the reacted ozone, their concentration in surface layer grows proportionally of time of treatment, of induction effect is not observed, the ratio of function groups is constant at initial period (Fig. 13).

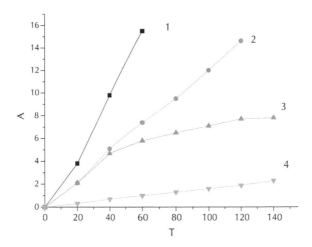

FIGURE 13   The kinetics of functional groups increasing at the action of O3 on polystyrene, A – concentrations, mol/l x 10; T – min; 1 – O3 absorbance; 2 - ~C(=O)~; 3 - -O-O-; 4 - ~C(=O)-OH.

In accordance with conversion increase it is possible to see as more reactive peroxide, hydroxyl and carbonyl groups react with ozone and transform up to carboxyl groups. The transient formation of peroxide radicals in the system was detected in direct experiments on their EPR-spectrum [64,67].

5.9   RISE OF PLASTIC ADHESION

The polyolefins and many other plastics are characterized by low adhesion to metals, paper and other surfaces [68,69]. Besides other reasons it is stipulated by absence in chains of macromolecules of function groups, capable to formation of hydrogen bridges, complexes with a charge transfer or with rather major electrostatic interaction. These interactions influence a free surface energy, energy of cohesion, interfacial tension and other related properties. Published data [70,71] allow us to see in what direction introduction of those or other function groups will change properties of a material. The outcomes of measurement of adhesion of dyes to a promoted surface of PE were compared to surface tension γ [72]. From the data given on Fig. 14 follows, that at $\gamma < \sim 2.8$ n/m² the adhesion is negligible, at small increase γ it grows fast, reaching maximal values at $\gamma \sim 3$–3.1 n/m².

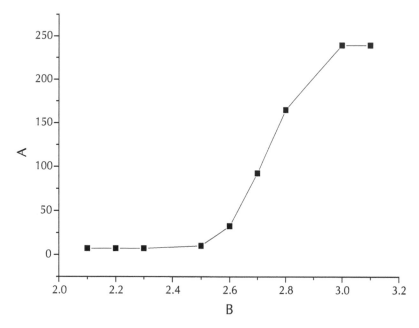

FIGURE 14 Influence of surface tension (B, N/m²) on printing dyes adhesion (A, N/m).

It is possible to raise affinity of a surface to other materials by introduction of polar units to a macromolecules by copolymerisation of ethylene with vinyl acetate [73,74], acrylonitrile [75], or methyl methacrylate [76].

Alongside with modification of polymers by synthetic procedures the wide spread was received the circumscribed above processes of oxidizing modification [77]. So, at plotting antirust coats on metals the method of flame spray coating simultaneously happen destructive modification of a surface of particles, in outcome the affinity of polymer to a surface increases. The increase of concentration of polar function groups in a contact zone is observed so as to increase of an amount of macromolecules strongly bound with metal [78]. As an example, Figure 15 is given dependence of a resistance to delamination of connection surface PE-steel samples from time of exposure at different temperatures.

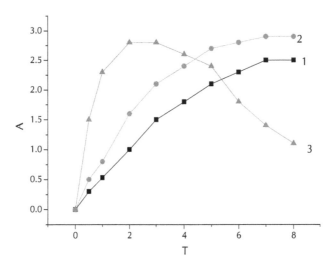

FIGURE 15   The dependence of resistance to delamination of connection surface PE-steel samples from time of exposure on air at different temperatures [78].

Polar groups appeared in a course thermo-oxidation in contact layer can often participate in formation of adhesion links due it the speed of reaching of maximal value of forces of affinity thus is incremented. Plotting of a decorative pictures, marking and art furnish of oil-cloth, packing materials, tare from PE or PETP are bound to major difficulties owing to poor moisten ability of films and feeble cohesion of a colorant with its [79]. Wide usage of treatment by electric discharge appreciably has solved this problem (Figure 11). When the treatment by discharge does not yield desirable outcome, will one use various roundabout ways. In particular, in a case PETP films the good outcomes were obtained at introduction to initial macromolecules of PETP-fragments unsaturated cyclic dicarboxylic acids with their subsequent demolishing by ozone [80]

$-\sim\sim$OCH$_2$-CH$_2$-O-C(=O)-C$_6$H$_4$-C(=O)-O-CH$_2$-CH$_2$-O-C(=O)C=C-C(=O)$\sim\sim$

$$\downarrow O_3$$

$$\begin{array}{c} \text{CH}_2 \\ /\backslash \end{array}$$

$-\sim\sim$OCH$_2$-CH$_2$-OC(=O)-C$_6$H$_4$-C(=O)O-CH$_2$-CH$_2$-O-C(=O)-C C-C(=O)$\sim$

$$\begin{array}{cc} | & | \\ \text{HO-C} & \text{C=O} \\ || & | \\ \text{O} & \text{H} \end{array}$$

It is allowed to form on PETP surface of given number of polar groups and to decrease electron affinity of surface. The mentioned above ways of obtaining of function groups are used in a method of "active points" for creation of reactive centers for a grafting [81]. Surface treatment by hot air [82] or ozone [83,84] can be applied for an addition of various vinyl monomers to polyamides, polyolefins, starch, and other substances.

5.10   MODIFICATION OF A PLASTIC SURFACE BY ITS OXIDIZING PICKLING

The wide spread was received by processes of pickling at platting of plastic [85,86]. The volume of production of goods from metallized plastics is large enough. Activation of a surface before vacuum spraying of decorative metal layer usually is carried out with the help of a flame or electric discharge [87,88], whereas before electrochemical plotting of metal on a surface of polymer it usually handle in solutions chromates in sulfuric acid or other reagents [88,89]. At manufacture of containers for long-lived storage of water from silvered PE it was recommended also to promoted surface of PE by chromic mixture for rise of water resistance of connections between sprayed silver and polymer [90]. The pickling of a surface of carbon filaments by nitric acid is allowed to raise the strength of carbonic plastics and considerably improve other physical-mechanical indexes [91]. Pickling in practice of scientific research has a wide application at research of defects in crystal units. The detection of defects in such units is based on more light permeability and more high solubility of reagent in less ordered zones of a partly crystallized material and, as a consequence, the greatest rate of amorphous zones degradations [92]. The pickling of cellulose, polyethers and polyamides proceeds easy under action of acids and alkalis.

For carbon-chain polymers the methods based on oxidizing destruction were designed. In particular, the optimal conditions of PE pickling by fuming nitric acid were developed [93]. Dicarboxylic acids were obtained as a main product after such treatment. Acids produced from high crystallized PE were rather homogeneous and have identical chain length [94].   At low temperatures most strongly change the molecular weight, while change of density and the degrees of crystallinity are insignificant. All this specifies that after breaking a chain the fragments further do not vary. The decreasing of chain length of stayed crystal oligomers frequently is stopped completely after particular time.

Other reagent for pickling is the ozone [93]. It is more of soft destroying agent in comparison with fuming nitric acid and it can be applied at room temperature, when even the poor crystallites are rather steady. Products of reaction are, mainly, the dicarboxylic acids, which are easy for deleting by extraction. The outcomes of disrupted polymers molecular weight determination was conclusion, that macromolecules tearing up on the border of crystallites, mostly on tucks and chain length of oligomer acids appropriate to single and double width of a crystallite [95]. Length of received acids at time of pickling permanently decreased, mirroring change of boundary between accessible and inaccessible layers of crystallites. The number of specific reagents for disrupt pickling PETP was offered, for example methylamine [96]. For pickling polyoxymethylene it was applied heating at high pressure [97].

In electronic - microscopic researches frequently one used pickling by gas discharge at pressures 10^{-3}–10^{-4} torr [98]. Atoms and the ions of oxygen generated in discharge, destroy and delete surface layers of a material. The rates of degradation depend on degree of order [99]. The nylon6,6 and PETP is formed after pickling by(with) characteristic pictures of lamellar structure [100]. The method of pickling in a gas discharge is usually more informative, than methods based on extractions by solvents [101,102].

KEYWORDS

- **Hydrolytic degradation**
- **Oxidizing destruction**
- **Plastic adhesion**
- **Polycarbonates film**
- **Polyethylenterephtalate films**

REFERENCES

1. N. N. Emanuel, N. N. *J. Pol.Sci..* **C51** 69 (1995).
2. Alkenis, A. F., Zaikov, G. E., and V. I. Karlivan. The Chemical Stability of Polyethers, *Zinantne Publ.,* (1998).
3. Moiseev, Yu. V., Zaikov, G. E. The Chemical Stability of Polymers at the aggressive media. *Chemistry,* (1998).
4. Moiseev, Yu. V., Markin, V. S., and Zaikov, G. E. *Uspekhi Chimii,* **45** 510 (1996).
5. Rudakova, T. E., Thesisis of PhD Diss. Moscow, *Inst.Chemical Physics* RAS, (1975).
6. Rudakova, T. E., Moiseev, Yu. V., Astrina, V. I., Razumova, L. L., Vlasov, S. V., and Zaikov, G. E. *Visokomolekularnie soedinenija,* **A17**, 1791 (Russ) (1995).
7. Kohler, F. J. and Krause, L. M.. *Kunstoffe,* **67**, 731 (1991).
8. US Pat. 3088173, (1961).
9. US Pat. 3142582, (1961).
10. US Pat. 3426754, (1964).
11. Zimin, Yu. B., Sagalaev, G. V., Vlasov, C. V., Surgenko, V. V., Markov, N. G.m Astrina, V. I., and Markov, A. V. *Plast. massi.,* **N4**, 60 (Russ.) (1997).
12. Kimoto, K. K. Jap. Pat. 20595, (1971).
13. GB Pat. 843850, (1958).
14. Shen, M. and Bever, M. *J. Mater. Sci.,* 7, 742 (1992).
15. Astrina, V. I., Razumova, L. L. Schatalova, O. V., Vlasov, S. V. Sagalaev, G. V., Ozerova, L. I., and Zaikov. G. E. *Visokomolekularnie soedinenija,* **A20**, 342 (Russ) (1998).
16. Astrina, V. I., Vlasov, S. V. Sagalaev, G. V., Gumen, R. G., Sokolskii, V. A., Ozerova, L. I., and Ovchinnikov, Yu. K. *Plast. massi.,* **N4**, 50 (Russ.) (1997).
17. Chochlov, A. A, Pavlov, N. N., and Sade, V. A. *Plast. massi.,* **N10**, 26. (Russ.) (1996).
18. Fisher, E. W., Wendorff, J. N., Dettenmater, M., Lieser, G., and Voight-Martin, J. *J. Macromolec. Sci.,* **B12**, 41 (1976).
19. Astrina, V. I., Vlasov, S. V., Sagalaev, G. V., Moiseeb, Yu. V., Gumen, R. G.. *Plast. massi.,* **N5**, 18 (Russ.) (1977).
20. Astrina, V. I., Razumova, L. L., Schatalova, O. V., Vlasov, S. V., and Zaikov, G. E. *Visokomolekularnie soedinenija,* **B21**, 505 (Russ) (1999).
21. Kozlov, P. V. In: Polymer Film Materials, *Chemistry,* Moscow, 18 (1976).
22. Askadskii, A. A. Polymer Deformations. *Chemistry,* 19 (1993).

23. Vlasov, S. V. Thesisis of PhD Diss., Moscow, *Inst. Thin Chemical Technology*, (1966).
24. Neverov, A. N., Perov, B. V., Zherdev, Yu. V. *Visokomolekularnie soedinenija*, **A11**, 1059 (Russ) (1999).
25. Sazhin, B. I. Electric Properties of Polymers. *Chemistry*, Leningrad, (1970).
26. Rudakova, T. E., Kuleva, S. S., Moiseev, Yu. V., Pashkjavechus, V. V., and Zaikov, G. E. *Trudi LITNIITP*, Kaunas, **3**, 265 (1974).
27. Pashkjavechus, V. V., Zaikov, G. E., Rudakova, T. E., Kuleva, S. S., and Moiseev, Yu. V. *Trudi LITNIITP*, Kaunas, **3**, 165 (1974).
28. Moiseev, Yu. V. Thesis's Dr.Sci. Diss., Moscow State Univ., (2012).
29. Pashkjavechus, V. V. Thesis's Dr.Sci. Diss., Polytehn. Institute, Kaunas, (2012).
30. Zaikov, G. E. *Uspekhi chimii*, **44**, 1805 (2012).
31. Emanuel, N. M. and Zaikov, G. E. *Visokomolekularnie soedinenija*, **A17**, 2122 (Russ) (1995).
32. Chemical Reactions of Polymers. Ed. E. M. Fettes, Interscience Publ., N.Y.-London, (1964).
33. Sirota, A. G. Structure Modification and Properties of Polyolefines. *Chemistry*, Moscow, 173 (1994).
34. Hara, K. and Imoto, T. Kolloid-Z und Z. *fur Polimere*, **229**, 4 (1999).
35. Moskovch, Yu. L., Pozamontir, A. G., Razumovskii, S. D. Freidin, B. G., Tsiskovskii, V. K., Shalun, I. M. Pat. USSR 191112, (1966).
36. Wittaker, D. GB Pat. 581279, (1946).
37. Polyethylene and other Polyolefines Ed. by V. P. Kozlov and N. A. Plate *Mir*, Moscow, 302 (1964).
38. Manyasek, Z. and Bellush, D. Polypropylene, Ed. V. I. Polinovskii and I. K. Yartseva, *Chemistry*, (1967).
39. Rikuo, T. and Nabuchiko, M. Jap. Pat. 46-24414, (1975).
40. Guillett, J. E. US Pat. 2828296, (1958).
41. Thompson, W. and Leeder, G US Pat. 291 384, (1959).
42. Knyazev, V. K. and Sidorov, N. A. Application of irradiated Polyethylene in Radioelectronic. *Energy*, Moscow, (1972).
43. Spitsin, V. I., Zubov, P. I., Kabanov, V. A. and Grozinskaya, Z. N. *Visokomolekularnie soedinenija*, **A8**, 604 (Russ) (1966).
44. Rossma, K. *J. Pol. Sci.*, **19**, 141 (1956).
45. Kreidl, W. H. and Hartman, F. *Plastics Technol.* **1**, 31 (1955).
46. Podgornii, Yu. M., Shulman, M. A., and Auzans, P. A. *Plast. massi.*, **N1**, 56 (Russ.) (1979).
47. Bloyer, S. *Mod. Plast.*, **32**, 105 (1968).
48. Evko, E. I. *Zh. Phyz. Chem.*, **42**, 3140 (1968).
49. Kefeli, A. A., Razumovskii, S. D., and Zaikov, G. E. *Visokomolekularnie soedinenija*, **A13**, 803 (Russ) (1971).
50. Knyazev, V. K. and Sidorov, N. A. Irradiated Polyethylene in Technique. *Chemistry*, Moscow, 374 (1974).
51. Livschiz, R. M. and Rogovin, Z. A. In: *Progress Polymer Chemistry*, Ed. by V. V. Korshak, *Nauka*, Moscow, 178 (1969).
52. Shapiro, A. *Chemistry and Technology of Polymers*, **N2**, 3 (1958).
53. Arthur, J. C. and Blouin, F. A. *J. Appl. Polymer. Sci.*, **8**, 2813 (1964).
54. Azizov, U., Usmanov, H. and Sadikov, M. I. *Visokomolekularnie soedinenija*, **A7**, 19 (Russ) (1965).
55. Emanuel, N. M. *Visokomolekularnie soedinenija*, **A21**, 2624 (Russ) (1979).
56. Leeds, Sh. *Tappi*, **44**, 244 (1961).

57. Bagirov, M. A., Abasov, S. A., Agaev, Ch. G. Abbasov, T. F., and Kabulov, U. A. *Plast. massi.*, **N4**, 52 (Russ.) (1972).

58. Bagirov, M. A., Abasov, S. A., Agaev, Ch. G., Abbasov, T. F., and Kabulov, U. A. *Visokomolekularnie soedinenija*, **B11**, 833 (Russ.) (1973).

59. Razumovskii, S. D., Karpuhin, O. N., Kefeli, A. A. Pocholok, T. V., and Zaikov, G.E. *Visokomolekularnie soedinenija*, **A13**, 782 (Russ) (1971).

60. Kefeli, A. A., Razumovskii, S. D., and Zaikov, G. E. *Eur. Polymer J.*, **7**, 275 (1971).

61. Kefeli, A. A., Razumovskii, S. D., Markin, V. S. and Zaikov, G. E. *Visokomolekularnie soedinenija*, **A14**, 137 (Russ.) (1972).

62. Kefeli, A. A., Razumovskii, S. D., and Zaikov, G. E. *Visokomolekularnie soedinenija*, **A18**, 609 (Russ.) (1976).

63. Gaponova, I. S., Golub, V. A., Kefeli, A. A., Zaikov, G. E., Pariiskii, G. B., Razumovskii, S. D., and Toptigin, D. Ya. *Visokomolekularnie soedinenija*, **A20**, 2038 (Russ.) (1978).

64. Razumovskii, S. D. and Zaikov, G. E. *Ozon and its Reactions with Organuc Compounds.* Elsevier, Amsterdam-N.-Y.-London, (1984).

65. Dogadkin, B. A., Tugov, I. I., Tutorskii, I. A., Altzitser, V. S., Krochina, A. S., and Shershnev, V. A. *Visokomolekularnie soedinenija*, **3**, 729 (Russ.) (1961).

66. Gul, V. E., Dogadkin, B. A., Tugov, I. I., Tutorskii, I. A., Altzitser, V. S., Shershnev, V. A., and Kaplunov, Ya.N. USSR. Pat. 13804, (1961).

67. Gaponova, I. S., Golberg, V. M., Kefeli, A. A., Zaikov, G. E., Pariiskii, G. B., Razumovskii, S. D., and Toptigin, D. Ya. *Visokomolekularnie soedinenija*, **B20**, 699 (Russ.) (1978).

68. Belii, V. A., Egorenkov, I. I., and Pleskachevskii, Yu. M. Adhesion of a Polymers to Metals. *Nauka i Technica*, p. 295 (1971).

69. Berlin, A. A. and Basin, V. E. The Foundations of Polymer Adhesion. *Chemistry*, Moscow, (1974).

70. Van Amerongen, G. I. *Elastomers and Plastomers.* Amsterdam, p. 17 (1950).

71. Kraus, G. In: *Adhesion and Adhesives.* London, p. 45 (1954).

72. Allan, A. J. *J. Polymer Sci.*, **38**, 297 (1959).

73. Okui, N. and Kawai, T. Macromolek. *Chem.*, **154**, 161 (1972).

74. Duntov, F. I., Teteryan, P. A., and Krendel, V. Ch. *Plast. massi.*, **N1**, 19 (Russ.) (1972).

75. Andreev, L. N., Krentsel, B. A., Litmanovitch, A. D., Polyak, L S., and Topchiev, A. V. Izvestia Acad. Sci. USSR, *ser. chem.*, p.1507 (1959).

76. Alexander, R., Anspon, Ch., Braun, F., Klampitt, B., and Hyus, R. Chemistry a. *Technology of Polymers.*, **N4**, 70 (1967).

77. Polymeric Film Materials Ed. V. E. Gul, *Chemistry* (1976).

78. Rekner, F. V., Rentse, L. K., and Kalnin, M. M. In: *Modifications of Polymeric Materials*, *Zinantne.*, **3**, 33 (1972).

79. Chozhevets, L. A., Katishonok, L. A., and Michelson, Yu. A. In: *Modifications of Polymeric Materials*, Polytechnic Inst. Publ., Riga, **2**, 116 (1969).

80. Volozhin, A. I., Vorobjeva, L. I., Krutko, E. T., Razumovskii, S. D., Gukalov, S. P. Zernov, P. P., and Levdanskii, V. A. USSR Pat. 563809, (1977).

81. Korshak, V. V. *Progress in Polymer Chemistry*, Nauka, p. 141 (1965).

82. Korshak, V. V., Mozgova, K. K., and Shkolina, M. A. *Visokomolekularnie soedinenija*, **2**, 957 (Russ.) (1960).

83. Kargin, V. A., Kozlov, P. V., Plate, N. A., and Konoreva, I. I. *Visokomolekularnie soedinenija*, **2**, 114 (Russ) (1959).

84. Korshak, V. V., Mozgova, K. K., and Shkolina, M. A. *Visokomolekularnie soedinenija*, **1**, 1573 (Russ) (1959).

85. Gezas, S. I. Decoration Treatment of Plastic Things. *Chemistry*, (Russ) (1978).

86. Shalkauskas, M. and Vashkalis, A. Chemical Metallization of Plastics. *Chemistry*, (Russ) (1977).

87. Rudic, T. A., Trofimova, A. A., Sharafutdinova, D. I., and Sibirjakova, I. V. *Plast. massi.*, **N1**, 58 (Russ.) (1979).

88. Piiroya, E. K., Viikna, A. Ch., Granat, A., Tiikma, L. V., and Ebber, A. V. *Plast. massi.*, **N3**, 24 (Russ.) (1979).

89. Melastchenko, N. F. and Ovdienko, A. P. *Plast. massi.*, **N3**, 40 (Russ.) (1979).

90. Charieva, E. E., Koryukin, A. V., Vinogradova, A. M., and Korolev, A. Ya. *Plast. massi.*, **N9**, 60 (Russ.) (1979).

91. *Plast. massi.*, **N2**, 22 (Russ.) (1979).

92. Battista, O. A., Cruz, M. M., and Ferraro, C. F. In: *Surface and Colloid Chemistry*, **3**, Ed. by E. Mattijevic, J. Wiley, N.-Y., (1971).

93. Palmer, R. P. and Cobold, A. J. Makromolek. *Chem.*, **74**, 174 (1964).

94. Keller, A. and Udagawa, Y. *J. Polymer Sci.*, **8**, A-2, 19; A-2, 1971 (1973).

95. Priest, D. *J. Polymer Sci.*, **9**, A-2, 1971 (1977).

96. Farrow, G., Ravens, D. A., and Ward, J. M. *Polymer*, **3**, 17 (1962).

97. Wanderlich, B. and Bopp, R. Makromolek. *Chem.*, **147**, 79 (1971).

98. Dietl, J. J. *Kunstoffe*, **59**, 792 (1969).

99. Spit, B. J. *Polymer*, **4**, 109 (1963).

100. Holland, V. F. and Anderson, F. R. *J. Appl. Phys.*, **31**, 1516 (1960).

101. Mackie, J. S. and Rudin, A. *J. Polymer Sci.*, **49**, 407 (1961).

102. Shen, Li Li and Kargin, V. A. *Visokomolekularnie soedinenija*, **3**, 1102 (Russ) (1961).

CHAPTER 6

POLYELECTROLYTE MICROSENSORS AS A NEW TOOL FOR METABOLITES' DETECTION

L. I. KAZAKOVA, G. B. SUKHORUKOV, and L. I. SHABARCHINA

CONTENTS

6.1   INTRODUCTION

Enzyme based micron sized sensing system with optical readout was fabricated by co-encapsulation of urease and dextran couple with pH sensitive dye SNARF-1 into polyelectrolyte multilayer capsules. The co-precipitation of calcium carbonate, urease, and dextran followed up by multilayer film coating and Ca- extracting by EDTA resulted in formation of 3.5–4 micron capsules, what enable the calibrated fluorescence response to urea in concentration range from 10^{-6} to 10^{-1} M. Sensitivity to urea in concentration range of 10^{-5} to 10^{-1} M was monitored on capsule assemblies (suspension) and on single capsule measurements. Urea presence can be monitored on single capsule level as illustrated by confocal fluorescent microscopy.

The design of micro and nanostructured systems for *in situ* and *in vivo* sensing become an interesting subject nowadays for biological and medical oriented research [1-3]. Miniaturization of sensing elements opens a possibility for non-invasive detection and monitoring of various analytes exploiting cell and tissues residing reporters. Typical design of such sensor is nano- and micro-particles loaded with sensing substances enable to report on presence of analyte by optical means [4-9]. The particles containing fluorescent dye can be use as a sensor for relevant analytes such as H^+, Na^+, K^+, Cl^-, and so on [10-13]. The fluorescent methods are most simple and handy among the possible ways of registration. They provide high sensitivity and relative simplicity of data read-out. For analysis of various metabolites it is necessary to use the enzymatic reactions to convert analyte to optically detectable compound [14,15]. In order to proper functioning all components of sensing elements (fluorescence dyes, peptides, and enzymes) are to be immobilized in close proximity of each other. That "tailoring" of several components in one sensing entity represents a challenge in developing of a generic tool for sensor construct. One approach to circumvent problem has been introduced by the Photonic Explorers for Bioanalyse with Biologically Localized Embedding (PEBBLE) system [5,16]. The PEBBLE is a generic term to describe use co-immobilization of sensitive components in inert polymers, substantially polyacrylamide, by the microemulsion polymerization technique [17] This technique is useful for fluorescent probe, but to our mind, is too harsh for peptides and enzymes capsulation due to organic solvents involved in particle processing. Multilayer polyelectrolyte microcapsules have not this shortcoming as they are operated fully in aqueous solution at mild condition. These capsules are fabricated using the Layer-by-Layer (LbL) technique based on the alternating adsorption of oppositely charged polyelectrolytes onto sacrificial colloidal templates [18,19] Immobilization of one or more enzymes within polyelectrolyte microcapsules can be accomplished by the co precipitation of these enzymes into the calcium carbonate particles, followed by particle dissolution in mild condition leaving a set of protein retained in capsule [15,20,21]. A fluorescence dye can be included in polyelectrolyte capsules as well. Thus, the multilayer polyelectrolyte encapsulation technique microcapsules allows in principle combining enzyme activity for selected metabolite and registration ability of dyes in one capsule. In this work we demonstrate urea detection using capsules containing urease and pH sensitive dye.

The concentration of urea in biological solutions (blood, urine) is a major characteristic of the condition of a human organism. Its value may suggest a number of acute and chronic diseases: Myocardial infarction, kidney, and liver dysfunction. The measurement of urea concentration is a routine procedure in clinical practice. Urease based enzymatic methods are most widely used for urea detection and use urease enzyme as a reactant. There are multiple urease-based methods which differ from each other by the manner of monitoring the enzymatic reaction. Despite of high specificity, reproducibility and extremely sensitivity for urea as urea is the only physiological substrate for urease, these methods are all laborious and time-consuming since freshly prepared chemical solutions and calibration are required daily. They all are lacking in-situ live monitoring what makes them inappropriate for analysis in vivo as residing sensors. Embedding of urease into polyelectrolytes microcapsules can help to solve these problems [22,23]. The encapsulated urease completely preserve its activity at least 5 days at the fridge storage [22].

Aim of this work was to demonstrate a particular example of a sensor system, which combines catalytic activity for urea and at the same time, enabling monitoring enzymatic reaction by optical recording. The proposed sensor system is based on multilayer polyelectrolyte microcapsules containing urease and a pH-sensitive fluorescent dye, which translates the enzymatic reaction into a fluorescently registered signal.

6.2 EXPERIMENTAL DETAILS

6.2.1 MATERIALS

Sodium poly(styrene sulfonate) (PSS, MW = 70,000) and poly(allylamine hydrochloride) (PAH, MW = 70,000), calcium chloride dihydrate, sodium carbonate, sodium chloride, ethylenediaminetetraacetic acid (EDTA), TRIS, maleic anhydride, sodium hydroxide (NaOH), and Bromocresol purple were purchased from Sigma-Aldrich (Munich, Germany). Urease (Jack bean, *Canavalia ensiformis*) was purchased from Fluka. SNARF-1 dextran (MW = 70,000) was obtained from Invitrogen GmbH (Molecular Probes #D3304, Karlsruhe, Germany). All chemicals were used as received. The bidistilleted water was used in all experiments.

6.2.2 PREPARATION OF SNARF-1 DEXTRAN AND SNARF-1 DEXTRAN/ UREASE CONTAINING $CACO_3$ MICRO-PARTICLES

The preparation of loaded $CaCO_3$ microspheres was carried out according to the co-precipitation method [20,21]. To prepare the $CaCO_3$ microspheres loaded with SNARF-1 dextran were used: 1.6 ml H_2O, 0.5 ml 1M $CaCl_2$, 0.5 ml 1M Na_2CO_3 and 0.4 ml SNARF-1 dextran solution (1 mg/ml).

To prepare the $CaCO_3$ microspheres contained different ratio of SNARF-1 dextran and urease were used:
- *Sample I*: 0.6 ml H_2O, 0.5 ml 1M $CaCl_2$, 0.5 ml 1M Na_2CO_3, 0.4 ml SNARF-1 dextran solution (1 mg/ml) and 1 ml urease (3 mg/ml);

- **Sample II**: 0.8 ml H_2O, 0.5 ml 1M $CaCl_2$, 0.5 ml 1M Na_2CO_3, 0.2 ml SNARF-1 dextran solution (1 mg/ml) and 1 ml urease (3 mg/ml).

The solutions were rapidly mixed and thoroughly agitated on a magnetic stirrer for 30 s at 4°C. After the agitation, the precipitate was separated from the supernatant by centrifugation (250 × g, 30 s) and washed three times with water. The procedure resulted in highly spherical microparticles containing SNARF-1 dextran or SNARF-1 dextran and urease with an average diameter ranging from 3.5-4 µm.

6.2.3   FABRICATION OF SNARF-1 DEXTRAN LOADED MICROCAPSULES

Microcapsules were prepared by alternate LbL deposition of oppositely charged polyelectrolytes poly(allylamine hydrochloride) (PAH, MW = 70,000) and poly(styrene sulfonate) (PSS, MW = 70,000) onto $CaCO_3$ particles containing SNARF-1 dextran or SNARF-1 dextran and urease to give the following shell architecture: (PSS/PAH)4PSS. Short ultrasound pulses were applied to the sample prior to the addition of each polyelectrolyte in order to prevent particle aggregation. The decomposition of the $CaCO_3$ core was achieved by treatment with EDTA (0.2 M, pH 7.0) followed by triple washing with water. The microcapsules were immediately subjected to further analysis or stored as suspension in water at 4°C.

6.2.4   SPECTROSCOPIC STUDY

All spectroscopic studies were carried out with UV–vis spectrophotometer Varian Cary 100 at constant agitation and thermostatic control at 20°C.

The SNARF-1 dextran concentrations in different capsules samples were estimated by matching absorption intensity of supernatant after coprecipitation of the dye with $CaCO_3$ to intensity of calibrated of SNARF-1 dextran concentrations in free solution. The average content of SNARF-1 dextran per capsule was calculated to be: 1) for SNARF-1 dextran $CaCO_3$ microparticales – 1 pg; 2) for SNARF-1 dextran/urease $CaCO_3$ microparticles: **Sample I** – 0.6 pg, **Sample II** – 0.2 pg.

The amount of active urease immobilized into the polyelectrolyte microcapsules was determined under assumption that the enzyme retain its activity while encapsulated. Free urease had 100 U/mg according to the data sheets. The activity of free and encapsulated enzyme were determined from the decomposition of urea into two ammonia molecules and CO_2 using a pH-sensitive dye Bromocresol purple [24]. The urease aliquot solutions were added to a reaction mixture contained a necessary amount of urea and 0.015 mM Bromocresol, whose pH was apriory brought up to 6.2. The reaction kinetics was recorded as a change in the optical absorption of the dye at 588 nm to obtain the linear calibration plot. Then, the known number of microcapsules containing urease and SNARF-1 dextran was added to the reaction solution. The revealed activity of enzyme was compared with amount of free urease.

6.2.5   SPECTROFLUORIMETRIC STUDY

All spectrofluorimetric studies of SNARF-1 dextran and SNARF-1 dextran/urease were carried out with the spectrofluorimeter *Varian Cary Eclipse*, at constant agitation, thermostatic control at 20°C, λ_{exc} = 540 nm, slit width: excitation at 10 nm and emission at 20 nm. The microcapsule suspensions were used at concentration 2 x 10^6 capsules/ml, which was estimated with the cytometer chamber. All solutions were prepared on bidistilleted water.

The TRIS-maleate buffer solutions for pH setting were prepared by adding appropriate quantity of 0.2 NaOH to 0.2 M TRIS and maleic anhydride mixture and diluted to 0.05 M concentration.

6.2.6   CONFOCAL LASER SCANNING MICROSCOPY

Confocal images were obtained by Leica Confocal Laser Scanning Microscope TCS SP. For capsules visualization 100 × oil immersion objective was used throughout. 10 μl of the SNARF-1 dextran/urease capsules suspension was placed on a coverslip. To this suspension 10 μl of 0.1 mol/L urea is added. After about 20 min confocal images were obtained. The red fluorescence emission was accumulated at 600-680 nm after excitation by the FITC-TRIC-TRANS laser at 543 nm.

6.3   DISCUSSION AND RESULTS

Degradation of urea ($CO(NH_2)_2$) is catalyzed by urease and results in the shift of the medium pH into the alkaline range.

$$CO(NH_2)_2 + H_2O = CO_2 + 2\ NH_3$$

Monitoring of the urea degradation can be done by using SNARF-1 as pH-sensitive dye to follow changes of the pH in the enzyme driven reaction. In order to fabricate the sensing microcapsule the urease and SNARF-1 bearing dextran were simultaneously co-precipitated to form $CaCO_3$ spherical particles 3.5-4 μm in size, containing both components urease and fluorescent dye SNARF-1 coupled dextran (MW=70,000) [15]. Then the particles were coated by standard LBL protocol with nine alternating layers of oppositely charged polyelectrolytes PSS and PAH. The formed shell had a $(PSS/PAH)_4PSS$ architecture. After the dissolution of $CaCO_3$ with the EDTA solution the obtained capsule samples contain an enzyme and a fluorescent dye in its cavity.

The dye and enzyme concentrations inside the polyelectrolyte capsule are predetermined essentially at the stage of formation of the $CaCO_3$/SNARF-1 dextran/urease conjugate microparticles. Obviously, the amount of both components of urease and SNARF-1 dextran in the capsules and their ration should play an important role while functioning of entire sensing system is concerned. However, the final composition of co-precipitated particles and later capsules in fabricated samples may be different, though the same initial concentration of components used while preparing co-precipitating particles. It depends on a number of factors: adsorption, capturing, and distribution of the components among the $CaCO_3$ particles, the size and the number of the particles yielded [25]. These parameters might vary from one experiments to another and therefore, it makes problematic to obtain two capsule samples with exactly the same content of encapsulated substances while relying on single capsule detection. Yet, it is imperative to observe this condition to reproduce the efficiency of any sensor. Thus, we always run experiments with at least two samples of capsules in parallel produced independent and having the same parameters at preparation. One of major problem on single particle/capsule detecting is deviation of fluorescent intensity from one particles to another due to uneven fluorescent distribution over population of capsules. To avoid this bottleneck and to obtain a sensor whose reliability and efficiency would not depend on the concentration of the reacting and registering substances in single capsule we opted the SNARF-1 fluorescent dye for the present study (Figure 1).

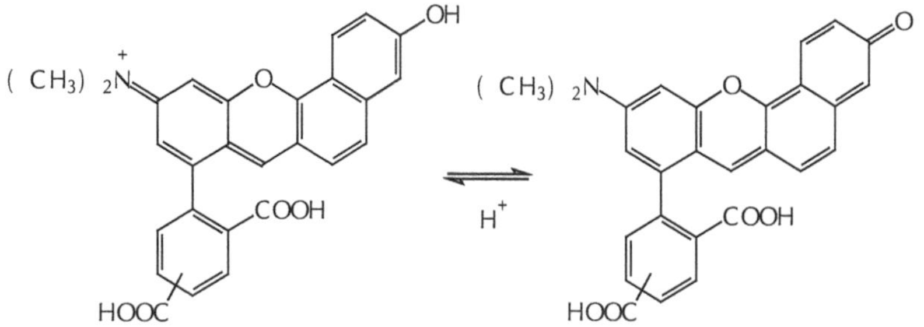

FIGURE 1   Structure of the protonated and deprotonated forms of the SNARF-1 dye.

The emission spectrum of SNARF-1 undergoes a pH-dependent wavelength shift from 580 nm in the acidic medium to 640 nm in alkaline environment. The ratio R = I_{580nm}/I_{640nm} of the fluorescence intensities from the dye at two emission wavelengths allows to determine the pH value according to the ratiometric method. Dual emission wavelength monitoring is well established method eliminating a number of fluorescence measurement artifacts, including photobleaching, sample's size thickness variation, measuring instrument stability, and non-uniform loading of the indicator [26]. It becomes very important particularly if one is using not an ensemble of capsules but only single or few capsules in the analysis, e.g. in the experiments with cells.

In order to verify a feasibility of fluorescence based urea sensing on two component co-encapsulation we fabricated two polyelectrolyte capsule samples of the (PSS/PAH)$_4$PSS shell architecture with different content of dye and urease. The first sample (*Sample I*) contained in average 0.6 pg SNARF-1 dextran per capsule, while the content of the SNARF-1 dextran in the other sample was 0.2 pg per capsule (*Sample II*). The concentration of active urease in samples was opposite 0.2 and 0.6 pg/capsule respectively what gives an average SNARF-1 dextran/urease ratio of 3:1 and 1:3 in these investigated samples.

Spectrofluoremetric studies were carried out to determine the correlation between the fluorescence intensity of the SNARF-1 dextran/urease capsules and the pH of the medium. Both the capsule samples were stored for 10 min in the 0.05 M TRIS-maleate buffer at pH in the range 5.5–9. The excitation wavelength was 540 nm. The capsules fluorescence spectra of the first sample are shown in Figure 2. The spectra obtained for both samples were similar. The encapsulated dye is capable to provide information of the medium acidity in a reasonably wide range of pH. It is seen that fluorescence spectra are characteristic for every pH value. This fact can be used for calibration regardless amount of dye per capsules in studied samples.

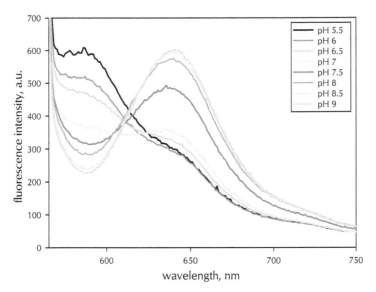

FIGURE 2   Fluorescence spectra of the SNARF-1 dextran/urease capsules in the 0.05 M TRIS-maleate buffer at pH in the range 5.5-9.

The ratio between fluorescence intensity and pH can be described by the following equation according to [26]:

$$pH = pK_a - \log\left(\frac{R - R_{\min}}{R_{\max} - R} \times \frac{I_{640nm}(B)}{I_{640nm}(A)}\right) \tag{1}$$

where $R = I_{580nm}/I_{640nm}$; $R_{\min}$ and $R_{\max}$ – are the minimal and maximal R values in the titration curve (Figure 3, curves 3,4); $I_{640nm}(A)$ and $I_{640nm}(B)$ - fluorescence intensities at 640 nm for the protonated and deprotonated forms of the dye, i.e. in the acidic and alkaline media, respectively. The $R_{\min}$ and $R_{\max}$ meanings depend on the experimental conditions. In this study they were determined to be:

   Sample 1: $R_{\min} = 0.41$, $R_{\max} = 1.96$, $I_{640nm}(B)/I_{640nm}(A)=2$;
   Sample 2: $R_{\min} = 1.06$, $R_{\max} = 2.14$, $I_{640nm}(B)/I_{640nm}(A)= 1.69$.

To yield of the pK_a value the data were plotted as the log of the $[H^+]$ versus the $\log\{(R-R_{\min})/(R_{\max}-R)*(I_{640nm}B)/I_{640nm}A\}$. In this form, the data gave a linear plot with an intercept equal to the pK_a (Figure 4, curves 3,4). As follows from this data pK_a for sample 1 is equal 7.15, for the sample 2–7.25. Thus, the pK_a value differed on 0.1 for different samples whereas the concentration of dye in them differed in 3 times. However dependence of fluorescence intensity for the 2 sample was linear in the smaller interval of values

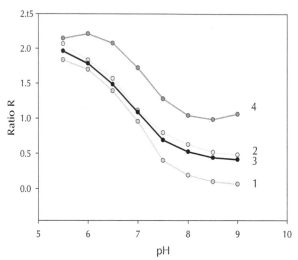

FIGURE 3   Change of fluorescence intensity ratio –R at 580 and 640 nm for: curve 1- SNARF-1 dextran water solution; curve 2 – containing SNARF-1 dextran capsules; curve 3 – SNARF-1 dextran/urease capsules (sample I, 0.6 pg dye/capsule; curve 4 – SNARF-1 dextran/urease capsules (sample II, 0.2 pg dye/capsule) in 0.05M TRIS-maleate buffer at pH in range 5.5–9.

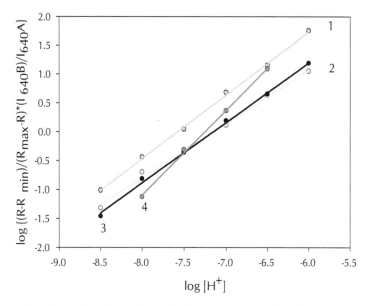

FIGURE 4   Experimental points and theoretical curves generated by the ratiometric method to determine the pK_a values: curve 1 – for SNARF-1 dextran water solution, $pK_a = 7.58$; curve 2 – for containing SNARF-1 dextran capsules, $pK_a = 7.15$; curve 3 – for SNARF-1 dextran/urease capsules (sample I, 0.6 pg dye/capsule), $pK_a = 7.15$; curve 4 – for SNARF-1 dextran/urease capsules (sample II, 0.2 pg dye/capsule), $pK_a = 7.25$.

In Figure 3, the curves 1, 2 shows the ratio R dependences on pH value for free fluorescent dye in comparison with SNARF-1 dextran capsules. The calculation of pK_a values (Figure 4, the curves 1, 2) has demonstrated that for encapsulated dye it is less, than for the free dye solution. It is reasonable to assume that this effect is the result of the interaction between SNARF-1 dextran and polyelectrolyte shell, notably with PAA, because they have opposite charges.

Figure 5 illustrates the effect the urea in concentration from 10^{-6} to 10^{-2} mol/L produces on fluorescence spectra of capsules containing SNARF-1 dextran and urease. Urea concentration dependence is reflected spectroscopically by apparent pH change in the course of enzymatic reaction inside the capsules. The ammonium ions generated via enzymatic reaction in capsule interior effect on pH shift what is recorded by SNARF-1 on the plot (Figure 5). The fluorescent spectra were measured at the 30 minute time point after adding urea solutions at these concentrations to the SNARF-1 dextran/urease capsule's samples. Our particular attention was paid to the kinetics of the change at the fluorescence intensity ratio at 580 nm to 640 nm R (Figure 6) and its relevance to amount of SNARF/urease. Parameter R was plotted versus time and as one can see on curves at high concentration of the urea substrate (10^{-3}M) level off at about 15–20 minutes after the beginning of the of the enzymatic reaction, while at low concentrations of urea (10^{-5} M) the time needed for flattening out spectral characteristics reaches 25–30 minutes. Remarkably, there are no substantial changes for samples at variable concentrations of the dye and urease inside the capsule at least at studied range of 0.2–0.6 pg per capsule. SNARF-1 dextran indicates only the course of the enzymatic reaction, therefore the time needed for the R parameter curve to level off correlates with the time needed to reach the equilibrium in the enzymatic reaction urea/urease occurring inside the capsule. Presumably, it takes about few minutes to equilibrate concentration of urea and its access to urease. We assume rather fast diffusion of urea through the multilayers due to small molecular size of the molecules.

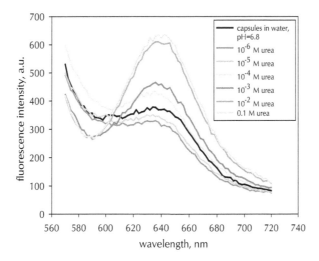

FIGURE 5   SNARF-1 dextran/urease capsule fluorescence spectra in water in the presence of urea from 10^{-6} to 0.1 M.

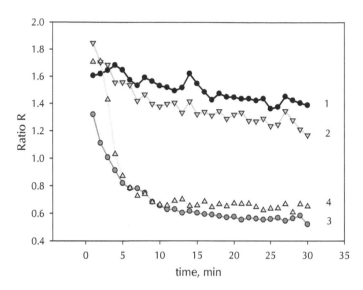

FIGURE 6 Kinetics of the change of the SNARF-1 dextran/urease capsule fluorescence intensity expressed through the fluorescence intensity ratio R at 580 and 640 nm in the presence of 10^{-5} M (curve 1 – sample I, curve 2 – sample II) and 10^{-3} M urea (curve 3 – sample I, curve 4 – sample II).

To calculate the apparent pH caused by urea concentration the values of R_{min}, R_{max}, pK_a and $I_{640nm}(B)/I_{640nm}(A)$ were used. R_{max} was determined as a fluorescence intensity ratio at 580 nm and 640 nm in the capsules stored in bidistilled water (pH = 6.4). R_{min} is the ratio of the fluorescence intensity spectrum related to the minimal R value at 580 nm and 640 nm in the SNARF-1 dextran/urease capsules when "high" concentration of urea (0.1 M) was added and 30 min after the enzymatic reaction begun. The pK_a value were assumed to be = 7.15 and 7.25 for sample 1 and 2 respectively, that was in accord with the experiment with buffer solutions *(Figure 3,4)*. From the values obtained a calibration curve of the pH dependence inside the capsules on the urea concentration present in the solution was plotted. The calibration curve is presented in Figure 7.

Thus, to determine the urea concentration in the solution it is necessary to obtain the following three spectra: 1) of the capsules without substrate (urea); 2) of the capsules at "high" concentration of urea in the solution, e.g. 0.1 M as used for our calibration plot; 3) of the capsules in the investigated sample studied. These data will suffice to calculate the values of R_{min}, R_{max} and $I_{640nm}(B)/I_{640nm}(A)$, which are characteristic of each particular sample of the sensing capsules. Using Equation 1 one then can calculate the pH as apparently reached in the capsules in the course of urea degradation and to compare its value with the calibration curve in Figure 7.

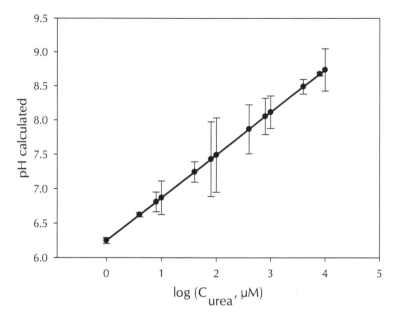

FIGURE 7 Calibration curve for detecting urea using the SNARF-1 dextran/urease capsules in water solutions.

It is worth to notice, that this calibration curve is obtained for SNARF-1 dextran/urease capsule in pure water without substantial contamination of any salt which could buffer the systems and spoil truly picture for urea detection. We carried out experiments to build a similar calibration curve in the presence of the 0.001 M TRIS-maleate buffer (were used solutions with the pH 6.5 and 7.5) but it resulted in overwhelming effect of pH buffering. Buffering the solution eliminates the pH change caused in a course of enzymatic reactions. Thus, it sets a limit for detection of urea concentration using SNARF-1-dextran/urease capsules. However the calibration in conditions of particular experimental system is reasonable at salt free solution assumption. Summarizing, one can state the presented in Figure 7 calibration curve as suitable for estimation of urea concentrations *in situ* in water solutions.

The feasibility studies on single capsule detection of urea presence were carried out using confocal fluorescent microscopy. CLSM image of SNARF-1 dextran/urease capsules in absence and at 0.1M urea added to the same capsules are presented on Figure 8. Small distinctions in the form and the sizes of capsule population are connected with non-uniformity of SNARF-1 dextran/urease $CaCO_3$ particles received in co-precipitation process what is rather often observed for calcium carbonate templated capsules containing proteins [20].

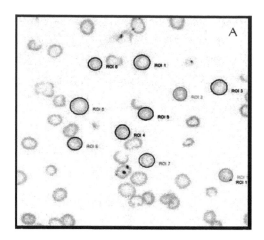

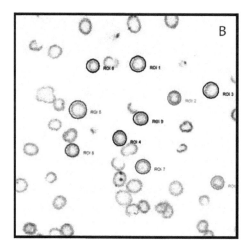

FIGURE 8 Confocal fluorescence microscopy images of (PSS/PAH)$_4$PSS capsules loaded with SNARF-1 dextran (MW = 70kDa) and urease enzyme in water (image A) and 0.1 M urea concentrations (image B). The Table 1 shows the increase of mean energy of individual capsules before (F_{low}) and after the addition 0.1 M urea solution (F_{high}) to water capsule suspension. The red fluorescence emission was accumulated at 600-680 nm after excitation by the FITC-TRIC-TRANS laser at 543 nm.

The images have been processed with Leica Confocal Laser Scanning Microscope TCS SP software to quantify the effect. The image areas corresponding to location of 10 selected capsules are set off in different colors - ROI$_{1-10}$ (Region Of Interest). The value of average intensity of a luminescence is defined as parameter mean energy by formula:

$$I^2_{Mean} = \frac{1}{N_{Pixel}} \sum_{Pixel} I^2_i$$

where I_{mean} – the average image energy of ROI areas;
N_{pixel} - the total number of pixels that are included in the calculation;
I_i – the energy correspond for particular pixel.

The energy meanings of individual capsules are presented in the Table 1.

TABLE 1 The change of mean fluorescence intensity of 10 selected capsules in presence of 0.1 M urea.

ROI	Mean Energy, I_{low}	Mean Energy, I_{high}	I_{high}/I_{low}
ROI 1	2947.95	6467.42	2.1939
ROI 2	4635.87	10316.34	2.2253
ROI 3	2424.13	5694.92	2.3493
ROI 4	4021.83	11161.52	2.7752
ROI 5	1666.76	3727.05	2.2361
ROI 6	3177.11	7237.68	2.2781
ROI 7	2237.61	5932.76	2.6514
ROI 8	3192.70	7158.54	2.2422
ROI 9	4100.76	8944.83	2.1813
ROI 10	2758.57	6127.32	2.2212

Although, value of fluorescence intensity is seen to be different for each capsule the more than double increase of integrated intensity is well pronounced in all monitored capsules upon addition of urea solution. The distribution of energy relation therefore remains almost constant for each capsule. These data on single capsules are in good agreement with data presented on Figure 5 obtained on entire capsule population. Indeed, integrated area under spectra for black (no urea) and light blue (0.1 M urea) with spectral range of 600–680 nm is about twice in difference. This fact demonstrates the principal applicability to use single capsule for carrying out analysis of urea presence.

6.4 CONCLUSION

In this study we have demonstrated a particular example of a sensor system, which combines catalytic activity for the substrate (urea) and at the same time enabling to monitor the enzymatic reaction by co-encapsulated pH sensitive dye. Substrate sensitive enzyme urease was co-encapsulated together with SNARF-1 coupled to dextran in multilayer microcapsules. Enzymatic activity was recorded by fluorescent changes caused by increasing of pH in course of enzymatic cleavage of urea as measured on population of capsules and on single capsule imaging by confocal fluorescent microscope. Suggested method can be used to measure the concentration of urea in solutions where the content of urea is fairly high (blood, urine) and also able to detect urea at concentration down to 10^{-5} M at non buffering solution. Spectroscopic parameters of microencapsulated sensors were found stable in regardless ratio between urease and SNARF-dextran in concentration range of 0.2-0.6 pg per capsules what encourages such as microencapsulated sensing system as robust. Although, that pH sensitivity of dye has a limitation to function in buffers the concept of co-encapsulation of metabolite active enzymes and dyes sensitive to product of enzymatic reactions is illustrated to be workable in reasonable concentration range and applicable for single capsule based detecting.

The presented results prove the concept of feasibility of microencapsulated enzyme/dye systems for local metabolite sensing and optical online recording. These co-encapsulating sensors have advantages over well-known PEBBLE systems as they could sense the substances via extra enzymatic reaction what is much more prospective in term of analytes to be monitored, especially in biological systems as cells and tissue. The micron size of the sensors will pave the way for producing and applicability of injectable and implantable sensing systems like 'a smart tattoo' or delivered to the cell or tissue and serving for on-line monitoring of various biological processes. Aspects of, considered here, urea sensing has a particular challenge to measure concentration of urea in vivo (in cells and tissue, e.g. in skin epithelium), what remains subject of further research.

KEYWORDS

- **Confocal fluorescent microscopy**
- **Confocal laser scanning microscopy**
- **Layer-by-Layer (LbL) technique**
- **Microcapsules**
- **Microemulsion polymerization technique**

REFERENCES

1. Arregui, F. J. *Sensors Based on Nanostructured Materials*, (2009).
2. Fehr, M., Okumoto, S., Deuschle, K., Lager, I., Looger, L. L., Persson, J., Kozhukh, L., Lalonde, S., and Frommer, W. B. *Biochem Soc Trans.*, **33(1)**, 287-290 (2005).
3. Vo-Dinh, T.,Griffin, G. D., Alarie, J. P., Cullum, B., Sumpter, B., and Noid, D. *Summary Nanomedicine*, **4(8)**, 967-979 (2009).
4. Sukhorukov, G. B., Rogach, A. L., Garstka, M., Springer, S., Parak, W. J., Muñoz-Javier, A., Kreft, O., Skirtach, A. G., Susha, A. S., Ramaye, Y., Palankar, R., and Winterhalter, M., *Small,* **3(6)**, 944-955.
5. Lee, Y. E., Smith, R., and Kopelman, R. *Annu Rev Anal Chem (Palo Alto Calif)*, **2**, 57–76 (2009).

6. Sukhorukov, G. B., Rogach, A. L., Zebli, B., Liedl, T., Skirtach, A. G., Köhler, K., An-
    tipov, A. A., Gaponik, N., Susha, A. S., Winterhalter, M., and Parak, W. J. *Small*, **1(2)**,
    194–200 (2005).
7. B. G. De Geest, S. De Koker, G. B. Sukhorukov, O. Kreft, W. J. Parak, A. G.
    Skirtach, J. Demeester, S. C. De Smedt and W. E. Hennink, *Soft Matter*, **5**, 282.
    J. Peteiro-Cartelle, M. Rodríguez-Pedreira, F. Zhang, P. Rivera Gil, L. L del Mercato and
    W. J. Parak, *Nanomedicine*, 2009, **4 (8)**, 967–979 (2009).
8. M. J. Sailor and E. C. Wu, *Advanced Functional Materials*, 2009, **19 (20)**, 3195–3208.
9. S. Nayak and M. J. McShane, *Sensor Letters*, **4**, 433–439 (2006).
10. O. Kreft, A. Muñoz Javier, G. B. Sukhorukov and W. J. Parak, *J. Mater. Chem.*, **(42)**,
    4471–4476.
11. J. Q.Brown and M. J. McShane, *IEEE Sensors Journal*, **5**, 1197–1205 (2005).
12. L, L. del Mercato, A. Z. Abbasi and W. J. Parak, *Small*, 2011, DOI: 10.1002/smll.201001144
13. J. Q.Brown and M. J. McShane, *Biosensors and Bioelectronics*, **21**, 1760–1769 (2005).
14. E. W. Stein, D. V. Volodkin, M. J. McShane and G. B. Sukhorukov, *Biomacromolecules*,
    7, 710–719 (2006)
15. M. Brasuel, J. W. Aylott, H. Clark, H. Xu, R. Kopelman M. Hoyer, T. J. Miller, R. Tjalkens
    and M. Philbert, *Sensors and Materials*, **14**, 309–338 (2002).
16. H. Xu, J. W. Aylott and R. Kopelman, *Analyst*, **127**, 1471–1477 (2002).
17. E. Donath, G. B. Sukhorukov, F. Caruso, S. A. Davis and H. Möhwald, *Angew. Chem.*, Int.
    Ed., **37**, 2202–2205 (1998).
18. G. B. Sukhorukov, E. Donath, S. Davis, H. Lichtenfeld, F. Caruso, V. I. Popov and H.
    Möhwald, *Polym. Adv. Technol.*, **9**, 759–767 (1998).
19. A.I. Petrov, D. V. Volodkin and G. B. Sukhorukov, *Biotechnol. Prog.*, **21**, 918–925 (2005).
20. G. B. Sukhorukov, D. V. Volodkin, A. M. Gunther, A. I. Petrov, D. B. Shenoy and H. Möh-
    wald, *Journal of Materials Chemistry*, **14**, 2073–2081 (2004).
21. Y. Lvov, A.A. Antipov, A. Mamedov, H. Möhwald, G. B. Sukhorukov, *Nano Lett.,*1, 125
    (2001).
22. Y. Lvov and F. Caruso, *Anal. Chem.*, **73**, 4212 (2001).
23. S. Paddeu, A. Fanigliulo, M. Lanzin, T. Dubrovsky and C. Nicolini, *Sens. Actuators*, **25**,
    876–882 (1995).
24. D. Halozana, U. Riebentanz, M. Brumen and E. Donath, Colloids and Surfaces A: Physi-
    cochem. Eng. Aspects, **342**, 115–121 (2009).
25. J. E. Whitaker, R. P. Haugland and F. G. Prendergast, *Anal Biochem.*, **194 (2)**, 330–44
    (1991).
26. A. A. Antipov, G. B. Sukhorukov, S. Leporatti, I. L. Radtchenko, E. Donath and
    H. Möhwald, *Colloids and Surfaces A: Physicochemical and Engineering Aspects*,
    198-200, 53–541 (2002).

CHAPTER 7

ACTIVITY OF LIPOSOMAL ANTIMICROBIC PREPARATIONS CONCERNING STAPHYLOCOCCUS AUREUS

N. N. IVANOVA, G. I. MAVROV, S. A. DERKACH, and E. V. KOTSAR

CONTENTS

7.1   INTRODUCTION

It has been found the antimicrobic activity of liposomal lincomycin depends on the composition and charge of liposomes. The antimicrobic activity of liposomal lincomycin, obtained on the basis of egg lecithin is higher than antimicrobic activity of the solution lincomycin in three times concerning planktonic cells of *Staphylococcus aureus*. Negatively charged liposomes received on the basis of polar lipids and lincomycins were more effective, than neutral liposomes. The minimum inhibitory concentration (MIC) of them is less of MIC of lincomycin in seven times concerning planktonic cells of *Staphylococcus aureus*. The antimicrobic activity of liposomal benzoyl peroxide (BP) obtained on the basis of egg lecithin was in 14 times higher than antimicrobic activity of the solution BP concerning of *Staphylococcus aureus*.

The skin of patients with atopic dermatitis planted on different microorganisms, whose number is much bigger than the skin of healthy people. For example, *Staphylococcus aureus* sow from the skin of patients with atopic dermatitis in 80–100% of cases. Skin diseases of microbial etiology in most cases basic drug treatment and prevention are antibiotics. Widespread use of antibiotics has negative consequences, one of which is the emergence of pathogens with resistance to penicillin, gentamicin, tetracycline, methicillin, lincomycin, and sulfonamides, as well as a new generation of antibiotics—quinolones and cephalosporins. Consequently, the problem of prevention and treatment of infectious diseases is urgent. One of the ways of its solution is the introduction of new chemotherapeutic agents into medical practice.

It is known that nanoparticles and liposomal forms of medicines allow significantly improving the efficacy, reducing toxicity, and therapeutic dose and qualitatively changing the nature of their actions. Thus, in the work [4] it was shown that, despite the low concentrations, the efficiency of liposomal benzyl penicillin to inhibit growth of bacterial biofilms of *Staphylococcus aureus* was higher than the intact benzyl penicillin. The purpose of the study was to investigate the efficiency of antimicrobial agents in liposomal form relatively to *Staphylococcus aureus*.

7.2   MATERIALS AND METHODS

The strain ATCC 25923 *Staphylococcos aureus* was taken from SE " Mechnicov Institute of Microbiology and Immunology AMSU". We have also used the following items: egg lecithin (Ukraine, "Biolek"), DMSO (Russia), the mixture of negatively charged lipids that were obtained by the original technology of Dr. Nina Ivanova, substance of BP ("Aldrich" USA), lincomycin (JSC "Darnitsa", Ukraine), Mueller–Hiton agar (HiMedia Laboratories Pvt. Limited (Индия)), meat-peptone broth (MPB).

7.2.1   THE RECEIVING OF LIPOSOMES

The substance of BP dissolved in chloroform due to its poor solubility in aqueous solutions and added to an alcohol or chloroform solution of lipids in the ratio of BP—Lipids 1:10 and 1:20. The liposomes were obtained by evaporating the lipids and antibiotics on a rotary vacuum evaporator (Switzerland). Next mixture was suspended in sterile buffered saline. Liposomes prepared in the extruder EmulsiFlex–C5 (Canada

"Avestin"), punching with compressed air (10 cycles) to achieve a constant optical density on spectrophotometer (DU–7Spectrophotometer Beckman, USA) at temperatures that above the phase transition temperature of any of the lipid components was present. The average size of liposomes was 160–180 nm, concentration of lipids in the liposomes was 2%, ratio LN—lipids and BP—lipids was 1:20 [2].

The cleaning of switched antibiotics in liposomes from those that are not involved in making liposomes using ultracentrifuge (MSE-Superspeed Centrifuge 65, England) for an hour at 105000g. The output of liposomes was determined spectrophotometrically at 450 nm.

The determination of MIC of antibiotics was performed by microtiter method. Antimicrobial agents were diluted by serial dilutions MPB in flat-bottomed plates, they were also added to the culture of *Staphylococcus aureus*, and were incubated for 24 hr at 34°C. The control was culture *Staphylococcus aureus* without antimicrobial agents. After that mixture sowed on solid nutrient medium Mueller-Hinton agar for calculation of amount of colony forming particles and MIC definitions. The MIC was considered to be the lowest concentration, which retards the growth of *Staphylococcus aureus* during the incubation period.

7.3   DISCUSSION AND RESULTS

Among the antibiotics we stopped on lincomycin. It has a bacteriostatic effect on a wide range of microorganisms, with increasing doses of lincomycin it has a bactericidal effect. Antimicrobial mechanism of action of lincomycin is the inhibition of protein synthesis in the cells of microorganisms. The drug is active with respect to grampositive aerobic and anaerobnih microorganisms, including *Staphylococcus* species.

The BP has a wide spectrum of antimicrobial activity. It is active against the bacteria, in the case it is also resistant to antibiotics [3,4].

In the first phase of study, the MICs of lincomycin and BP in liposomes on the basis of egg lecithin were used, which is a soft lipid and traditionally used in the creation of liposomal forms of drugs. As a result, the definition MIC of lincomycin and its liposomal preparations were found and also liposomes received on the basis of egg lecithin and lincomycin are more effective than epy solution of lincomycin under the action of *Staphylococcus aureus* planktonic cells (Figure 1, №2). The MIC of liposomes this composition decreased in three times in comparison with the MIC of the lincomycin solution.

Negatively charged liposomes received on the basis of polar lipids and lincomycins were the most effective. The using of negatively charged liposomes that contained lincomycin reduced the MIC of the lincomycin solution in seven times concerning planktonic cells of *Staphylococcus aureus*. (Figure 1, No3).

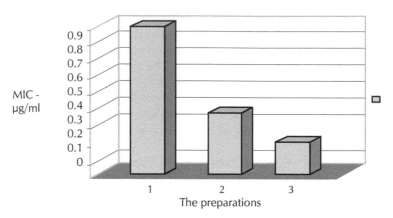

FIGURE 1 The MIC of the liposomal lincomycin №1—the control (lincomycin solution), №2—liposomal lincomycin on the basis of egg lecithin, №3—negatively charged liposomal lincomycin.

Liposomal BP on the basis of egg lecithin was also more effective against *Staphylococcus aureus* in comparison with the MIC of intact BP. At definition MIC of liposomal forms BP on the basis of egg lecithin, it has been found that its MIC made 31 µg/ml that was in 14 times less MIC of BP dissolved in dimethyl sulfoxide (DMSO).

7.4   CONCLUSION

1. It is found that minimally inhibitory concentration of liposomal antibiotic solution with the neutral charge (lecithin liposomes) decreased in three times in comparison with minimally inhibitory concentration of the lincomycin solution concerning *Staphylococcus aureus*.
2. The using of negatively charged liposomes on the basis of polar lipids with the antibiotic, strengthens efficiency of antibiotic action in greater degree than using lecithin liposomes. Negatively charged liposomes containing lincomycin reduced its minimally inhibitory concentration in seven times concerning *Staphylococcus aureus*.
3. The minimally inhibitory concentration of the liposomal antimicrobic preparation BP with the neutral charge (lecithin liposomes) decreased in 14 times in comparison with minimally inhibitory concentration of BP concerning *Staphylococcus aureus*.
4. The received results enable to predict the using of liposomal forms of antimicrobic substances for increase of pharmacological efficiency in treatment of *Staphylococcus* infections.

KEYWORDS

- **Antimicrobic activity**
- **Benzoyl peroxide**
- **Liposomes**
- **Minimum inhibitory concentration (MIC)**
- *Staphylococcus aureus*

REFERENCES

1. Kim H. J. and Jones, M. N., *J Liposome Res.*, №14.-C, pp. 123–139 (2004).
2. Sorokoumova, G. M., Selishcheva, A. A., and Kaplun, A. P., *Educational tools in bioorganic chemistry*. M., **105** P. (in Russian) (2000).
3. Tanghetti, E. A. and Popp, K. F. *Dermatol Clin.*, **27**, 17–24 (2009).
4. Tanghetti, E. A., *Cutis.* **82**, 5–11 (2008).

CHAPTER 8

POLYELECTROLYTE ENZYME-BEARING MICRODIAGNOSTICUM—A NEW STEP IN CLINICAL-BIOCHEMISTRY ANALYSIS

SERGEY A. TIKHONENKO, ALEXEI V. DUBROVSKY,
EKATERINA A. SABUROVA, and LYUDMILA I. SHABARCHINA

CONTENTS

8.1 INTRODUCTION

Clinical and biochemical analyzes are among the most common methods used to diagnose human diseases. Investigations of this kind are known to include general blood and urine tests, the study of a number of other biological fluids [1-3]. Until recently, these tests were carried out by chemical methods, but, due to the toxicity of many of them, the low sensitivity and other shortcomings, enzymological methods widespread received today. However, along with the obvious advantages of this approach, there are some drawbacks: The ambiguity of the analysis in the presence of aggressive high-molecular compounds to the enzyme, in particular, proteases and other intracellular components; one-time use of the enzyme, and so on. Thus, there is a need to protect the enzyme from the adverse effects, while maintaining access to it of the substrate, to increase its stability during prolonged storage, as well as to develop the reusability of the enzyme [4]. One type of such a defense is encapsulation of enzymes in polyelectrolyte microcapsules (PMC) [5,6].

The PMC are the product of a new field of polymer nanotechnology. At present this area is booming around the world: in the U.S., EU, China, Australia, and another countries. This, along with pure fundamental research of structure, physico-chemical and biological properties of polyelectrolyte microcapsules, increasing emphasis on applied research aimed at practical use of PMC, particularly in medicine, chemical engineering, biotechnology and many other areas [7]. The combination of unique properties and relatively simple technology of preparation for a wide range of PMC with the given parameters (structural, mechanical, and functional), simplicity of inclusion of a wide variety of substances, and the ability to control membrane permeability PMC makes promising their use as tools for targeted drug delivery to the organs and tissues, [8-13], depot [8] and the therapeutic effect, as microreactors, microcontainers.

The PMC containing enzyme can be used as a microdiagnosticum to detect a substrate, or inhibitor, or activator of the encapsulated enzyme in native biological fluids and in wasted water.

8.2 EXPERIMENTAL DETAILS

8.2.1 REAGENTS

Pig skeletal muscle lactate dehydrogenase (EC 1.1.1.27) (M4 isoform) was isolated as in [14]. Urease (EC 3.5.1.5, jack bean) and proteinase K (EC 3.4.21.64) were from Fluka (no. 94 285) and Sigma (P8044), respectively. The initial reagents used in the preparation of the samples were as follows: Sodium poly(styrenesulfonate) (70 kDa), poly(diallyldimethylammonium chloride) (70 kDa), and poly(allylamine hydrochloride) (70 kDa) from Aldrich (Germany); dextran sulfate (10 kDa) and EDTA from Sigma (Germany); and calcium chloride, sodium carbonate, and sodium chloride from Reakhim (Russia). The polyelectrolyte solutions were brought to a specified pH value with concentrated solutions of NaOH or HCl.

The $CaCO_3$ microspherulites containing enzymes were prepared as in [6]. One volume of a 1 M $CaCl_2$ aqueous solution was added to two volumes of enzyme solution (1.0 mg/ml) with vigorous stirring. Then an equal volume of aqueous 0.33 M Na_2CO_3

was rapidly added. After stirring for 30 s, the suspension of particles was allowed to stand for 15 min at room temperature without stirring. The formation of microspherulites was controlled with a light microscope. These composite microspherulites with inclusions of urease or LDH had a narrow size distribution with a mean diameter of ~4.5 mm.

Preparation of PEMC involved adsorption of alternating layers of polyanions (PSS, DS) and polycations (PAA, PDADMA) on the surface of the $CaCO_3$–enzyme composite microspherulites in solutions containing 0.5 M NaCl and polyelectrolytes at 2 mg/ml. After each adsorption stage, the system was triply washed in order to remove unbound polymer molecules. After depositing a necessary number of polyelectrolyte layers, the carbonate core was dissolved in 0.2 M EDTA for 12 h. The microcapsules thus prepared were triply washed with water; the enzyme remained inside the microcapsules. All procedures were performed at 20°C.

Urease activity was determined from the decomposition of urea into ammonia and carbon dioxide CO_2 with a pH-sensitive dye Bromocresol Violet. The reaction mixture contained a necessary amount of urea and 0.015 mM Bromocresol Violet adjusted to pH 6.2. The known number of PEMC containing urease was added to the solution. Then, the kinetics of the reaction was measured from a change in absorption of the dye at 588 nm [15]. The activity of the free enzyme in the presence of the polyelectrolyte was measured by a similar method, a necessary amount of the polyelectrolyte with pH adjusted to 6.2 was added to the reaction solution.

Lactate dehydrogenase activity was determined from a change in NADH absorption at 340 nm with a Specord M-40 spectrophotometer (Carl Zeiss, Germany) [15]. The reaction was initiated by introducing 20 ml of the enzyme (0.05–0.10 mg/ml) or a necessary amount of microcapsules with the enzyme into the reaction mixture containing pyruvate at a specified concentration, 0.2 mM NADH, and, in a number of cases, the polyelectrolyte.

8.2.2   STABILITY OF ENCAPSULATED ENZYMES AGAINST PROTEINASE K

We prepared two samples, each containing an aqueous suspension (0.5 ml) of microcapsules with urease (7.4×10^7 capsules/ml) in 0.01 M Tris-HCl (pH 8.0). A solution (0.5 ml) of proteinase K at 4 mg/ml in the same buffer was added to the first sample, and the same amount of the buffer was added to the second sample. Then, the solutions were incubated at 37°C. Every 10 min, 50-ml portions were taken and added to the urease assay mixture (1.95 ml) with 125 mM urea. Free urease at 20 µg/ml instead of microcapsules was used as a reference.

8.3   DISCUSSION AND RESULTS

8.3.1   SHELL FORMATION WITH A SELECTED PAIR OF POLYELECTROLYTES

Since incorporation of enzymes into capsules of any origin usually alters the enzyme activity, it is most important to select a pair of oppositely charged polyelectrolytes that would be optimal for the operation of the projected microdiagnostic. Figure 1 shows the dependence of the enzyme activity on the concentration of different polyelectrolytes.

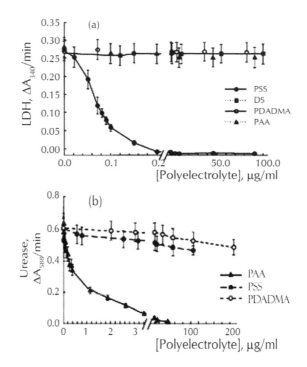

FIGURE 1 Dependence of (a) LDH and (b) urease activity on the concentration of polyelectrolytes (legend). LDH assay: LDH, 0.5 mg/ml; pyruvate, 1 mM; NADH, 0.2 mM; Tris-HCl, 0.05 M; pH 6.2. Urease assay: urease, 0.5 mg/ml; urea, 125 mM; Bromocresol Violet, 0.015 mM; pH 6.2.

The PSS polyanion with a hydrophobic backbone is a strong inhibitor for LDH, whereas the PAA polycation with a highly hydrophobic backbone is a strong inhibitor for urease. A 50% inhibition of LDH and urease activity is observed at low concentrations of the polyelectrolytes (fractions of a microgram per milliliter). An important factor responsible for the inhibitory effect of the polyelectrolytes is the presence of the hydrophobic polymer backbone. Polyelectrolytes with a hydrophilic backbone (e.g., DS) have virtually no effect on the activity of the enzymes (Figure 1a). However, the capsules prepared from these polyelectrolytes are not stable. The activity of LDH remains unchanged even in the presence of the studied polyelectrolytes with positively charged ionogenic groups up to very high concentrations. Therefore, polyelectrolytes that can be used for the preparation of polyelectrolyte microcapsules containing enzymes must satisfy the following criteria: first, these polyelectrolytes must not inactivate the enzyme at concentrations suitable for the formation of the capsule shell (1–2 mg/ml), and, second, they must have a hydrophobic backbone. It is these two criteria (sometimes incompatible) that determine the proper choice of a pair of oppositely charged polyelectrolytes for use in the design of polyelectrolyte enzymatic micro-diagnostics.

The experiments and calculations have demonstrated that the best pairs of poly-electrolytes for the shells of microcapsules are PAA/DS and PAA/PSS for LDH, and PSS/PAA and PSS/PDADMA for urease in the given sequence of layer deposition. On this basis, we designed and prepared PEMC containing LDH and urease with different polyelectrolyte compositions and different numbers of layers.

8.3.2 THE DEPENDENCE OF PROTEINS' DISTRIBUTION WITHIN POLYELECTROLYTE MICROCAPSULES ON PH OF THE MEDIUM

Transmission electron microscopy of ultrathin sections and confocal laser scanning microscopy were used to study the distribution of proteins within polyelectrolyte mi-crocapsules. Since obtaining quality images using transmission electron microscopy is possible in case studies of electron dense objects, the iron containing protein, fer-ritin, with a value of pI 4.7, was selected for encapsulation. Polyelectrolytes poly allyl amine (PAA) and polystyrene (PSS) that were used have values of ionization constants of functional groups of 10.5 and 1.9 respectively, which removes the question of their charge in the range of pH investigated. Photos of ultrathin sections of samples contain-ing polyelectrolyte microcapsules ferritin at pH values below or near the isoelectric point are shown in Figure 2. It was found that the distribution of proteins inside the capsule depends on the protein isoelectric point and the charge of the internal poly-electrolyte layer. The distribution of protein in the interior of the capsule is available in two versions: a uniform distribution of protein throughout the volume and concentra-tion of its aggregates in wall space (Figure 2).

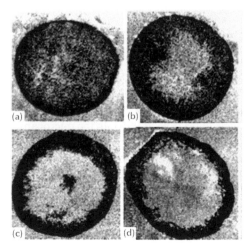

FIGURE 2 Electron micrograph of ultrathin sections of polyelectrolyte microcapsules containing ferritin at pH 2 (a), 3 (b), 4 (c), 5 (d). The polyelectrolyte shell of the composition is (PAA/PSS)3.

Thus, if the pH is less than the isoelectric point (p*I* 4.7), a protein is positively charged, while the inner layer of polyelectrolyte microcapsules is presented as a poly-cation, the protein molecules are distributed throughout its volume. Protein molecules lose their charge values near the isoelectric point and are concentrated in the wall space of the capsule due to hydrophobic interactions with a polyelectrolyte shell. If the polyanion PSS was used as the first layer in the formation of a shell, the protein at all pH values in the range studied was located in the wall space (Figure 3). We attribute this to the electrostatic interaction between protein molecules and the polyelectrolyte at low pH and hydrophobic interactions in the region of the isoelectric point.

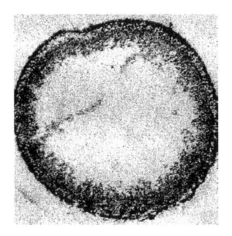

FIGURE 3   Electron micrograph of ultrathin sections of polyelectrolyte microcapsules containing ferritin at pH 2. The polyelectrolyte shell of the composition is (PSS/PAA)3.

8.3.3   ACTIVITY AND CATALYTIC CHARACTERISTICS OF ENCAPSULATED ENZYMES

According to the results presented above, the microcapsules containing LDH were prepared so that the first layer of their shell was a positively charged polyelectrolyte, PAA, whereas the second and subsequent even layers were composed of different neg-atively charged polyelectrolytes (DS or PSS). In the microcapsules containing urease, the first layer of their shell was a negatively charged polyelectrolyte, PSS, whereas the second and subsequent even layers were composed of different positively charged polyelectrolytes (PAA or PDADMA). This was done to preventing or diminish direct contact of the inactivating polyelectrolytes with the enzyme. It can be seen in Figure 4a that LDH activity in the (PAA/PSS)3 microcapsule is relatively low. Therefore, the inhibitory effect of PSS on the activity of LDH manifests itself even in the case where the PSS layer is formed after the PAA layer. Microcapsules composed of different bilayers, e.g., (PAA/DS)2(PAA/PSS), exhibit relatively high activity of the enzyme. However, the microcapsules in which the negatively charged polyelectrolyte in the shell is represented only by DS have zero activity. The reason for this is that DS has a polar sugar backbone; moreover, although DS does not inhibit LDH, in the pair with PAA it forms a loose unstable shell. Eventually, a shell of the composition (PAA/DS)2(PAA/PSS), which involves polyelectrolytes with polar and hydrophobic backbones, was found to be optimal for encapsulation of LDH. Similarly, the inhibitory effect of the positively charged polyelectrolyte on the activity of urease was taken into account in the design and preparation of the microcapsule shell for the urease microdiagnostic. The urease-containing microcapsules (Figure 4b) in which the first layer is a "noninhibitory" polyelectrolyte, PSS, exhibit sufficiently high activity.

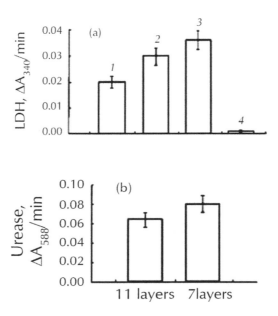

FIGURE 4 Activity of (a) LDH and (b) urease in microcapsules with different composition and different number of layers. LDH capsules: (*1*) (PAA/PSS)3, (*2*) (PAA/DS)(PAA/PSS)2, (*3*) (PAA/DS)2(PAA/PSS), and (*4*) (PAA/DS)3. Urease capsules: (PSS/PAA)3PSS (seven layers) and (PSS/PAA)5PSS (eleven layers). Assay conditions as in Figure 2, with 1.2·106 capsules/ml.

In order to determine the catalytic characteristics of the encapsulated enzymes, we obtained the dependences of the stationary rate of substrate conversion on the substrate concentration. As an example, the curves of saturation of urease with urea in the reaction of urea decomposition are depicted in Figure 5. It can be seen that the dependences for urease in microcapsules, are generally similar to those for free enzyme, except small differences in the affinity constants. In particular, the Michaelis constant K_M with respect to urea is 7.1 ☐2.2 mM for urease in microcapsules of eleven and seven layers, whereas the K_M for free urease is 2.5 ☐0.7 mM. The maximal rate V_{max} for urease in microcapsules of eleven layers is 20% lower than that for urease in microcapsules of seven layers. The K_M with respect to pyruvate for microcapsules containing LDH was not different from for free enzyme.

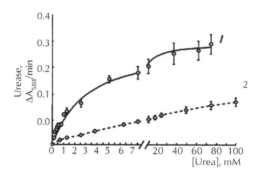

FIGURE 5   Dependences of the activity of (*1*) free urease and (*2*) urease in (PSS/PAA)3PSS microcapsules on the urea concentration.

8.3.4 *ACTIVITY OF FREE AND ENCAPSULATED ENZYMES DURING LONG-TERM STORAGE AND UNDER THE ACTION OF PROTEINASE K*

It is known that enzymes, especially at low concentrations, are inactivated during storage in solution. For oligomeric enzymes, this is associated with their dissociation into subunits and structural distortion of the active site. In the general case, inactivation of different enzymes in long-term storage can be caused by spontaneous thermal denaturation; chemical modification as a result of hydrolysis, oxidation, and other processes; and bacterial contamination. We investigated the stability of encapsulated LDH and urease, in the course of long-term storage. The dependences of the residual activity of the enzymes on the incubation time are plotted in Figure 6. It can be seen that, unlike free enzymes in solution, the encapsulated enzymes retain their activity for several months. The investigation into the nature of the factors responsible for the high stability of an enzyme in a PEMC during long-term storage will be reported in a separate paper.

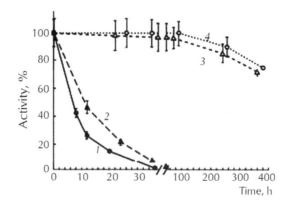

FIGURE 6   Variations in the activity of (*1*) free LDH, (*2*) free urease, (*3*) LDH in (PAA/DS)2(PAA/PSS) microcapsules, and (*4*) urease in (PSS/PAA)3PSS microcapsules during long-term storage in H$_2$O, $T = 21°C$. LDH and urease, 50 mg/ml; microcapsules, 6.2$10^7$ ml^{-1}.

We also investigated the influence of proteolytic. Figure 6 shows the dependences of the activity of free (curve *1**) and encapsulated (curve *2**) urease on the time of incubation with proteinase K at 37°C. For comparison, the dependences obtained in the absence of proteinase K (curves *1, 2*) are also shown in Figure 6. Urease was encapsulated into the (PSS/PAA)3PSS shell. As can be seen in Figure 7, the activity of free urease in the presence of the proteolytic enzyme steeply decreases to zero, whereas encapsulated urease retains the ability to decompose urea in the presence of proteinase. Since, the enzymes in PEMC are not degraded by proteinase K, these microcapsules can be used for quantitative analysis of low-molecular compounds in biological fluids containing proteolytic enzymes. This obviates the need for removing proteinases from the biological fluid to be analyzed, which is a laborious process.

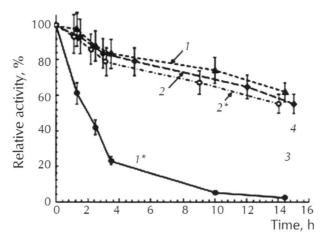

FIGURE 7   Dynamics of the activity of (*1, 1**) free urease and (*2, 2**) urease in (PSS/PAA) 3PSS microcapsules at 37°C: (*1**, *2**) in the presence and (*1, 2*) in the absence of proteinase K.

CONCLUSIONS

Analysis of the data obtained for two enzymes belonging to different classes allows the following conclusions:

1. When the shell of the capsule is prepared from a properly chosen pair of polyelectrolytes, the encapsulated enzyme retains high affinity to the substrate.
2. The enzyme incorporated into a multilayered polyelectrolyte capsule retains activity for several months, whereas, the activity of the free enzyme in solution drops nearly to zero in a few days.
3. The encapsulated enzyme completely retains activity in the presence of proteinase. This obviates the need for their removal from the biological fluid to be analyzed.
4. The proposed micro-diagnostic based on encapsulated enzymes can be repeatedly used for analyzing biological fluids, thus further reducing the enzyme expenditure.

5. Such enzymatic micro-diagnostics represent a new class of biosensors with significant advantages over the existing clinical/biological enzymatic methods of analysis.
6. The distribution of proteins inside polyelectrolyte microcapsules depend on charges of protein and of inner polyelectrolyte layer.

REFERENCES

1. Dolgov, V. V. and Selivanova, A. V. Biochemical studies in clinical diagnostic laboratories, primary care health services. *Vital Diagnostics SPb*, St. Petersburg., p. 231 (2006).
2. Dolgov, V. V., Shevchenko, O. P., Sharyshev, A. A., and Bondar, V. A. Turbodimetriya in laboratory practice. *M. Reaform*, p. 176 (2007).
3. Menshikova, V. V. *Methods of clinical laboratory tests*. Moscow: Labora., p. 304 (2009).
4. Caruso, F., Trau, D., Mohwald, H., and Renneberg, R. *Enzyme Encapsulation in Layer-by-Layer Engineered Polymer Multilayer Capsules*. Langmuir, **16**, 1485-1488 (2000).
5. Decher, G., Hong, J. D., and Schmitt, J. Buildup of ultrathin multilayer films by a self- assembly process. Alternating adsorption of anionic and cationic polyelectrolytes on charged surfaces. *Thin Solid Films*, **210(1-2)**. 831–835 (1992).
6. Petrov, A. I., Volodkin, D. V., and Sukhorukov, G. B. Protein calcium carbonate co-precipitation: a tool of protein encapsulation. *Biotechnol. Prog.* **21(3)** 918–925 (2005).
7. Skirtach, A. G., et al., Nanoparticles distribution control by polymers: Aggregates versus nonaggregates. *Journal of Physical Chemistry C*, **111(2)** 555–564 (2007).
8. Rivera-Gil, P., et al., Intracellular processing of proteins mediated by biodegradable polyelcctrolyte capsules. *Nano Lett*, **9(12)** 4398–402 (2009).
9. Sabini, E., et al., Structural basis for substrate promiscuity of dCK. *J Mol Biol*, **378(3)** 607–21 (2008).
10. Sieker, F., et al., Differential tapasin dependence of MHC class I molecules correlates with conformational changes upon peptide dissociation: a molecular dynamics simulation study. *Mol Immunol*, **45(14)** 3714–22 (2008).
11. Sieker, F., Springer, S., and Zacharias, M. Comparative molecular dynamics analysis of tapasin-dependent and -independent MHC class I alleles. *Protein Sci*, **16(2)** 299–308 (2007).
12. Borodina, T. N., Rumsh, L. D., Kunizhev, S. M., Sukhorukov, G. B., Vorozhtsov, G. N., Feldman, B. M., Rusanova, A. V., Vasilyeva, T. V., Strukova, C. M., and Markvicheva, E. A. The inclusion of extracts of medicinal plants in biodegradable microcapsules. *Biomedical Chemistry*. **53(6)** 662–671 (2007).
13. Borodina, T. N., Rumsh, L. D., Kunizhev, S. M., Sukhorukov, G. B., Vorozhtsov, G. N., Feldman, B. M., and Markvicheva, E. A. Polyelectrolyte microcapsules as delivery systems of bioactive substances. *Biomedical Chemistry*. **53(5)** 557–565 (2007).
14. Saburova, E. A., Bobreshova, M. E., Elfimova, L. I., and Sukhorukov, B. I. Biokhimiya (Moscow) **65(8)**, 1151 [*Biochemistry* (Moscow) **65(8)**, 976 (2000)].
15. *Medical Laboratory Technologies: A Handbook*, Ed. by Karpishchenko (Intermedika, St. Petersburg, 2002) [in Russian].

CHAPTER 9

CNT/POLYMER NANOCOMPOSITES: A DETAILED REVIEW ON MATHEMATICAL MODELING AND EXPERIMENTAL CASE STUDIES

M. ESMAEILI, R. ANSARI, V. MOTTAGHITALAB, A. K. HAGHI, B. XI, G. M. SPINKS, G. G. WALLACE, and A. I. MINETT

CONTENTS

SECTION I: MATHEMATHICAL MODELING

9.1 INTRODUCTION

In this chapter, the modeling of mechanical properties of carbon nanotube (CNT)/ polymer nanocomposites is reviewed. It starts with the structural and intrinsic mechanical properties of CNTs. Then we introduce some computational methods that have been applied to polymer nanocomposites, covering from molecular scale (e.g., molecular dynamics (MD) and Monte Carlo (MC)) and microscale (e.g., Brownian dynamics (BD), dissipative particle dynamics (DPD), lattice Boltzmann (LB), time-dependent Ginzburg–Landau method, and dynamic density functional theory (DFT) method) to mesoscale and macroscale (e.g., micromechanics, equivalent-continuum, and self-similar approaches, and finite element method (FEM)). Hence, the knowledge and understanding of the nature and mechanics of length and orientation of nano-tube and load transfer between nano-tube and polymer is critical for manufacturing of enhanced CNT-polymer composites and will enable in tailoring of the interface for specific applications or superior mechanical properties. So, in this review a state of these parameters in mechanics of CNT-polymer composites will be discussed along with some directions for future research in this field.

The CNTs were first observed by Iijima, almost two decades ago [1], and since then, extensive work has been carried out to characterize their properties [2-4]. A wide range of characteristic parameters has been reported for CNT nanocomposites. There are contradictory reports that show the influence of CNTs on a particular property (e.g., Young's modulus) to be improving, indifferent or even deteriorating [5]. However, from the experimental point of view, it is a great challenge to characterize the structure and to manipulate the fabrication of polymer nanocomposites. The development of such materials is still largely empirical and a finer degree of control of their properties cannot be achieved so far. Therefore, computer modeling and simulation will play an ever increasing role in predicting and designing material properties, and guiding such experimental work as synthesis and characterization, For polymer nanocomposites, computer modeling and simulation are especially useful in the hierarchical characteristics of the structure and dynamics of polymer nanocomposites ranging from molecular scale and microscale to mesoscale and macroscale, in particular, the molecular structures and dynamics at the interface between nanoparticles and polymer matrix. The purpose of this review is to discuss the application of modeling and simulation techniques to polymer nanocomposites. This includes a broad subject covering methodologies at various length and time scales and many aspects of polymer nanocomposites. We organize the review as follows. In section I we will discuss about CNTs and nano composite properties. In Section II, we introduce briefly the computational methods used so far for the systems of polymer nanocomposites which can be roughly divided into three types—molecular scale methods (e.g., MD and MC), microscale methods (e.g., BD, DPD, LB, time dependent Ginzburg–Lanau method, and DFT method), and mesoscale and macroscale methods (e.g., micromechanics, equivalent-continuum and self-similar approaches, and FEM) [6]. Many researchers used this method for determine the mechanical properties of nanocomposite that in section

III will be discussed. In section IV modeling of interfacial load transfer between CNT and polymer in nanocomposite will be introduced and finally we conclude the review by emphasizing the current challenges and future research directions.

9.2 CARBON NANOTUBES'S (CNTS) AND NANOCOMPOSITE PROPERTIES

9.2.1 INTRODUCTION TO CNTS

The CNTs are one dimensional carbon materials with aspect ratio greater than 1000. They are cylinders composed of rolled-up graphite planes with diameters in nanometer scale [7-10]. The cylindrical nanotube usually has at least one end capped with a hemisphere of fullerene structure. Depending on the process for CNT fabrication, there are two types of CNTs [8-11]—single-walled CNTs (SWCNTs) and multi-walled CNTs (MWCNTs). The SWCNTs consist of a single graphene layer rolled up into a seamless cylinder whereas MWCNTs consist of two or more concentric cylindrical shells of graphene sheets coaxially arranged around a central hollow core with van der Waals (vdW) forces between adjacent layers. According to the rolling angle of the graphene sheet, CNTs have three chiralities—armchair, zigzag, and chiral one. The tube chirality is defined by the chiral vector, Ch = na_1 + ma_2 (Figure 1), where the integers (n, m) are the number of steps along the unit vectors (a1 and a2) of the hexagonal lattice [9, 10]. Using this (n, m) naming scheme, the three types of orientation of the carbon atoms around the nanotube circumference are specified. If n = m, the nanotubes are called "armchair". If m = 0, the nanotubes are called "zigzag". Otherwise, they are called "chiral". The chirality of nanotubes has significant impact on their transport properties, particularly the electronic properties. For a given (n, m) nanotube, if (2n + m) is a multiple of 3, then the nanotube is metallic, otherwise the nanotube is a semiconductor. Each MWCNT contains a multi-layer of graphene, and each layercan have different chiralities, so the prediction of its physical properties is more complicated than that of SWCNT. Figure 1 shows the CNT with different chiralities.

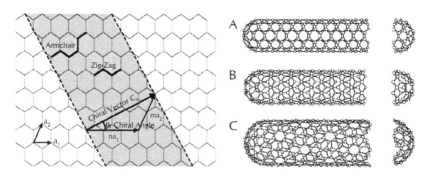

FIGURE 1 Schematic diagram showing that how a hexagonal sheet of graphene is rolled to form a CNT with different chiralities (A—armchair, B—zigzag, and C—chiral).

9.2.2 CLASSIFICATION OF CNT/POLYMER NANOCOMPOSITES

Polymer composites, consisting of additives and polymer matrices, including thermoplastics, thermosets, and elastomers, are considered to be an important group of relatively inexpensive materials for many engineering applications. Two or more materials are combined to produce composites that possess properties that are unique and cannot be obtained each material acting alone. For example, high modulus carbon fibers or silica particles are added into a polymer to produce reinforced polymer composites that exhibit significantly enhanced mechanical properties including strength, modulus, and fracture toughness. However, there are some bottlenecks in optimizing the properties of polymer composites by employing traditional micron-scale fillers. The conventional filler content in polymer composites is generally in the range of 10–70 wt%, which in turn results in a composite with a high density and high material cost. In addition, the modulus and strength of composites are often traded for high fracture toughness [12]. Unlike traditional polymer composites containing micron-scale fillers, the incorporation of nanoscale CNTs into a polymer system results in very short distance between the fillers, thus the properties of composites can be largely modified even at an extremely low content of filler. For example, the electrical conductivity of CNT/ epoxy nanocomposites can be enhanced several orders of magnitude with less than 0.5 wt% of CNTs [13]. The CNTs are amongst the strongest and stiffest fibers ever known. These excellent mechanical properties combined with other physical properties of CNTs exemplify huge potential applications of CNT/polymer nanocomposites. Ongoing experimental works in this area have shown some exciting results, although the much-anticipated commercial success has yet to be realized in the years ahead. In addition, CNT/polymer nanocomposites are one of the most studied systems because of the fact that polymer matrix can be easily fabricated without damaging CNTs based on conventional manufacturing techniques, a potential advantage of reduced cost for mass production of nanocomposites in the future. Following the first report on the preparation of a CNT/polymer nanocomposite in 1994 [14], many research efforts have been made to understand their structure–property relationship and find useful applications in different fields, and these efforts have become more pronounced after the realization of CNT fabrication in industrial scale with lower costs in the beginning of the 21st Century [15]. According to the specific application, CNT/polymer nanocomposites can be classified as structural or functional composites [16]. For the structural composites, the unique mechanical properties of CNTs, such as the high modulus, tensile strength and strain to fracture, are explored to obtain structural materials with much improved mechanical properties. As for CNT/polymer functional composites, many other unique properties of CNTs, such as electrical, thermal, optical, and damping properties along with their excellent mechanical properties, are utilized to develop multi-functional composites for applications in the fields of heat resistance, chemical sensing, electrical and thermal management, photoemission, electromagnetic absorbing and energy storage performances, and so on.

9.3   MODELING AND SIMULATION TECHNIQUES

9.3.1   MOLECULAR SCALE METHODS

The modeling and simulation methods at molecular level usually employ atoms, molecules or their clusters as the basic units considered. The most popular methods include molecular mechanics (MM), MD, and MC simulation. Modeling of polymer nanocomposites at this scale is predominantly directed toward the thermodynamics and kinetics of the formation, molecular structure, and interactions. The diagram in Figure 1 describes the equation of motion for each method and the typical properties predicted from each of them [17–22]. We introduce here the two widely used molecular scale methods—MD and MC.

MOLECULAR DYNAMICS (MD)

The MD is a computer simulation technique that allows one to predict the time evolution of a system of interacting particles (e.g., atoms, molecules, granules, etc.) and estimate the relevant physical properties [23, 24]. Specifically, it generates such information as atomic positions, velocities, and forces from which the macroscopic properties (e.g., pressure, energy, and heat capacities) can be derived by means of statistical mechanics.

The MD simulation usually consists of three constituents:
1.   A set of initial conditions (e.g., initial positions and velocities of all particles in the system),
2.   The interaction potentials to represent the forces among all the particles, and
3.   The evolution of the system in time by solving a set of classical Newtonian equations of motion for all particles in the system. The equation of motion is generally given by:

$$\vec{F}_1(t) = m_i \frac{d^2 \vec{r}_1}{dt^2} \tag{1}$$

where $\vec{F}_i(t)$ is the force acting on the i th atom or particle at time t which is obtained as the negative gradient of the interaction potential U, mi is the atomic mass, and $\vec{F}_i$ the atomic position. A physical simulation involves the proper selection of interaction potentials, numerical integration, periodic boundary conditions, and the controls of pressure and temperature to mimic physically meaningful thermodynamic ensembles. The interaction potentials together with their parameters, that is, so-called force field, describe in detail how the particles in a system interact with each other, that is, how the potential energy of a system depends on the particle coordinates. Such a force field may be obtained by quantum method (e.g., *ab initio*), empirical method (e.g., Lennard, Mores, and Mayer) or quantum-empirical method (e.g., embedded atom model, glue model, and bondorder potential). The criteria for selecting a force field include the accuracy, transferability, and computational speed. A typical interaction potential U may consist of a number of bonded and nonbonded interaction terms:

$$U(\vec{r_1},\vec{r_2},\vec{r_3},...,\vec{r_n}) = \sum_{i_{bond}}^{N_{bond}} U_{bond}(i_{bond},\vec{r_a},\vec{r_b}) + \sum_{i_{angle}}^{N_{angle}} U_{angle}(i_{angle},\vec{r_a},\vec{r_b},\vec{r_c}) + \sum_{i_{torsion}}^{N_{torsion}} U_{torsion}(i_{torsion},\vec{r_a},\vec{r_b},\vec{r_c},\vec{r_d})$$

$$+ \sum_{i_{inversion}}^{N_{inversion}} U_{inversion}(i_{inversion},\vec{r_a},\vec{r_b},\vec{r_c},\vec{r_d}) + \sum_{i=1}^{N-1}\sum_{j>i}^{N} U_{vdw}(i,j,\vec{r_a},\vec{r_b}) + \sum_{i=1}^{n-1}\sum_{j>i}^{N} U_{electrostatic}(i,j,\vec{r_a},\vec{r_b}) \qquad (2)$$

The first four terms represent bonded interactions, that is, bond stretching U_{bond}, bond-angle bend U_{angle}, dihedral angle torsion $U_{torsion}$, and inversion interaction $U_{inversion}$, while the last two terms are non-bonded interactions, that is, vdW energy U_{vdw} and electrostatic energy $U_{electrostatic}$. In the equation, $\vec{r_a},\vec{r_b},\vec{r_c},\vec{r_d}$ are the positions of the atoms or particles specifically involved in a given interaction; N_{bond}, N_{angle}, $N_{torsion}$, and $N_{inversion}$ stand for the total numbers of these respective interactions in the simulated system; i_{bond}, i_{angle}, $i_{torsion}$, and $i_{inversion}$ uniquely specify an individual interaction of each type; i and j in the vdW and electrostatic terms indicate the atoms involved in the interaction. There are many algorithms for integrating the equation of motion using finite difference methods. The algorithms of varlet, velocity varlet, leap-frog, and Beeman, are commonly used in MD simulations [23]. All algorithms assume that the atomic position $\vec{r}$, velocities $\vec{v}$, and accelerations $\vec{a}$, can be approximated by a Taylor series expansion:

$$\vec{r}(t+\delta t) = \vec{r}(t) + \vec{v}(t)\delta t + \frac{1}{2}\vec{a}(t)\delta^2 t + ... \qquad (3)$$

$$\vec{v}(t+\delta t) = \vec{v}(t)\delta t + \frac{1}{2}\vec{b}(t)\delta^2 t + ... \qquad (4)$$

$$\vec{a}(t+\delta t) = \vec{a}(t) + \vec{b}(t)\delta t + ... \qquad (5)$$

On the whole, a good integration algorithm should conserve the total energy and momentum and be time-reversible. It should also be easy to implement and computationally efficient, and permit a relatively long time step. The Verlet algorithm is probably the most widely used method. It uses the positions $\vec{r}(t)$ and accelerations $\vec{a}(t)$ at time t, and the positions $\vec{r}(t-\delta t)$ from the previous step (t – δ) to calculate the new positions $\vec{r}(t+\delta t)$ at (t + δt), we have:

$$\vec{r}(t+\delta t) = \vec{r}(t) + \vec{v}(t)\delta t + \frac{1}{2}\vec{a}(t)\delta t^2 + \dots \tag{6}$$

$$\vec{r}(t-\delta t) = \vec{r}(t) - \vec{v}(t)\delta t + \frac{1}{2}\vec{a}(t)\delta t^2 + \dots \tag{7}$$

$$\vec{r}(t+\delta t) = 2\vec{r}(t)\delta t - \vec{r}(t-\delta t) + \vec{a}(t)\delta t^2 + \dots \tag{8}$$

The velocities at time t and $t + \frac{1}{2\delta t}$ can be respectively estimated:

$$\vec{v}(t) = \left[\vec{r}(t+\delta t) - \vec{r}(t-\delta t)\right]/2\delta t \tag{9}$$

$$\vec{v}(t+1/2\delta t) = \left[\vec{r}(t+\delta t) - \vec{r}(t-\delta t)\right]/\delta t \tag{10}$$

The MD simulations can be performed in many different ensembles, such as grand canonical (μVT), microcanonical (NVE), canonical (NVT) and isothermal–isobaric (NPT). The constant temperature and pressure can be controlled by adding an appropriate thermostat (e.g., Berendsen, Nose, Nose–Hoover, and Nose–Poincare) and barostat (e.g., Andersen, Hoover, and Berendsen), respectively. Applying MD into polymer composites allows us to investigate into the effects of fillers on polymer structure and dynamics in the vicinity of polymer–filler interface and also to probe the effects of polymer–filler interactions on the materials properties.

MONTE CARLO (MC)

The MC technique, also called Metropolis method [24], is a stochastic method that uses random numbers to generate a sample population of the system from which one can calculate the properties of interest. A MC simulation usually consists of three typical steps. In the first step, the physical problem under investigation is translated into an analogous probabilistic or statistical model. In the second step, the probabilistic model is solved by a numerical stochastic sampling experiment. In the third step, the obtained data are analyzed by using statistical methods. The MC provides only the information on equilibrium properties (e.g., free energy and phase equilibrium), different from MD which gives non-equilibrium as well as equilibrium properties. In a NVT ensemble with N atoms, one hypothesizes a new configuration by arbitrarily or systematically moving one atom from position $i\,j$. Due to such atomic movement, one can compute the change in the system Hamiltonian ΔH:

$$\Delta H = H(j) - H(i) \tag{11}$$

where $H_{(i)}$ and $H_{(j)}$ are the Hamiltonian associated with the original and new configuration, respectively.

This new configuration is then evaluated according to the following rules. If $\Delta H = 0$, then the atomic movement would bring the system to a state of lower energy. Hence, the movement is immediately accepted and the displaced atom remains in its new position. If $\Delta H \geq 0$, the move is accepted only with a certain probability $P\ i \to j$ $i \to j$ which is given by:

$$P i \to j \propto \exp(-\frac{\Delta H}{K_B T}) \tag{12}$$

where K_B K_B is the Boltzmann constant. According to Metropolis et al. [25], one can generate a random number ζ between 0 and 1 and determine the new configuration according to the following rule:

$$\xi \leq \exp(-\frac{\Delta H}{K_B T}) \ ; \text{ the move is accepted.} \tag{13}$$

$$\xi \rangle \exp(-\frac{\Delta H}{K_B T}) \ ; \text{ the move is not accepted.} \tag{14}$$

If the new configuration is rejected, one counts the original position as a new one and repeats the process by using other arbitrarily chosen atoms. In a μVT ensemble, one hypothesizes a new configuration j by arbitrarily choosing one atom and proposing that it can be exchanged by an atom of a different kind. This procedure affects the chemical composition of the system. Also, the move is accepted with a certain probability. However, one computes the energy change ΔU associated with the change in composition. The new configuration is examined according to the following rules. If $\Delta U = 0$, then move of compositional change is accepted.

However, if $\Delta U \geq 0$, the move is accepted with a certain probability which is given by:

$$P i \to j \propto \exp(-\frac{\Delta U}{K_B T}) \tag{15}$$

where ∆U is the change in the sum of the mixing energy and the chemical potential of the mixture. If the new configuration is rejected one counts the original configuration as a new one and repeats the process by using some other arbitrarily or systematically chosen atoms. In polymer nanocomposites, the MC methods have been used to investigate the molecular structure at nanoparticle surface and evaluate the effects of various factors.

9.3.2   MICROSCALE METHODS

The modeling and simulation at microscale aim to bridge molecular methods and continuum methods and avoid their shortcomings. Specifically, in nanoparticle–polymer systems, the study of structural evolution (i.e., dynamics of phase separation) involves the description of bulk flow (i.e., hydrodynamic behavior) and the interactions between nanoparticle and polymer components. Note that hydrodynamic behavior is relatively straightforward to handle by continuum methods but is very difficult and expensive to treat by atomistic methods. In contrast, the interactions between components can be examined at an atomistic level but are usually not straightforward to incorporate at the continuum level. Therefore, various simulation methods have been evaluated and extended to study the microscopic structure and phase separation of these polymer nanocomposites, including BD, DPD, LB, time-dependent Ginsburg–Landau (TDGL) theory, and dynamic DFT. In these methods, a polymer system is usually treated with a field description or microscopic particles that incorporate molecular details implicitly. Therefore, they are able to simulate the phenomena on length and time scales currently inaccessible by the classical MD methods.

BROWNIAN DYNAMICS (BD)

The BD simulation is similar to MD simulations [26]. However, it introduces a few new approximations that allow one to perform simulations on the microsecond timescale whereas MD simulation is known up to a few nanoseconds. In BD the explicit description of solvent molecules used in MD is replaced with an implicit continuum solvent description. Besides, the internal motions of molecules are typically ignored, allowing a much larger time step than that of MD. Therefore, BD is particularly useful for systems where there is a large gap of time scale governing the motion of different components. For example, in polymer–solvent mixture, a short time-step is required to resolve the fast motion of the solvent molecules, whereas the evolution of the slower modes of the system requires a larger time step. However, if the detailed motion of the solvent molecules is concerned, they may be removed from the simulation and their effects on the polymer are represented by dissipative (-γP) and random (σ ζ   (t)) force terms. Thus, the force in the governing Equation (16) is replaced by a Langevin equation:

$$F_i(t) = \sum_{i \neq j} F_{ij}^c - \gamma P_i + \sigma \zeta_i(t) \qquad (16)$$

where $F^c_{i \, j}$ is the conservative force of particle j acting on particle i, γ, and σ are constants depending on the system, P_i the momentum of particle i, and $\varsigma(t)$ a Gaussian random noise term. One consequence of this approximation of the fast degrees of freedom by fluctuating forces is that the energy and momentum are no longer conserved, which implies that the macroscopic behavior of the system will not be hydrodynamic. In addition, the effect of one solute molecule on another through the flow of solvent molecules is neglected. Thus, BD can only reproduce the diffusion properties but not the hydrodynamic flow properties since the simulation does not obey the Navier–Stokes equations.

DISSIPATIVE PARTICLE DYNAMICS (DPD)

The DPD was originally developed by Hoogerbrugge and Koelman [27]. It can simulate both Newtonian and non-Newtonian fluids, including polymer melts and blends, on microscopic length and time scales. Like MD and BD, DPD is a particle-based method. However, its basic unit is not a single atom or molecule but a molecular assembly (i.e., a particle). The DPD particles are defined by their mass M_i, position r_i, and momentum P_i. The interaction force between two DPD particles i and j can be described by a sum of conservative F^C_{ij}, dissipative F^D_{ij}, and random forces F^R_{ij} [28-30]:

$$ F_j = F_j{}^C + F_j{}^D + F^R{}_j \tag{17} $$

while the interaction potentials in MD are high-order polynomials of the distance r_{ij} between two particles, in DPD the potentials are softened so as to approximate the effective potential at microscopic length scales. The form of the conservative force in particular is chosen to decrease linearly with increasing r_{ij}. Beyond a certain cut-off separation r_c, the weight functions and thus the forces are all zero. Because the forces are pair wise and momentum is conserved, the macroscopic behavior directly incorporates Navier–Stokes hydrodynamics. However, energy is not conserved because of the presence of the dissipative and random force terms which are similar to those of BD, but incorporate the effects of Brownian motion on larger length scales. DPD has several advantages over MD, for example, the hydrodynamic behavior is observed with far fewer particles than required in a MD simulation because of its larger particle size. Besides, its force forms allow larger time steps to be taken than those in MD.

LATTICE BOLTZMANN (LB)

The LB [31] is another microscale method that is suited for the efficient treatment of polymer solution dynamics. It has recently been used to investigate the phase separation of binary fluids in the presence of solid particles. The LB method is originated from lattice gas automaton which is constructed as a simplified, fictitious molecular dynamic in which space, time and particle velocities are all discrete. A typical lattice gas automaton consists of a regular lattice with particles residing on the nodes. The main feature of the LB method is to replace the particle occupation variables (Boolean variables), by single-particle distribution functions (real variables) and neglect individual particle motion and particle–particle correlations in the kinetic equation. There are several ways to obtain the LB equation from either the discrete velocity model or the Boltzmann kinetic equation, and to derive the macroscopic Navier–Stokes equations from the LB equation. An important advantage of the LB method is that microscopic physical interactions of the fluid particles can be conveniently incorporated into the numerical model. Compared with the Navier–Stokes equations, the LB method can handle the interactions among fluid particles and reproduce the microscale mechanism of hydrodynamic behavior. Therefore it belongs to the MD in nature and bridges the gap between the molecular level and macroscopic level. However, its main disadvantage is that it is typically not guaranteed to be numerically stable and may lead to physically unreasonable results, for instance, in the case of high forcing rate or high interparticle interaction strength.

TIME-DEPENDENT GINZBURG–LANDAU METHOD (TDGL)

The TDGL is a microscale method for simulating the structural evolution of phase-separation in polymer blends and block copolymers. It is based on the Cahn–Hilliard–Cook (CHC) nonlinear diffusion equation for a binary blend and falls under the more general phase-field and reaction-diffusion models [32-34]. In the TDGL method, a free-energy function is minimized to simulate a temperature quench from the miscible region of the phase diagram to the immiscible region. Thus, the resulting time-dependent structural evolution of the polymer blend can be investigated by solving the TDGL/CHC equation for the time dependence of the local blend concentration. Glotzer and co-workers have discussed and applied this method to polymer blends and particle-filled polymer systems [35]. This model reproduces the growth kinetics of the TDGL model, demonstrating that such quantities are insensitive to the precise form of the double-well potential of the bulk free-energy term. The TDGL and CDM methods have recently been used to investigate the phase-separation of polymer nanocomposites and polymer blends in the presence of nanoparticles [36-40].

DYNAMIC DFT METHOD

Dynamic DFT method is usually used to model the dynamic behavior of polymer systems and has been implemented in the software package Mesodyn TM from Accelrys [41]. The DFT models the behavior of polymer fluids by combining Gaussian mean-field statistics with a TDGL model for the time evolution of conserved order parameters. However, in contrast to traditional phenomenological free-energy expansion methods employed in the TDGL approach, the free energy is not truncated at a certain level, and instead retains the full polymer path integral numerically. At the expense of a more challenging computation, this allows detailed information about a specific polymer system beyond simply the Flory–Huggins parameter and mobilities to be included in the simulation. In addition, viscoelasticity, which is not included in TDGL approaches, is included at the level of the Gaussian chains. A similar DFT approach has been developed by Doi and coworkers [42, 43] and forms the basis for their new software tool Simulation Utilities for Soft and Hard Interfaces (SUSHI), one of a suite of molecular and mesoscale modeling tools (called OCTA) developed for the simulation of polymer materials [44]. The essence of dynamic DFT method is that the instantaneous unique conformation distribution can be obtained from the off-equilibrium density profile by coupling a fictitious external potential to the Hamiltonian. Once such distribution is known, the free energy is then calculated by standard statistical thermodynamics. The driving force for diffusion is obtained from the spatial gradient of the first functional derivative of the free energy with respect to the density. Here, we describe briefly the equations for both polymer and particle in the diblock polymer–particle composites [38].

9.3.3 MESOSCALE AND MACROSCALE METHODS

Despite the importance of understanding the molecular structure and nature of materials, their behavior can be homogenized with respect to different aspects which can be at different scales. Typically, the observed macroscopic behavior is usually explained by ignoring the discrete atomic and molecular structure and assuming that the material is continuously distributed throughout its volume. The continuum material is thus assumed to have an average density and can be subjected to body forces such as gravity and surface forces. Generally speaking, the macroscale methods (or called continuum methods hereafter) obey the fundamental laws of—(i) continuity, derived from the conservation of mass, (ii) equilibrium, derived from momentum considerations and Newton's second law, (iii) the moment of momentum principle, based on the model that the time rate of change of angular momentum with respect to an arbitrary point is equal to the resultant moment, (iv) conservation of energy, based on the first law of thermodynamics, and (v) conservation of entropy, based on the second law of thermodynamics. These laws provide the basis for the continuum model and must be coupled with the appropriate constitutive equations and the equations of state to provide all the equations necessary for solving a continuum problem. The continuum method relates the deformation of a continuous medium to the external forces acting on the medium and the resulting internal stress and strain. Computational approaches range from simple closed-form analytical expressions to micromechanics and complex structural mechanics calculations based on beam and shell theory. Some continuum methods that have been used in polymer nanocomposites, including micromechanics models (e.g., Halpin–Tsai model and Mori–Tanaka model), equivalent-continuum model, self-consistent model, and finite element analysis.

9.3.4   MICROMECHANICS

Since the assumption of uniformity in continuum mechanics may not hold at the microscale level, micromechanics methods are used to express the continuum quantities associated with an infinitesimal material element in terms of structure and properties of the micro constituents. Thus, a central theme of micromechanics models is the development of a representative volume element (RVE) to statistically represent the local continuum properties. The RVE is constructed to ensure that the length scale is consistent with the smallest constituent that has a first-order effect on the macroscopic behavior. The RVE is then used in a repeating or periodic nature in the full-scale model. The micromechanics method can account for interfaces between constituents, discontinuities, and coupled mechanical and non-mechanical properties. Their purpose is to review the micromechanics methods used for polymer nanocomposites. Thus, we only discuss here some important concepts of micromechanics as well as the Halpin–Tsai model and Mori–Tanaka model.

BASIC CONCEPTS

When applied to particle reinforced polymer composites, micromechanics models usually follow such basic assumptions as:
1.   Linear elasticity of fillers and polymer matrix,
2.   The fillers are axisymmetric, identical in shape and size, and can be characterized by parameters such as aspect ratio,
3.   Well-bonded filler–polymer interface and the ignorance of interfacial slip, filler–polymer debonding or matrix cracking.

The first concept is the linear elasticity, that is, the linear relationship between the total stress and infinitesimal strain tensors for the filler and matrix as expressed by the following constitutive equations:

For filler,
$$\sigma^f = C^f \varepsilon^f \tag{18}$$

For matrix,
$$\sigma^m = C^m \varepsilon^m \tag{19}$$

where C is the stiffness tensor. The second concept is the average stress and strain. Since the point wise stress field $\sigma(x)$ and the corresponding strain field $\varepsilon(x)$ are usually non-uniform in polymer composites, the volume–average stress $\bar{\sigma}$ and strain $\bar{\varepsilon}$ are then defined over the representative averaging volume V, respectively:

$$\bar{\sigma} = \frac{1}{V} \int \sigma(x) dv \tag{20}$$

$$\bar{\varepsilon} = \frac{1}{V}\int \varepsilon(x)dv \tag{21}$$

Therefore, the average filler and matrix stresses are the averages over the corresponding volumes V_f and V_m, respectively:

$$\bar{\sigma}_f = \frac{1}{V_f}\int \sigma(x)dv \tag{22}$$

$$\bar{\sigma}_m = \frac{1}{V_m}\int \sigma(x)dv \tag{23}$$

The average strains for the fillers and matrix are defined, respectively, as

$$\bar{\varepsilon}_f = \frac{1}{V_f}\int \varepsilon(x)dv \tag{24}$$

$$\bar{\varepsilon}_m = \frac{1}{V_m}\int \varepsilon(x)dv \tag{25}$$

Based on the above definitions, the relationships between the filler and matrix averages and the overall averages can be derived as follows:

$$\bar{\sigma} = \bar{\sigma}_f v_f + \bar{\sigma}_m v_m \tag{26}$$

$$\bar{\varepsilon} = \bar{\varepsilon}_f v_f + \bar{\varepsilon}_m v_m \tag{27}$$

where v_f and v_m are the volume fractions of the fillers and matrix, respectively.

The third concept is the average properties of composites which are actually the main goal of a micromechanics model. The average stiffness of the composite is the tensor C that maps the uniform strain to the average stress:

$$\bar{\sigma} = \bar{\varepsilon}C \tag{28}$$

The average compliance S is defined in the same way:

$$\bar{\varepsilon} = \bar{\sigma}S \tag{29}$$

Another important concept is the strain–concentration and stress–concentration tensors A and B which are basically the ratios between the average filler strain (or stress) and the corresponding average of the composites.

$$\overline{\varepsilon_f} = \overline{\varepsilon}A \tag{30}$$

$$\overline{\sigma_f} = \overline{\sigma}B \tag{31}$$

Using the above concepts and equations, the average composite stiffness can be obtained from the strain concentration tensor A and the filler and matrix properties:

$$C = C_m + v_f(C_f - C_m)A \tag{32}$$

HALPIN–TSAI MODEL

The Halpin–Tsai model is a well-known composite theory to predict the stiffness of unidirectional composites as a functional of aspect ratio. In this model, the longitudinal E_{11} and transverse E_{22} engineering moduli are expressed in the following general form:

$$\frac{E}{E_m} = \frac{1 + \zeta \eta v_f}{1 - \eta v_f} \tag{33}$$

where E and E_m represent the Young's modulus of the composite and matrix, respectively, v_f is the volume fraction of filler, and η is given by:

$$\eta = \frac{\dfrac{E}{E_m} - 1}{\dfrac{E_f}{E_m} + \zeta_f} \tag{34}$$

where E_f represents the Young's modulus of the filler and ζ_f the shape parameter depending on the filler geometry and loading direction. When calculating longitudinal modulus E_{11}, ζ_f is equal to l/t, and when calculating transverse modulus E_{22}, ζ_f is equal to w/t. Here, the parameters of l, w, and t are the length, width, and thickness of the dispersed fillers, respectively. If $\zeta_f \rightarrow 0$, the Halpin–Tsai theory converges to the inverse rule of mixture (lower bound):

$$\frac{1}{E} = \frac{v_f}{E_f} + \frac{1-v_f}{E_m}$$

(35)

Conversely, if $\zeta_f \rightarrow \infty$, the theory reduces to the rule of mixtures (upper bound):

$$E = E_f v_f + E_m (1 - v_f)$$

(36)

MORI–TANAKA MODEL

The Mori–Tanaka model is derived based on the principles of Eshelby's inclusion model for predicting an elastic stress field in and around ellipsoidal filler in an infinite matrix. The complete analytical solutions for longitudinal E_{11} and transverse E_{22} elastic moduli of an isotropic matrix filled with aligned spherical inclusion are [45, 46]:

$$\frac{E_{11}}{E_m} = \frac{A_0}{A_0 + v_f (A_1 + 2v_0 A_2)}$$

(37)

$$\frac{E_{22}}{E_m} = \frac{2A_0}{2A_0 + v_f (-2A_3 + (1 - v_0 A_4) + (1 + v_0) A_5 A_0)}$$

(38)

where E_m represents the Young's modulus of the matrix, v_f the volume fraction of filler, v_0 the Poisson's ratio of the matrix, parameters, A0, A1,...,A5 are functions of the Eshelby's tensor and the properties of the filler and the matrix, including Young's modulus, Poisson's ratio, filler concentration and filler aspect ratio [45].

EQUIVALENT-CONTINUUM AND SELF-SIMILAR APPROACHES

Numerous micromechanical models have been successfully used to predict the macroscopic behavior of fiber-reinforced composites. However, the direct use of these models for nanotube-reinforced composites is doubtful due to the significant scale difference between nanotube and typical carbon fiber. Recently, two methods have been proposed for modeling the mechanical behavior of SWCNT composites—equivalent-continuum approach and self-similar approach [47]. The equivalent-continuum approach was proposed by Odegard et al. [48]. In this approach, the MD was used to model the molecular interactions between SWCNT–polymer and a homogeneous equivalent-continuum reinforcing element (e.g., a SWCNT surrounded by polymer) was constructed as shown in Figure 2. Then, micromechanics are used to determine the effective bulk properties of the equivalent-continuum reinforcing element embedded in a continuous polymer. The equivalent-continuum approach consists of four major steps. Step 1–the MD simulation is used to generate the equilibrium structure of a SWCNT–polymer composite and then to establish the RVE of the molecular model and the equivalent-continuum model. Step 2—the potential energies of deformation for the molecular model and effective fiber are derived and equated for identical loading conditions. The bonded and non-bonded interactions within a polymer molecule are quantitatively described by MM. For the SWCNT/polymer system, the total potential energy U^m of the molecular model is:

$$U^m = \sum U^r(K_r) + \sum U^\theta(K_\theta) + \sum U^{vdw}(K_{vdw}) \tag{39}$$

Where U^r, U^θ, and U^{vdw} are the energies associated with covalent bond stretching, bond-angle bending, and vdW interactions, respectively. An equivalent-truss model of the RVE is used as an intermediate step to link the molecular and equivalent-continuum models. Each atom in the molecular model is represented by a pin-joint, and each truss element represents an atomic bonded or non-bonded interaction.

The potential energy of the truss model is:

$$U^t = \sum U^a(E^a) + \sum U^b(E^b) + \sum U^c(E^c) \tag{40}$$

Where U^a, U^b, and U^c are the energies associated with truss elements that represent covalent bond stretching, bond-angle bending, and vdW interactions, respectively. The energies of each truss element are a function of the Young's modulus, E. Step 3—a constitutive equation for the effective fiber is established. Since the values of the elastic stiffness tensor components are not known a priori, a set of loading conditions are chosen such that each component is uniquely determined from:

$$U^f = U^t = U^m \tag{41}$$

Step 4—overall constitutive properties of the dilute and unidirectional SWCNT/polymer composite are determined with Mori–Tanaka model with the mechanical properties of the effective fiber and the bulk polymer. The layer of polymer molecules that are near the polymer/nanotube interface (Figure 2) is included in the effective fiber, and it is assumed that the matrix polymer surrounding the effective fiber has mechanical properties equal to those of the bulk polymer. The self-similar approach was proposed by Pipes and Hubert [49] which consists of three major steps—first, a helical array of SWCNs is assembled. This array is termed as the SWCNT nanoarray where 91 SWCNTs make up the cross-section of the helical nanoarray. Then, the SWCNT nanoarrays is surrounded by a polymer matrix and assembled into a second twisted array, termed as the SWCNT nanowire Finally, the SWCNT nanowires are further impregnated with a polymer matrix and assembled into the final helical array—the SWCNT microfiber. The self-similar geometries described in the nanoarray, nanowire, and microfiber (Figure 3) allows the use of the same mathematical and geometric model for all three geometries [49].

FINITE ELEMENT METHOD (FEM)

The FEM is a general numerical method for obtaining approximate solutions in space to initial-value and boundary-value problems including time-dependent processes. It employs preprocessed mesh generation, which enables the model to fully capture the spatial discontinuities of highly inhomogeneous materials. It also allows complex, nonlinear tensile relationships to be incorporated into the analysis. Thus, it has been widely used in mechanical, biological and geological systems. In FEM, the entire domain of interest is spatially discretized into an assembly of simply shaped subdomains (e.g., hexahedra or tetrahedral in three dimensions, and rectangles or triangles in two dimensions) without gaps and without overlaps. The subdomains are interconnected at joints (i.e., nodes). The implementation of FEM includes the important steps shown in Figure 4. The energy in FEM is taken from the theory of linear elasticity and thus the input parameters are simply the elastic moduli and the density of the material. Since these parameters are in agreement with the values computed by MD, the simulation is consistent across the scales. More specifically, the total elastic energy in the absence of tractions and body forces within the continuum model is given by [50]:

$$U = U_v + U_k \tag{42}$$

$$U_k = 1/2 \int dr p(r) \left| \dot{U}_r \right|^2 \tag{43}$$

$$U_v = \frac{1}{2} \int dr \sum_{\mu,\nu,\lambda,\sigma=1}^{3} \varepsilon_{\mu\nu}(r) C_{\mu\nu\lambda\sigma} \lambda\sigma(r) \tag{44}$$

Where U_v is the Hookian potential energy term which is quadratic in the symmetric strain tensor e, contracted with the elastic constant tensor C. The Greek indices (i.e., m, n, l, and s) denote Cartesian directions. The kinetic energy U_k involves the time rate of change of the displacement field $\dot{U}$, and the mass density ρ.

These are fields defined throughout space in the continuum theory. Thus, the total energy of the system is an integral of these quantities over the volume of the sample dυ. The FEM has been incorporated in some commercial software packages and open source codes (e.g., ABAQUS, ANSYS, Palmyra, and OOF) and widely used to evaluate the mechanical properties of polymer composites. Some attempts have recently been made to apply the FEM to nanoparticle-reinforced polymer nanocomposites. In order to capture the multiscale material behaviors, efforts are also underway to combine the multiscale models spanning from molecular to macroscopic levels [51,52].

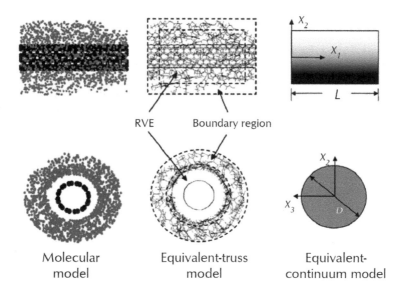

FIGURE 2   Equivalent-continuum modeling of effective fiber .

9.4   MULTI SCALE MODELING OF MECHANICAL PROPERTIES

In Odegard's study [48], a method has been presented for linking atomistic simulations of nano-structured materials to continuum models of the corresponding bulk material. For a polymer composite system reinforced with SWCNTs, the method provides the steps whereby the nanotube, the local polymer near the nanotube, and the nanotube/polymer interface can be modeled as an effective continuum fiber by using an equivalent-continuum model. The effective fiber retains the local molecular structure and bonding information, as defined by MD, and serves as a means for linking the equivalent-continuum and micromechanics models. The micromechanics method is then available for the prediction of bulk mechanical properties of SWCNT/polymer composites as a function of nanotube size, orientation, and volume fraction. The utility of this method was examined by modeling tow composites that both having a interface. The elastic stiffness constants of the composites were determined for both aligned and three-dimensional randomly oriented nanotubes, as a function of nanotube length and volume fraction. They used Mori–Tanaka model [53] for random and oriented fibers position and compare their model with mechanical properties, the interface between fiber and matrix was assumed perfect. Motivated by micrographs showing that embedded nanotubes often exhibit significant curvature within the polymer, Fisher et al. [54] have developed a model combining finite element results and micromechanical methods (Mori–tanaka) to determine the effective reinforcing modulus (ERM) of a wavy embedded nanotube with perfect bonding and random fiber orientation assumption. This ERM is then used within a multiphase micromechanics model to predict the effective modulus of a polymer reinforced with a distribution of wavy nanotubes. We found that even slight nanotube curvature significantly reduces the effective reinforcement when compared to straight nanotubes. These results suggest that nanotube waviness may be an additional mechanism limiting the modulus enhancement of nanotube-reinforced polymers. Bradshaw et al. [55] investigated the degree to which the characteristic waviness of nanotubes embedded in polymers can impact the effective stiffness of these materials. A 3D finite element model of a single infinitely long sinusoidal fiber within an infinite matrix is used to numerically compute the dilute strain concentration tensor. A Mori–Tanaka model utilizes this tensor to predict the effective modulus of the material with aligned or randomly oriented inclusions. This hybrid finite element micromechanical modeling technique is a powerful extension of general micromechanics modeling and can be applied to any composite microstructure containing non-ellipsoidal inclusions. The results demonstrate that nanotube waviness results in a reduction of the effective modulus of the composite relative to straight nanotube reinforcement. The degree of reduction is dependent on the ratio of the sinusoidal wavelength to the nanotube diameter. As this wavelength ratio increases, the effective stiffness of a composite with randomly oriented wavy nanotubes converges to the result obtained with straight nanotube inclusions.

The effective mechanical properties of CNT-based composites are evaluated by Liu and Chen [56] using a 3D nanoscale RVE based on 3D elasticity theory and solved by the FEM. Formulas to extract the material constants from solutions for the RVE under three loading cases are established using the elasticity. An extended rule of mixtures, which can be used to estimate the Young's modulus in the axial direction of the RVE and to validate the numerical solutions for short CNTs, is also derived using the strength of materials theory. Numerical examples using the FEM to evaluate the effective material constants of a CNT-based composites are presented, which demonstrate that the reinforcing capabilities of the CNTs in a matrix are significant. With only about 2% and 5% volume fractions of the CNTs in a matrix, the stiffness of the composite in the CNT axial direction can increase as many as 0.7 and 9.7 times for the cases of short and long CNT fibers, respectively. These simulation results, which are believed to be the first of its kind for CNT-based composites, are consistent with the experimental results reported in the literature Schadler et al. [57], Wagner et al. [58], and Qian et al.[59]. The developed extended rule of mixtures is also found to be quite effective in evaluating the stiffness of the CNT-based composites in the CNT axial direction. Many research issues need to be addressed in the modeling and simulations of CNTs in a matrix material for the development of nanocomposites. Analytical methods and simulation models to extract the mechanical properties of the CNT-based nanocomposites need to be further developed and verified with experimental results. The analytical method and simulation approach developed in this paper are only a preliminary study. Different type of RVEs, load cases and different solution methods should be investigated. Different interface conditions, other than perfect bonding, need to be investigated using different models to more accurately account for the interactions of the CNTs in a matrix material at the nanoscale. Nanoscale interface cracks can be analyzed using simulations to investigate the failure mechanism in nanomaterials. Interactions among a large number of CNTs in a matrix can be simulated if the computing power is available. The SWCNTs and MWCNTs as reinforcing fibers in a matrix can be studied by simulations to find out their advantages and disadvantages. Finally, large multiscale simulation models for CNT-based composites, which can link the models at the nano, micro, and macro scales, need to be developed, with the help of analytical and experimental work [56]. The three RVEs proposed in [60] and shown in Figure 3 are relatively simple regarding the models and scales and pictures in Figure 4 are Three loading cases for the cylindrical RVE. However, this is only the first step toward more sophisticated and large scale simulations of CNT-based composites. As the computing power and confidence in simulations of CNT-based composites increase, large scale 3D models containing hundreds or even more CNTs, behaving linearly or nonlinearly, with coatings or of different sizes, distributed evenly or randomly, can be employed to investigate the interactions among the CNTs in a matrix and to evaluate the effective material properties. Other numerical methods can also be attempted for the modeling and simulations of CNT-based composites, which may offer some advantages over the FEM approach. For example, the boundary element method, Liu et al. [60] and Chen and Liu [61], accelerated with the fast multipole techniques , Fu et al. [62], Nishimura et al. [63], and the mesh free methods may enable one to model an RVE with thousands of CNTs in a matrix on a desktop computer [64]. Analysis of the CNT-based composites using the boundary element method is already underway and will be reported subsequently.

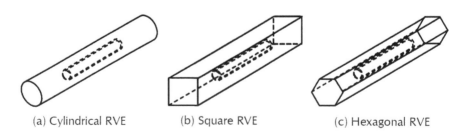

(a) Cylindrical RVE   (b) Square RVE   (c) Hexagonal RVE

FIGURE 3   Three nanoscale RVEs for the analysis of CNT-based nanocomposites [56].

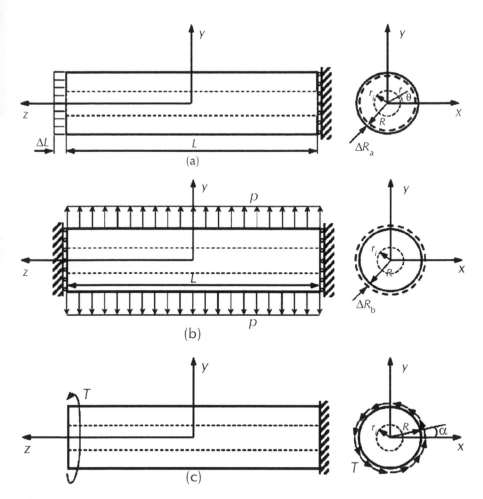

FIGURE 4   Three loading cases for the cylindrical RVE used to evaluate the effective material properties of the CNT-based composites. (a) under axial stretch DL, (b) under lateral uniform load P, and (c) under torsional load T [56].

The effective mechanical properties of CNT based composites are evaluated using square RVEs based on 3D elasticity theory and solved by the FEM. Formulas to extract the effective material constants from solutions for the square RVEs under two loading cases are established based on elasticity. Square RVEs with multiple CNTs are also investigated in evaluating the Young's modulus and Poisson's ratios in the transverse plane. Numerical examples using the FEM are presented, which demonstrate that the load-carrying capabilities of the CNTs in a matrix are significant. With the addition of only about 3.6% volume fraction of the CNTs in a matrix, the stiffness of the composite in the CNT axial direction can increase as much as 33% for the case of long CNT fibers [65]. These simulation results are consistent with both the experimental ones reported in the literature [56–59,66]. It is also found that cylindrical RVEs tend to overestimate the effective Young's moduli due to the fact that they overestimate the volume fractions of the CNTs in a matrix. The square RVEs, although more demanding in modeling and computing, may be the preferred model in future simulations for estimating the effective material constants, especially when multiple CNTs need to be considered. Finally, the rules of mixtures, for both long and short CNT cases, are found to be quite accurate in estimating the effective Young's moduli in the CNT axial direction. This may suggest that 3D FEM modeling may not be necessary in obtaining the effective material constants in the CNT direction, as in the studies of the conventional fiber reinforced composites. Efforts in comparing the results presented in this paper using the continuum approach directly with the MD simulations are underway. This is feasible now only for a smaller RVE of one CNT embedded in a matrix. In future research, the MD and continuum approach should be integrated in a multiscale modeling and simulation environment for analyzing the CNT-based composites. More efficient models of the CNTs in a matrix also need to be developed, so that a large number of CNTs, in different shapes and forms (curved or twisted), or randomly distributed in a matrix, can be modeled. The ultimate validation of the simulation results should be done with the nanoscale or microscale experiments on the CNT reinforced composites [64].

Griebel and Hamaekers [67] reviewed the basic tools utilized in computational nanomechanics and materials, including the relevant underlying principles and concepts. These tools range from subatomic *ab initio* methods to classical MD and multiple-scale approaches. The energetic link between the quantum mechanical and classical systems has been discussed, and limitations of the standing alone MD simulations have been shown on a series of illustrative examples. The need for multi-scale simulation methods to tackle nanoscale aspects of material behavior was therefore emphasized, that was followed by a review and classification of the mainstream and emerging multi-scale methods. These simulation methods include the broad areas of quantum mechanics, the MD and multiple-scale approaches, based on coupling the atomistic and continuum models. They summarize the strengths and limitations of currently available multiple-scale techniques, where the emphasis is made on the latest perspective approaches, such as the bridging scale method, multi-scale boundary conditions, and multi-scale fluidics. Example problems, in which multiple-scale simulation methods yield equivalent results to full atomistic simulations at fractions of the computational cost, were shown. They compare their results with Odegard, et al. [48], the micromechanic method was BEM Halpin–Tsai Equation [68] with aligned fiber by perfect bonding.

The solutions of the strain-energy-changes due to a SWCNT embedded in an infinite matrix with imperfect fiber bonding are obtained through numerical method by Wan, et al. [69]. A "critical" SWCNT fiber length is defined for full load transfer between the SWCNT and the matrix, through the evaluation of the strain-energy-changes for different fiber lengths The strain-energy-change is also used to derive the effective longitudinal Young's modulus and effective bulk modulus of the composite, using a dilute solution. The main goal of their research was investigation of strain-energy-change due to inclusion of SWCNT using FEM. To achieve full load transfer between the SWCNT and the matrix, the length of SWCNT fibers should be longer than a 'critical' length if no weak interphase exists between the SWCNT and the matrix [69].

A hybrid atomistic/continuum mechanics method is established in the Feng, et al. study [70] the deformation and fracture behaviors of CNTs in composites. The unit cell containing a CNT embedded in a matrix is divided in three regions, which are simulated by the atomic-potential method, the continuum method based on the modified Cauchy–Born rule, and the classical continuum mechanics, respectively. The effect of CNT interaction is taken into account *via* the Mori–Tanaka effective field method of micromechanics. This method not only can predict the formation of Stone–Wales (5-7-7-5) defects, but also simulate the subsequent deformation and fracture process of CNTs. It is found that the critical strain of defect nucleation in a CNT is sensitive to its chiral angle but not to its diameter. The critical strain of Stone–Wales defect formation of zigzag CNTs is nearly twice that of armchair CNTs. Due to the constraint effect of matrix, the CNTs embedded in a composite are easier to fracture in comparison with those not embedded. With the increase in the Young's modulus of the matrix, the critical breaking strain of CNTs decreases.

Estimation of effective elastic moduli of nanocomposites was performed by the version of effective field method developed in the framework of quasi-crystalline approximation when the spatial correlations of inclusion location take particular ellipsoidal forms [71]. The independent justified choice of shapes of inclusions and correlation holes provide the formulae of effective moduli which are symmetric, completely explicit, and easily to use. The parametric numerical analyses revealed the most sensitive parameters influencing the effective moduli which are defined by the axial elastic moduli of nanofibers rather than their transversal moduli as well as by the justified choice of correlation holes, concentration, and prescribed random orientation of nanofibers [72].

Li and Chou [73,74] have reported a multiscale modeling of the compressive behavior of CNT/polymer composites. The nanotube is modeled at the atomistic scale, and the matrix deformation is analyzed by the continuum FEM. The nanotube and polymer matrix are assumed to be bonded by vdW interactions at the interface. The stress distributions at the nanotube/polymer interface under isostrain and isostress loading conditions have been examined. They have used beam elements for SWCNT using molecular structural mechanics, truss rod for vdW links and cubic elements for matrix. The rule of mixture was used as for comparision in this research. The buckling forces of nanotube/polymer composites for different nanotube lengths and diameters are computed. The results indicate that continuous nanotubes can most effectively enhance the composite buckling resistance.

Anumandla and Gibson [75] describe an approximate, yet comprehensive, closed form micromechanics model for estimating the effective elastic modulus of CNT-reinforced composites. The model incorporates the typically observed nanotube curvature, the nanotube length, and both 1D and 3D random arrangement of the nanotubes. The analytical results obtained from the closed form micromechanics model for nanoscale RVEs and results from an equivalent finite element model for ERM of the nanotube reveal that the reinforcing modulus is strongly dependent on the waviness, wherein, even a slight change in the nanotube curvature can induce a prominent change in the effective reinforcement provided. The micromechanics model is also seen to produce reasonable agreement with experimental data for the effective tensile modulus of composites reinforced with MWCNTs and having different MWCNT volume fractions.

Effective elastic properties for CNT reinforced composites are obtained through a variety of micromechanics techniques [76]. Using the in-plane elastic properties of graphene, the effective properties of CNTs are calculated utilizing a composite cylinders micromechanics technique as a first step in a two-step process. These effective properties are then used in the self-consistent and Mori–Tanaka methods to obtain effective elastic properties of composites consisting of aligned SWCNTs or MWCNTs embedded in a polymer matrix. Effective composite properties from these averaging methods are compared to a direct composite cylinders approach extended from the work of Hashin and Rosen [77] and Christensen and Lo [78]. Comparisons with finite element simulations are also performed. The effects of an interphase layer between the nanotubes and the polymer matrix as result of functionalization is also investigated using a multi-layer composite cylinders approach. Finally, the modeling of the clustering of nanotubes into bundles due to interatomic forces is accomplished herein using a tessellation method in conjunction with a multi-phase Mori–Tanaka technique. In addition to aligned nanotube composites, modeling of the effective elastic properties of randomly dispersed nanotubes into a matrix is performed using the Mori–Tanaka method, and comparisons with experimental data are made.

Selmi, et al. [79] deal with the prediction of the elastic properties of polymer composites reinforced with SWCNTs. Their contribution is the investigation of several micromechanical models, while most of the papers on the subject deal with only one approach. They implemented four homogenization schemes, a sequential one and three others based on various extensions of the Mori–Tanaka (M–T) mean-field homogenization model—two-level (M–T/M–T), two-step (M–T/M–T), and two-step (M–T/Voigt). Several composite systems are studied, with various properties of the matrix and the graphene, short or long nanotubes, fully aligned or randomly oriented in 3D or 2D. Validation targets are experimental data or finite element results, either based on a 2D periodic unit cell or a 3D RVE. The comparative study showed that there are cases where all micromechanical models give adequate predictions, while for some composite materials and some properties, certain models fail in a rather spectacular fashion. It was found that the two-level (M–T/M–T) homogenization model gives the best predictions in most cases. After the characterization of the discrete nanotube structure using a homogenization method based on energy equivalence, the sequential, the two-step (M–T/M–T), the two-step (M–T/Voigt), the two-level (M–T/M–T), and finite element models were used to predict the elastic properties of SWCNT/polymer composites. The data delivered by the micromechanical models are compared against those obtained by finite element analyzes or experiments. For fully aligned, long nanotube polymer composite, it is the sequential and the two-level (M–T/M–T) models which delivered good predictions. For all composite morphologies (fully aligned, two-dimensional in-plane random orientation, and three-dimensional random orientation), it is the two-level (M–T/M–T) model which gave good predictions compared to finite element and experimental results in most situations. There are cases where other micromechanical models failed in a spectacular way.

Luo, et al. [80] have used multi-scale homogenization (MH) and FEM for wavy and straight SWCNTs, they have compare their results with Mori–Tanaka, Cox, and Halpin–Tsai, Fu, et al. [81] and Lauke [82]. Trespass, et al. [83] used 3D elastic beam for C-C bond, 3D space frame for CNT, and progressive fracture model for prediction of elastic modulus, they used rule of mixture for compression of their results. Their assumption was embedded a single SWCNT in polymer with perfect bonding. The multi-scale modeling, MC, FEM, and using equivalent continuum method was used by Spanos and Kontsos [84] and compared with Zhu, et al. [85] and Paiva, et al. [86] results.

Bhuiyan et al. [87], the effective modulus of CNT/PP composites is evaluated using FEA of a 3D RVE which includes the PP matrix, multiple CNTs and CNT/PP interphase and accounts for poor dispersion and nonhomogeneous distribution of CNTs within the polymer matrix, weak CNT/polymer interactions, CNT agglomerates of various sizes, and CNTs orientation and waviness. Currently, there is no other model, theoretical or numerical, that accounts for all these experimentally observed phenomena and captures their individual and combined effect on the effective modulus of nanocomposites. The model is developed using input obtained from experiments and validated against experimental data. The CNT reinforced PP composites manufactured by extrusion and injection molding are characterized in terms of tensile modulus, thickness and stiffness of CNT/PP interphase, size of CNT agglomerates and CNT distribution using tensile testing, and AFM and SEM, respectively. It is concluded that CNT agglomeration and waviness are the two dominant factors that hinder the great potential of CNTs as polymer reinforcement. The proposed model provides the upper and lower limit of the modulus of the CNT/PP composites and can be used to guide the manufacturing of composites with engineered properties for targeted applications. The CNT agglomeration can be avoided by employing processing techniques such as sonication of CNTs, stirring, calendaring, and so on, whereas CNT waviness can be eliminated by increasing the injection pressure during molding and mainly by using CNTs with smaller aspect ratio. Increased pressure during molding can also promote the alignment of CNTs along the applied load direction. The 3D modeling capability presented in this study gives an insight on the upper and lower bound of the CNT/PP composites modulus quantitatively by accurately capturing the effect of various processing parameters. It is observed that when all the experimentally observed factors are considered together in the FEA the modulus prediction is in good agreement with the modulus obtained from the experiment. Therefore, it can be concluded that the FEM models proposed in this study by systematically incorporating experimentally observed characteristics can be effectively used for the determination of mechanical properties of nanocomposite materials. Their result is in agreement with the results reported in [88], the theoretical micromechanical models, shown in Figure 5, are used to confirm that our FEM model predictions follow the same trend with the one predicted by the models as expected.

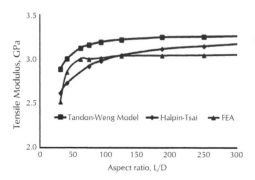

FIGURE 5 Effective modulus of 5 wt% CNT/PP composites—theoretical models vs. FEA.

For reasons of simplicity and in order to minimize the mesh dependency on the results the hollow CNTs are considered as solid cylinders of circular cross-sectional area with an equivalent average diameter, shown in Figure 6, calculated by equating the volume of the hollow CNT to the solid one.[87-88].

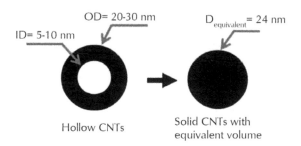

FIGURE 6 Schematic of the CNTs considered for the FEA.

The micromechanical models used for the comparison was Halpin–Tsai (H–T) [89] and Tandon–Weng (T–W) [90] model and the comparison was performed for 5 wt% CNT/PP. It was noted that the H–T model results to lower modulus compared to FEA because H–T equation does not account for maximum packing fraction and the arrangement of the reinforcement in the composite. A modified H–T model that account for this has been proposed in the literature [91]. The effect of maximum packing fraction and the arrangement of the reinforcement within the composite become less significant at higher aspect ratios [92].

A finite element model of CNT, inter-phase and its surrounding polymer is constructed to study the tensile behavior of embedded short CNTs in polymer matrix in presence of vdW interactions in inter-phase region by Shokrieh and Rafiee [93].The inter-phase is modeled using non-linear spring elements capturing the force-distance curve of vdW interactions. The constructed model is subjected to tensile loading to extract longitudinal Young's modulus. The obtained results of this work have been compared with the results of previous research of the same authors [94] on long embedded CNT in polymer matrix. It shows that the capped short CNTs reinforce polymer matrix less efficient than long CNTs.

Despite the fact that researches have succeeded to grow the length of CNTs up to 4 cm as a world record in US Department of Energy Los Alamos National Laboratory [95] and also there are some evidences on producing CNTs with lengths up to millimeters [96,97], the CNTs are commercially available in different lengths ranging from 100 nm to approximately 30 lm in the market based on employed process of growth [98-101]. Chemists at Rice University have identified a chemical process to cut CNTs into short segments [102]. As a consequent, it can be concluded that the SWCNTs with lengths smaller than 1000 nm do not contribute significantly in reinforcing polymer matrix. On the other hand, the efficient length of reinforcement for a CNT with (10, 10) index is about 1.2 lm and short CNT with length of 10.8 lm can play the same role as long CNT reflecting the uppermost value reported in our previous research [94]. Finally, it is shown that the direct use of Halpin–Tsai equation to predict the modulus of SWCNT/composites overestimates the results. It is also observed that application of previously developed long equivalent fiber stiffness [94] is a good candidate to be used in Halpin–Tsai equations instead of Young's modulus of CNT. Halpin–Tsai equation is not an appropriate model for smaller lengths, since there is not any reinforcement at all for very small lengths.

Earlier, a nanomechanical model has been developed by Chowdhury et, al. [103] to calculate the tensile modulus and the tensile strength of randomly oriented short CNTs reinforced nanocomposites, considering the statistical variations of diameter and length of the CNTs. According to this model, the entire composite is divided into several composite segments which contain CNTs of almost the same diameter and length. The tensile modulus and tensile strength of the composite are then calculated by the weighted sum of the corresponding modulus and strength of each composite segment. The existing micromechanical approach for modeling the short fiber composites is modified to account for the structure of the CNTs, to calculate the modulus and the strength of each segmented CNT reinforced composites. The MWCNTs with and without inter-tube bridging (see Figure 7) have been considered. Statistical variations of the diameter and length of the CNTs are modeled by a normal distribution. Simulation results show that CNTs inter-tube bridging, length, and diameter affect the nanocomposites modulus and strength. Simulation results have been compared with the available experimental results and the comparison concludes that the developed model can be effectively used to predict tensile modulus and tensile strength of CNTs reinforced composites.

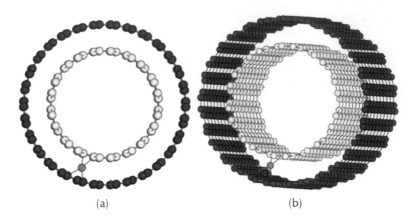

(a)   (b)

FIGURE 7   Schematic of MWCNT with inter-tube bridging, (a) top view and (b) oblique view [103].

The effective elastic properties of CNT-reinforced polymers have been evaluated by Tserpes and Chanteli [104] as functions of material and geometrical parameters using a homogenized RVE. The RVE consists of the polymer matrix, a MWCNT embedded into the matrix and the interface between them. The parameters considered are the nanotube aspect ratio, the nanotube volume fraction as well as the interface stiffness and thickness. For the MWCNT, both isotropic and orthotropic material properties have been considered. Analyses have been performed by means of a 3D FE model of the RVE. The results indicate a significant effect of nanotube volume fraction. The effect of nanotube aspect ratio appears mainly at low values and diminishes after the value of 20. The interface mostly affects the effective elastic properties at the transverse direction. Having evaluated the effective elastic properties of the MWCNT–polymer at the microscale, the RVE has been used to predict the tensile modulus of a polystyrene specimen reinforced by randomly aligned MWCNTs for which experimental data exist in the literature. A very good agreement is obtained between the predicted and experimental tensile moduli of the specimen. The effect of nanotube alignment on the specimen's tensile modulus has been also examined and found to be significant since as misalignment increases the effective tensile modulus decreases radically. The proposed model can be used for the virtual design and optimization of CNT-polymer composites since it has proven capable of assessing the effects of different material and geometrical parameters on the elastic properties of the composite and predicting the tensile modulus of CNT-reinforced polymer specimens.

9.5   MODELING OF THE INTERFACE

9.5 1   INTRODUCTION

The superior mechanical properties of the nanotubes alone do not ensure mechanically superior composites because the composite properties are strongly influenced by the mechanics that govern the nanotube-polymer interface. Typically in composites, the constituents do not dissolve or merge completely and therefore, normally, exhibit an interface between one another, which can be considered as a different material with different mechanical properties. The structural strength characteristics of composites greatly depend on the nature of bonding at the interface, the mechanical load transfer from the matrix (polymer) to the nanotube and the yielding of the interface. As an example, if the composite is subjected to tensile loading and there exists perfect bonding between the nanotube and polymer and/or a strong interface then the load (stress) is transferred to the nanotube, since the tensile strength of the nanotube (or the interface) is very high the composite can withstand high loads. However, if the interface is weak or the bonding is poor, on application of high loading either the interface fails or the load is not transferred to the nanotube and the polymer fails due to their lower tensile strengths. Consider another example of transverse crack propagation. When the crack reaches the interface, it will tend to propagate along the interface, since the interface is relatively weaker (generally) than the nanotube (with respect to resistance to crack propagation). If the interface is weak, the crack will cause the interface to fracture and result in failure of the composite. In this aspect, CNTs are better than traditional fibers (glass and carbon) due to their ability to inhibit nano and micro cracks. Hence, the knowledge and understanding of the nature and mechanics of load (stress) transfer between the nanotube and polymer and properties of the interface is critical for manufacturing of mechanically enhanced CNT-polymer composites and will enable in tailoring of the interface for specific applications or superior mechanical properties. Broadly, the interfacial mechanics of CNT-polymer composites is appealing from three aspects: mechanics, chemistry, and physics.

From a mechanics point of view, the important questions are:

1.   The relationship between the mechanical properties of individual constituents, that is nanotube and polymer, and the properties of the interface and the composite overall.
2.   The effect of the unique length scale and structure of the nanotube on the property and behavior of the interface.
3.   Ability of the mechanics modeling to estimate the properties of the composites for the design process for structural applications.

From a chemistry point of view, the interesting issues are:

1.   The chemistry of the bonding between polymer and nanotubes, especially the nature of bonding (e.g. covalent or non-covalent and electrostatic).
2.   The relationship between the composite processing and fabrication conditions and the resulting chemistry of the interface.
3.   The effect of functionalization (treatment of the polymer with special molecular groups like hydroxyl or halogens) on the nature and strength of the bonding at the interface.

From the physics point of view, researchers are interested in

1.   The CNT-polymer interface serves as a model nanomechanical or a lower dimensional system (1D) and physicists are interested in the nature of forces dominating at the nanoscale and the effect of surface forces (which are expected to be significant due to the large surface to volume ratio).

2. The length scale effects on the interface and the differences between the phenomena of mechanics at the macro (or meso) and the nanoscale.

9.5.2 SOME MODELING METHOD IN INTERFACE MODELING

Computational techniques have extensively been used to study the interfacial mechanics and nature of bonding in CNT-polymer composites. The computational studies can be broadly classified as atomistic simulations and continuum methods. The atomistic simulations are primarily based on MD simulations and DFT [105-110]. The main focus of these techniques was to understand and study the effect of bonding between the polymer and nanotube (covalent, electrostatic or vdW forces) and the effect of friction on the interface. The continuum methods extend the continuum theories of micromechanics modeling and fiber-reinforced composites (elaborated in the next section) to CNT-polymer composites [111-114] and explain the behavior of the composite from a mechanics point of view.

On the experimental side, the main types of studies that can be found in literature are as follows:

1. Researchers have performed experiments on CNT-polymer bulk composites at the macroscale and observed the enhancements in mechanical properties (like elastic modulus and tensile strength) and tried to correlate the experimental results and phenomena with continuum theories like micro-mechanics of composites or Kelly Tyson shear lag model [105,115-120].
2. Raman spectroscopy has been used to study the reinforcement provided by CNTs to the polymer, by straining the CNT-polymer composite and observing the shifts in Raman peaks [121-125].
3. *In situ* TEM straining has also been used to understand the mechanics, fracture and failure processes of the interface. In these techniques, the CNT-polymer composite (an electron transparent thin specimen) is strained inside a TEM and simultaneously imaged to get real-time and spatially resolved (1 nm) information [110,126].

9.5.3 NUMERICAL APPROACH

A MD model may serve as a useful guide, but its relevance for a covalent-bonded system of only a few atoms in diameter is far from obvious. Because of this, the phenomenological multiple column models that considers the interlayer radial displacements coupled through the vdW forces is used. It should also be mentioned the special features of load transfer, in tension and in compression, in MWCNT-epoxy composites studied by Schadler et al. [57] who detected that load transfer in tension was poor in comparison to load transfer in compression, implying that during load transfer to MWCNTs, only the outer layers are stressed in tension due to the telescopic inner wall sliding (reaching at the shear stress 0.5 MPa [127]), whereas all the layers respond in compression. It should be mentioned that NTCMs usually contain not individual, separated SWCNTs, but rather bundles of closest-packed SWCNTs [128], where the twisting of the CNTs produces the radial force component giving the rope structure more stable than wires in parallel. Without strong chemically bonding, load transfer between the CNTs and the polymer matrix mainly comes from weak electrostatic and vdW interactions, as well as stress/deformation arising from mismatch in the coefficients of thermal expansion [129]. Numerous researchers [130] have attributed lower than-predicted CNT-polymer composite properties to the availability of only a weak interfacial bonding. So Frankland et al. [106] demonstrated by MD simulation that the shear strength of a polymer/nanotube interface with only vdW interactions could be increased by over an order of magnitude at the occurrence of covalent bonding for only 1% of the nanotubes carbon atoms to the polymer matrix. The recent force-field-based molecular-mechanics calculations [131] demonstrated that the binding energies and frictional forces play only a minor role in determining the strength of the interface. The key factor in forming a strong bond at the interface is having a helical conformation of the polymer around the nanotube, polymer wrapping around nanotube improves the polymer-nanotube interfacial strength, although configurationally thermodynamic considerations do not necessarily support these architectures for all polymer chains [132]. Thus, the strength of the interface may result from molecular-level entanglement of the two phases and forced long-range ordering of the polymer. To ensure the robustness of data reduction schemes that are based on continuum mechanics, a careful analysis of continuum approximations used in macromolecular models and possible limitations of these approaches at the nanoscale are additionally required that can be done by the fitting of the results obtained by the use of the proposed phenomenological interface model with the experimental data of measurement of the stress distribution in the vicinity of a nanotube.

Meguid et.al. [133] investigated the interfacial properties of CNT reinforced polymer composites by simulating a nanotube pull-out experiment. An atomistic description of the problem was achieved by implementing constitutive relations that are derived solely from interatomic potentials. Specifically, they adopt the Lennard–Jones (LJ) interatomic potential to simulate a non-bonded interface, where only the vdW interactions between the CNT and surrounding polymer matrix was assumed to exist. The effects of such parameters as the CNT embedded length, the number of vdW interactions, the thickness of the interface, the CNT diameter and the cut-off distance of the LJ potential on the interfacial shear strength (ISS) are investigated and discussed. The problem is formulated for both a generic thermoset polymer and a specific two-component epoxy based on diglycidyl ether of bisphenol A (DGEBA) and triethylenetetramine (TETA) formulation. The study further illustrated that by accounting for different CNT capping scenarios and polymer morphologies around the embedded end of the CNT, the qualitative correlation between simulation and experimental pull-out profiles can be improved. Only vdW interactions were considered between the atoms in the CNT and the polymer implying a non-bonded system. The vdW interactions were simulated using the LJ potential, while the CNT was described using the modified Morse potential. The results reveal that the ISS shows a linear dependence on the vdW interaction density and decays significantly with increasing nanotube embedded length. The thickness of the interface was also varied and our results reveal that lower interfacial thicknesses favor higher ISS. When incorporating a 2.5Ψ cut-off distance to the LJ potential, the predicted ISS shows an error of approximately 25.7% relative to a solution incorporating an infinite cut-off distance. Increasing the diameter of the

CNT was found to increase the peak pull-out force approximately linearly. Finally, an examination of polymeric and CNT capping conditions showed that incorporating an end cap in the simulation yielded high initial pull-out peaks that better correlate with experimental findings. These findings have a direct bearing on the design and fabrication of CNT reinforced epoxy composites.

Fiber pull-out tests have been well recognized as the standard method for evaluating the interfacial bonding properties of composite materials. The output of these tests is the force required to pullout the nanotube from the surrounding polymer matrix and the corresponding interfacial shear stresses involved. The problem is formulated using a RVE which consists of the reinforcing CNT, the surrounding polymer matrix, and the CNT/polymer interface as depicted in Figure 8 (a and b) shows a schematic of the pull-out process, where x is the pullout distance and L is the embedded length of the nanotube. The atomistic-based continuum (ABC) multiscale modeling technique is used to model the RVE. The approach adopted here extends the earlier work of Wernik and Meguid [134].

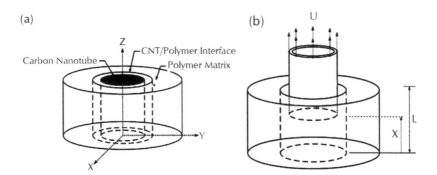

FIGURE 8   Schematic depictions of (a) the RVE and (b) the pull-out process [133].

The new features of the current work relate to the approach adopted in the modeling of the polymer matrix and the investigation of the CNT polymer interfacial properties as appose to the effective mechanical properties of the RVE. The idea behind the ABC technique is to incorporate atomistic interatomic potentials into a continuum framework. In this way, the interatomic potentials introduced in the model capture the underlying atomistic behavior of the different phases considered. Thus, the influence of the nanophase is taken into account *via* appropriate atomistic constitutive formulations. Consequently, these measures are fundamentally different from those in the classical continuum theory. For the sake of completeness, Wernik and Meguid provided a brief outline of the method detailed in their earlier work [133-134].

The cumulative effect of the vdW interactions acting on each CNT atom is applied as a resultant force on the respective node which is then resolved into its three Cartesian components. These processes is depicted in Figure 9, during each iteration of the pull-out process, the expression is re-evaluated for each vdW interaction and the cumulative resultant force and its three Cartesian components are updated to correspond to the latest pull-out configuration. Figure 10 shows a segment of the CNT with the cumulative resultant vdW force vectors as they are applied to the CNT atoms.

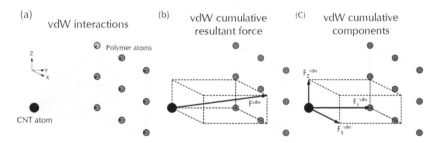

FIGURE 9   The process of nodal vdW force application, (a) vdW interactions on an individual CNT atom, (b) the cumulative resultant vdW force, and (c) the cumulative vdW Cartesian components.

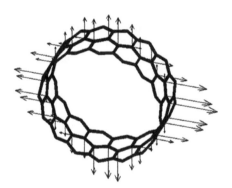

FIGURE 10   Segment of CNT with cumulative resultant vdW force vectors.

Yang et al. [135] investigated the CNT size effect and weakened bonding effect between an embedded CNT and surrounding matrix were characterized using MD simulations. Assuming that the equivalent continuum model of the CNT atomistic structure is a solid cylinder, the transversely isotropic elastic constants of the CNT decreased as the CNT radius increased. Regarding the elastic stiffness of the nanocomposite unit cell, the same CNT size dependency was observed in all independent components, and only the longitudinal Young's modulus showed a positive reinforcing effect whereas other elastic moduli demonstrated negative reinforcing effects as a result of poor load transfer at the interface. To describe the size effect and weakened bonding effect at the interface, a modified multi-inclusion model was derived using the concepts of an effective CNT and effective matrix. During the scale bridging process incorporating the MD simulation results and modified multi-inclusion model, we found that both the elastic modulus of the CNT and the adsorption layer near the CNT contributed to the size-dependent elastic modulus of the nanocomposites. Using the proposed multiscale bridging model, the elastic modulus for nanocomposites at various volume fractions and CNT sizes could be estimated. Among three major factors (CNT waviness, the dispersion state, and adhesion between the CNT and matrix), the proposed model considered only the weakened bonding effect. However, the present multiscale framework can be easily applied in considering the aforementioned factors and describing the real nanocomposite microstructures. In addition, by considering chemically grafted molecules (covalent or non-covalent bonds) to enhance the interfacial load transfer mechanism in MD simulations, the proposed multiscale approach can offer a deeper understanding of the reinforcing mechanism, and a more practical analytical tool with which to analyze and design functional nanocomposites. The analytical estimation reproduced from the proposed multiscale model can also provide useful information in modeling finite element-based RVEs of nanocomposite microstructures for use in multifunctional design.

The effects of the interphase and RVE configuration on the tensile, bending, and torsional properties of the suggested nanocomposite were investigated by Ayatollahi et al. [136]. It was found that the stiffness of the nanocomposite could be affected by a strong interphase much more than by a weaker interphase. In addition, the stiffness of the interphase had the maximum effect on the stiffness of the nanocomposite in the bending loading conditions. Furthermore, it was revealed that the ratio of Le/Ln in RVE can dramatically affect the stiffness of the nanocomposite especially in the axial loading conditions.

For CNTs not well bonded to polymers, Jiang et al. [137] established a cohesive law for CNT/polymer interfaces. The cohesive law and its properties (e.g. cohesive strength and cohesive energy) are obtained directly from the Lennard–Jones potential from the vdW interactions. Such a cohesive law is incorporated in the micromechanics model to study the mechanical behavior of CNT-reinforced composite materials. CNTs indeed improves the mechanical behavior of composite at the small strain. However, such improvement disappears at relatively large strain because the completely debonded nanotubes behave like voids in the matrix and may even weaken the composite. The increase of interface adhesion between CNTs and polymer matrix may significantly improve the composite behavior at the large strain [138].

Zalamea et al. [139] employed the shear transfer model as well as the shear lag model to explore the stress transfer from the outermost layer to the interior layers in MWCNTs. Basically, the interlayer properties between graphene layers were designated by scaling the parameter of shear transfer efficiency with respect to the perfect bonding. Zalamea et al. pointed out that as the number of layers in MWCNTs increases, the stress transfer efficiency decreases correspondingly. Shen et al. [140] examined load transfer between adjacent walls of DWCNTs using MD simulation, indicating that the tensile loading on the outermost wall of MWCNTs cannot be effectively transferred into the inner walls. However, when chemical bonding between the walls is established, the effectiveness can be dramatically enhanced. It is noted that in the above investigations, the loadings were applied directly on the outermost layers of MWCNTs, the stresses in the inner layers were then calculated either from the continuum mechanics approach [139] or MD simulation [140]. Shokrieh and Rafiee [93, 94] examined the mechanical properties of nanocomposites with capped SWCNTs embedded in a polymer matrix. The load transfer efficiency in terms of different CNTs lengths was the main concern in their examination. By introducing an interphase to represent the vdW interactions between SWCNTs and the surrounding matrix, Shokrieh and Rafiee [93,94] converted the atomistic SWCNTs into an equivalent continuum fiber in finite element analysis. The idea of an equivalent solid fiber was also proposed by Gao and Li [141] to replace the atomistic structure of capped SWCNTs in the nanocomposites cylindrical unit cell. The modulus of the equivalent solid was determined based on the atomistic structure of SWCNTs through molecular structure mechanics [142]. Subsequently, the continuum-based shear lag analysis was carried out to evaluate the axial stress distribution in CNTs. In addition, the influence of end caps in SWCNTs on the stress distribution of nanocomposites was also taken into account in their analysis. Tsai and Lu [143] characterized the effects of the layer number, inter-graphic layers interaction, and aspect ratio of MWCNTs on the load transfer efficiency using the conventional shear lag model and finite element analysis. However, in their analysis, the interatomistic characteristics of the adjacent graphene layers associated with different degrees of interactions were simplified by a thin interphase with different moduli. The atomistic interaction between the grapheme layers was not taken into account in their modeling of MWCNTs. In light of the forgoing investigations, the equivalent solid of SWCNTs was developed by several researchers and then implemented as reinforcement in continuum-based nanocomposite models. Nevertheless, for MWCNTs, the subjects concerning the development of equivalent continuum solid are seldom explored in the literature. In fact, how to introduce the atomistic characteristics, that is, the interfacial properties of neighboring graphene layers in MWCNTs, into the equivalent continuum solid is a challenging task as the length scales used to describe the physical phenomenon are distinct. Thus, a multi-scale based simulation is required to account for the atomistic attribute of MWCNTs into an equivalent continuum solid. In Lu and Tsai's study [144], the multi-scale approach was utilized to investigate the load transfer efficiency from surrounding matrix to DWCNTs. The analysis consisted of two stages. First, a cylindrical DWCNTs equivalent continuum was proposed based on MD simulation where the pullout extension on the outer layer was performed in an attempt to characterize the atomistic behaviors between neighboring graphite layers. Subsequently, the cylindrical continuum (denoting the DWCNTs) was embedded in a unit cell of nanocomposites, and the axial stress distribution as well as the load transfer efficiency of the DWCNTs was evaluated from finite element analysis. Both SWCNTs and DWCNTs were considered in the simulation and the results were compared with each other.

An equivalent cylindrical solid to represent the atomistic attributes of DWCNTs was proposed in this study. The atomistic interaction of adjacent graphite layers in DWCNTs was characterized using MD simulation based on which a spring element was introduced in the continuum equivalent solid to demonstrate the interfacial properties of DWCNTs. Subsequently, the proposed continuum solid (denotes DWCNTs) was embedded in the matrix to form DWCNTs nanocomposites (continuum model), and the load transfer efficiency within the DWCNTs was determined from FEM analysis. For the demonstration purpose, the DWCNTs with four different lengths were considered in the investigation. Analysis results illustrate that the increment of CNTs' length can effectively improve the load transfer efficiency in the outermost layers, nevertheless, for the inner layers, the enhancement is miniature. On the other hand, when the covalent bonds between the adjacent graphene layers are crafted, the load carrying capacity in the inner layer increases as so does the load transfer efficiency of DWCNTs. As compared to SWCNTs, the DWCNTs still possess the less capacity of load transfer efficiency even though there are covalent bonds generated in the DWCNTs.

9.6 CONCLUSION

Many traditional simulation techniques (e.g., MC, MD, BD, LB, Ginzburg–Landau theory, micromechanics, and FEM) have been employed, and some novel simulation techniques (e.g., DPD and equivalent-continuum and self-similar approaches) have been developed to study polymer nanocomposites. These techniques indeed represent approaches at various time and length scales from molecular scale (e.g., atoms), to microscale (e.g., coarse-grains, particles, and monomers) and then to macroscale (e.g, domains), and have shown success to various degrees in addressing many aspects of polymer nanocomposites. The simulation techniques developed thus far have different strengths and weaknesses, depending on the need of research. For example, molecular simulations can be used to investigate molecular interactions and structure on the scale of 0.1–10 nm. The resulting information is very useful to understanding the interaction strength at nanoparticle–polymer interfaces and the molecular origin of mechanical improvement. However, molecular simulations are computationally very demanding, thus not so applicable to the prediction of mesoscopic structure and properties defined on the scale of 0.1–10 mm, for example, the dispersion of nanoparticles in polymer matrix and the morphology of polymer nanocomposites. To explore the morphology on these scales, mesoscopic simulations such as coarse-grained methods, DPD, and dynamic mean field theory are more effective. On the other hand, the macroscopic properties of materials are usually studied by the use of mesoscale or macroscale techniques such as micromechanics and FEM. But these techniques may have limitations when applied to polymer nanocomposites because of the difficulty to deal with the interfacial nanoparticle-polymer interaction and the morphology, which are considered crucial to the mechanical improvement of nanoparticle-filled polymer nanocomposites. Therefore, despite the progress over the past years, there are a number of challenges in computer modeling and simulation. In general, these challenges represent the work in two directions. First, there is a need to develop new and improved simulation techniques at individual time and length scales. Secondly, it is important to integrate the developed methods at wider range of time and length scales, spanning from quantum mechanical domain (a few atoms) to molecular domain (many atoms), to mesoscopic domain (many monomers or chains), and finally to macroscopic domain (many domains or structures), to form a useful tool for exploring the structural, dynamic, and mechanical properties, as well as optimizing design and processing control of polymer nanocomposites. The need for the second development is obvious. For example, the morphology is usually determined from the mesoscale techniques whose implementation requires information about the interactions between various components (e.g., nanoparticle-nanoparticle and nanoparticle-polymer) that should be derived from molecular simulations. Developing such a multiscale method is very challenging but indeed represents the future of computer simulation and modeling, not only in polymer nanocomposites but also other fields. New concepts, theories and computational tools should be developed in the future to make truly seamless multiscale modeling a reality. Such development is crucial in order to achieve the longstanding goal of predicting particle-structure property relationships in material design and optimization.

The strength of the interface and the nature of interaction between the polymer and CNT are the most important factors governing the ability of nanotubes to improve the performance of the composite. Extensive research has been performed on studying and understanding CNT-polymer composites from chemistry, mechanics, and physics aspects. However, there exist various issues like processing of composites and experimental challenges, which need to be addressed to gain further insights into the interfacial processes.

SECTION TWO: EXPERIMENTAL CASE STUDIES

9.7   CASE STUDY I

In this case study, polyaniline (PAni)-SWCNT composite fibers have been produced wherein small loadings of nanotubes (< 1%) significantly increased the elastic modulus and tensile strength. Raman spectroscopy was employed to quantify the degree of nanotube alignment along the fiber axis, and to study the efficiency of load transfer between the polymer matrix and CNTs. The Herman's orientation factor for the nanotubes increased from 0.02 to 0.43 after drawing the as spun fiber. The drawing process was also shown to orient most of the nanotubes within about $\pm$ 30° of the fiber axis. Load transfer between the polymer matrix and the nanotubes was demonstrated by a shift of 90-130 cm^{-1}/strain in the nanotube D* band when a load was applied to the fiber indicating a strong interaction between the nanotubes and the PAni matrix.

9.7.1   INTRODUCTION

The SWCNTs are an attractive reinforcing and conducting filler for polymers due to their exceptional mechanical properties, high aspect ratio, and excellent electrical conductivity [1-6]. Numerous examples showing reinforcement of many polymers using SWCNTs have been reported and include poly(methyl methacrylate) [7] and polyvinyl alcohol [8] or epoxy [9] composites as films, and poly(p-phenylene benzobisoxazole) [10] and polyacrylonitrile [11] or polyvinyl alcohol [12] composites as fibers. Composite fibers of PAni containing SWCNTs have been recently reported and showed significant improvement in tensile strength and elastic modulus compared to neat PAni [13-15]. Fibers containing 0.76% (w/w) SWCNTs demonstrated a 50% increase in tensile stress (σ_b), a 120% increase in Young's modulus (E) and a 40% decrease in elongation at break (ε_b) compared with the neat PAni fibre [15]. The mechanical properties observed are dependent on the CNTs being aligned and on a strong interaction with the host matrix [16]. Here we use Raman spectroscopy to further explore this behaviur.

For polymer composite materials, a strong interfacial adhesion between reinforcing filler and the matrix material is required for successful load transfer and to increase the Young's modulus and tensile strength. Consequently, strong chemical or electrostatic bonds between the filler and the matrix will enhance this effect [17-19]. However, SWCNTs have a much higher aspect ratio than conventional fillers, hence substantial load transfer may be achieved by the much smaller, but significant vdW interactions [18]. Interfacial adhesion by non-covalent bonding of polymer to SWCNT can be promoted by favorable physical interactions between the polymer and the SW-CNT [20-21] and/or a donor-acceptor interaction through charge transfer [22]. These interactions result in better adhesion than possible with vdW interactions alone.

Raman spectroscopy has been used to probe interactions occurring in PAni nanotube [23-24] composites, the orientation of nanotube bundles within a matrix [25, 26], and the efficiency of load transfer from the host matrix to SWCNTs [27,28]. Unlike X-ray diffraction (XRD) methods [12], Raman spectroscopy can detect very low concentrations of SWCNTs in a polymer matrix [29,30]. The degree of orientation of aligned nanotubes can be estimated by polarized Raman spectroscopy due to the presence of a strong resonance Raman scattering effect [31,32]. Polarized Raman spectroscopy in combination with a mathematical model [33] has been employed to characterize the orientational order of nanotubes in polymers [34]. Using this model, the polarized Raman intensity of nanotubes is correlated with the orientation order parameters of SWCNTs in a uniaxially oriented system. An orientation distribution function can then be obtained.

The strain in the nanotubes within the composite can be determined from the shift in the D* band [27-29] observed using Raman spectroscopy. For example, thermal loading of a SWCNT-epoxy composite gave a 21 cm^{-1} shift per unit strain [35], while another study showed a 2 cm^{-1} shift per unit strain for a SWCNT-epoxy composite loaded in tension [27]. The band shift can be correlated to the change in C-C bond length [17], which in turn is proportional to overall SWCNT deformation [17].

The present study uses Raman spectroscopy techniques to investigate the origin of the reinforcing effect of SWCNTs in PAni fibers. This is achieved through evaluation of the degree of orientation of the tubes and determination of load transfer from the PAni matrix to the SWCNTs.

9.7.2   EXPERIMENTAL DETAILS

FABRICATION OF PANI FIBRES

The PAni spinning solution was prepared and a continuous mono filamentefiber was then spun and thermally stretched five times from the initial length [15,36]. PAni (Santa Fe Science and Technology, 280000g mol^{-1}), dichloroacetic acid (DCAA,

Merck, 98%), and 2-acrylamido-2-methyl-1-propanesulfonic acid (AMPSA, Aldrich, 99%) were all used without any further purification. The SWCNTs (HiPCO, CNI) were used as purchased and contained 5% w/w iron residue as determined by elemental analysis. The AMPSA (0.4 g) was sonicated for a few seconds in 20 g DCAA to produce a colorless solution. SWCNTs (18 mg) were then added to this solution, and it was sonicated for 30 min. Separately, the 0.9 g of AMPSA was added to 1.0 g of PAni eEmeraldine base) powder, and ground using a mortar and pestle to obtain a grey powder. The resultant powder was added to the prepared solution containing SWCNTs and AMPSA in DCAA. The solution was continuously stirred over a period of 30 min. under a N_2 atmosphere. This spinning solution was stirred for another 30 min. at 2000 revolutions per minute at °C. Next, bubbles in the viscous solution were removed using a dynamic vacuum over a period of 1 hr. The viscous, bubble-free solution was transferred to a N_2 pressure vessel to drive the spinning solution through a filter (200 μm, Millipore), then through a single-hole spinneret with a length/diameter ratio, L/D = 4 and D = 250 mm, and finally to the acetone coagulation bath at room temperature. In the coagulation bath, the solution solidified and the emergingefiber was taken up on the first bobbin (D = 2.5 cm) using a linear velocity of 3 m/min. Then, the semi-solidefiber was passed through warm air immediately above a hot plate at 10°C and collected with the second bobbin (D = 5 cm) using a linear velocity of 6 m/min. Thesfibers were then left at room temperature for at least 30 min. in preparation for the hot-drawing process. The drawing process involved stretching the as-spunefiber five times across a soldering iron wrapped with Teflon tape and heated to 10°C. Thesfibers were then dried at room temperature for 48 hr before characterization.

IINSTRUMENTATION

Raman spectra were obtained with the JOBIN Yvon Horiba Raman spectrometer model HR800. This Raman spectrometer employed a micro-Raman configuration where the incident laser and the Raman induced wavelength both pass through the same microscope lens. The spectra were recorded using a polarized 632.8 nm line of a Helium/Neon laser and a holographic diffraction grating, providing a spectral resolution of 1.5 cm^{-1}. A Gaussian Lorentzian-fitting function was used to obtain band position and intensity. The incident laser beam was focused onto the specimen surface through a 100 × objective lens, forming a laser spot of approximately 1 μm in diameter, using a capture time of 50 sec. Orientation Raman studies were performed when theefiber was positioned at 0°, 30°, 45°, 60°, 75°, and 90° to the plane of polarization of the incident laser. At each angle, spectra were collected at both the VV (parallel polarization of the incoming and scattered light) and VH (vertical polarization of the incoming and scattered light) configurations. In the VV configuration, the analyzer plate was parallel to the plane of polarization of the laser; and in the VH configuration, the analyzer plate was perpendicular to the plane of polarization of laser.

For Raman load transfer experiments one end of a 10 mm fiber sample was fixed to the Raman specimen stage, whilst the other end was attached to a metallic wire which was clamped in a fixed position. The Raman stage was then displaced by a known distance from which theefiber strain was calculated. For each strain increment, the shift in the D* band was determined.

9.7.3 RESULTS AND DISCUSSION

RAMAN SPECTROSCOPY OF PANI-CNTS FIBER

Initial investigations involved the collection of Raman spectra for thesfibers produced using the wet spinning techniques. Figure 1 shows the Raman spectra from "as received" nanotube powder and the PAniefiber (5X drawn) with no tubes or containing 0.76% w/w SWCNTs. The as received SWCNTs showed typical Raman resonance for SWCNTs [37]. Four prominent sets of peaks were observed– ~15—300 cm^{-1}(RBM), ~1600 cm^{-1} (G band), ~1300 cm^{-1} (D band), and ~2600 cm^{-1} (D* band) [38]. The G band is an intrinsic feature of sp2 carbon [39] and is common for both metallic and semiconducting nanotubes [40]. The D band can be attributed to defect induced sp2 bonded carbon material [40] and the origin of this is the double resonance mechanism which is also operative for graphite [44]. The D* band is the second order overtone of the D band [40]. The RBM band is related to the in-phase radial displacement [40, 42] and occurs over a range of frequencies dependent on nanotube diameter and chirality [45].

The G band may be used to study the degree of orientation of SWCNTs in the PAni matrix as it results from a symmetric normal mode of vibration, and exhibits a low depolarization ratio [43]. The D* band is useful in determining the load transfer from the polymer matrix to the SWCNTs [44]. The Raman spectra in Figure 1 (a and b) indicates that thesfibers produced here were in theeEmeraldine salt (ES) form with bands observed at 1337 cm^{-1}, 1496 cm^{-1}, and 1592 cm^{-1} as described [45,46]. The peak at 1337 cm^{-1} corresponds to the formation of the (C-$^{+}$) bond and this response was enhanced by addition of SWCNTs. The increase in peak intensity may be attributed to an increased level of doping [14,47]. With increased amounts of SWCNTs a decrease in the ratio of intensity in the 1496 cm^{-1} band (N-H in plane bending of benzoid ring) and the 1169 cm^{-1} band (C-H stretching of benzoid ring) was observed. These spectral changes can be attributed to the interaction of the bipolaronic structure of PAni and the SWCNTs [23,49].

SWCNT ORIENTATION

The analysis of SWCNT orientation usingdpolarized Raman spectroscopy was carried out based on the coordination system definen Figure 2.

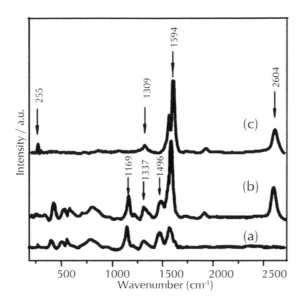

FIGURE 1   Raman spectra of (a) PAni-ES/AMPSAsfibers containing (a) % w/w SWCNT, (b) 0.76% wt SWCNT and (c) as received SWCNT Hipco powder.

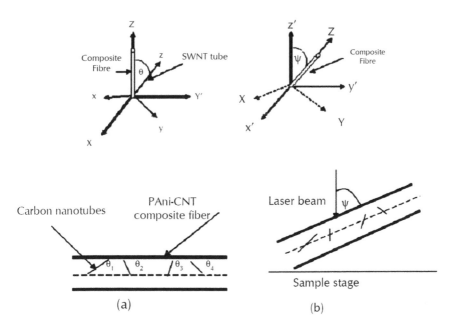

FIGURE 2   Schematic showing (a) the orientation of SWCNTs fiber (θ is angle of θ_1, θ_2, θ_3, θ_4 etc) with respect to the composite fiber axis (b) the orientation of the PAniefiber (ψ) with respect to the Raman sample stage.

The fiber axis is defined as Z and SWCNTs are oriented at angle θ with respect to the Z axis. The efiber was mounted on the stage of the Raman microscope such that the incident laser comes in along the x axis. The angle between the polarization plane and the efiber axis is ψ.

ALTERNATIVE

The θ is the angle at which the CNTs lie within the fiber with respect to the fiber axis. ψ is the angle of the nanocomposite fiber with respect to the laser beam during Raman spectroscopy.

Figures 3, 4, and 5 show the results of polarized Raman spectroscopy using the VV or VH configuration for the neat PAni and for fibees containing SWNT (0.76 %w/w) at different angles (ψ). For the neat PAni-ES/AMPSA (5X) Raman spectra were obtained in two directions, parallel (ψ = 0°) and vertical direction (ψ = 90°) with respect to then-polarization plane (Figure 3). The SWNTs are known to show high polarizability in all Raman modes, especially in the G (1594 cm^{-1}) and RBM (150–250) bands [48]. In contrast, a smaller orientation dependence in the Raman bands was observed for neat PAni fibres with only a small increase in intensity of the 1592 cm^{-1} band (C-C stretching of benzoid ring [48]) observed when thenpolarization plane was parallel to the fibee axis (ψ = 0°) compared with the vertical configuration. This result indicates some alignment of the PAni chains along the fibee axis. However, for 5x drawn fibees containing SWNTs (0.76% w/w) the G band (1594 cm^{-1}) intensity in the Raman spectra was critically dependent on orientation and monotonically decreases with increasing ψ (Figure 4 a). This can be compared to the band at 1169 cm^{-1} (C-H stretching of benzoid ring) which was independent of the fibee angle. These results are in agreement with previous studies that showed similar behavior for SWNT when used as reinforcing filler in other composites [4,51]. The intensity ratio of Ivv (ψ = 90°)/ Ivv (ψ = 0°) at 1592 cm^{-1} for the neat PAni fibee (5X drawn) is 0.85 when normalized to the peak at 1169 cm^{-1}. However, this ratio shows a significant decrease to 0.19 after addition of SWNTs and the introduction of the CNT G band in the 5X drawnefiber. This difference shows the much lower sensitivity of the polarised Raman scattering intensity to ψ for neat PAni fibees compared with theefiber containing SWNTs. In addition, Raman spectra show a much stronger orientation effect in the 5X drawn compositeefibers (Figure 4 a) compared with the as spun compositeefibers (Figure 5). For the as spunefibers, the change in intensity of the peak at 1592 cm^{-1} was insignificant with respect to ψ from the parallel to the perpendicular direction. It is clear that the drawing process is mostly responsible for the alignment of SWNTs in the fiber direction.

When the 5x drawn composite fibers were examined using the cross polarized VH configuration, the intensity variation of the G band with change in ψ showed a different trend (Figure 4 b). The intensity increased with increasing ψ and goes through a maximum at 45° and then decreased as ψ was increased to 90°. This observation is in good agreement with previous results obtained from polarized Raman spectroscopy of isolated SWNTs and their fibers [51,52]. The measurement of Raman intensity using the VH configuration is necessary to obtain the parameters required to determine the degree of orientation.

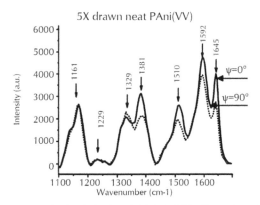

FIGURE 3   Raman spectra obtained for PAni-ES/AMPSAefibers (5x drawnefiber) obtained using the VV configuration for ψ=0° and 90° [ψ = the angle betweenefiber axis and polarization plane – see Figure 2].

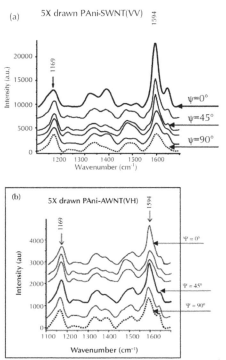

FIGURE   4   Raman   spectra   obtained   for   PAni-ES/AMPSA-SWNT   (0.76%   w/w) compositeefibers (5X drawn) obtained using (a) the VV configuration, (b) VH configuration. From top to bottom, the angle betweenefiber axis and polarization plane (ψ) is 0°, 30°, 45°, 60°, 75°, and 90°.

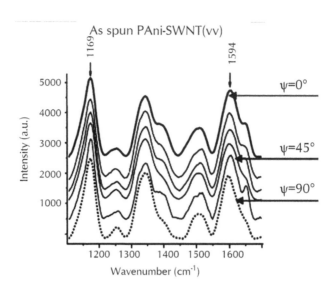

FIGURE 5 Raman spectra obtained for PAni-ES/AMPSA-SWNT (0.76% w/w) compositeefibers (as spun) obtained using the VV configuration. From top to bottom, the angle betweenefiber axis and polarization plane (ψ) is 0°, 30°, 45°, 60°, 75°, and 90°.

It has previously been shown that all Raman modes observed for SWNTs (e.g. the RBM and G band) exhibit the two fold symmetry in the VV and VH configurations [52-54]. Figure 6 shows the variation of intensity of the RBM band in VV (Figure 6 a) and VH (Figure 6 b) configuration at different angles in relation to thenpolarization plane. Compared to the G band, a lower RBM band (224 cm⁻¹) intensity ratio of Ivv (ψ = 90° Ivv (ψ = 0°) was observed due to absence of adpolarized peak of PAni in this region (Figure 6 a). Similar behavior was also observed for the RBM band compared to the G band in the VH configuration. The intensity increases with increasing ψ and reaches a maximum value at 45° and then decreases dramatically at 90°. The intensity pattern observed for both RBM and G bands in VH configuration are in good agreement with previously reported results for ψ dependence of RBM and G band in SWNT fiber with a maximum intensity at 45° [51,53].

It is acknowledged that the interference of the C-C peak of PAni and the G band may have a small effect on the analysis of the intensity of the G band for measurement of the degree of orientation of SWNTs and the distribution function. The intensity ratio data (I_{VV} (ψ°) / I_{VV} (ψ = 0°)) versus ψ for composite fibers are represented in Figure 7, for G and RBM bands. A small difference in intensity ratio at each ψ was observed for the G and RBM bands in the drawn PAni-ES/AMPSA-SWNTefibers. However, a comparison of data obtained for G band and RBM intensity ratios extracted from references [48,50]and for SWNT fibers exhibit a similar discrepancy. It can, therefore, be assumed that the interference between the PAni C-C peak and the G band has little effect on the intensity ratio. It can be seen from Figure 7 that the intensity ratio obtained in the present work is similar to previously published data [50,57].

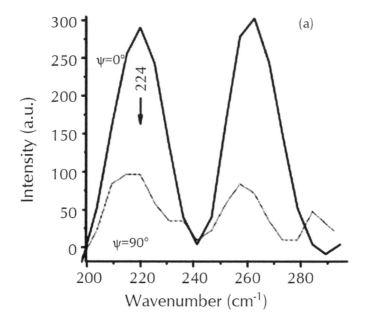

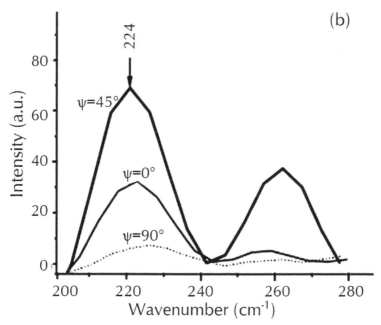

FIGURE 6 Raman spectra obtained using the VV or VH configuration for the RBM region, ψ = the angle between fiber axis and polarization plane.(a) PAni fiber (5x drawn fiber) for ψ=0° and 90°. VV configuration (b) PAni SWNT (0.76% w/w) composite fiber (5x drawn fiber) VH configuration.

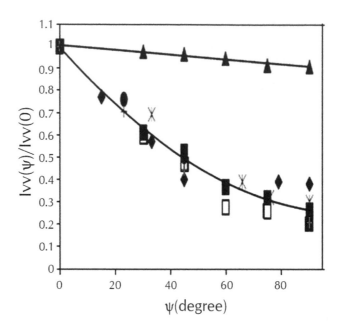

FIGURE 7   A plot of Ivv(ψ°)/Ivv 0° for different values of ψ for the G band and RBM regions. Data obtained here for this work and from the information given by literature (■)G band, PAni-ES/AMPSA-SWNT fiber (5x),(□) RBM, PAni-ES /AMPSA-SWNT fiber, (▲) G band, PAni-ES/AMPSA-SWNT fiber (as spun),(◆) G band SWNT fiber [47],(✳)RBM, SWNT fiber [47], (✹) G band, SWNT fiber [48],(+) RBM SWNT fiber [48].

The previous literature reported results from SWNT fibers containing small amounts of polymer (< 40%). Qualitatively, the results shown in Figure 7 suggest that the degree of alignment achieved in the drawn PAni-ES/AMPSA-SWNT fibers of the present study is similar to that reached in the SWNT fibers reported previously.

The variation indpolarized Raman band intensity can be used to quantitatively study the orientation of nanotube bundles along the compositeefiber axis [47]. The depolarized Raman scattering intensity can be correlated to the angle between the nanotube axis and the fiber axis (θ) and the angle between theefiber axis and thenpolarization plane(ψ) [35]. A simplified uniaxial model was used for quantitative evaluation of the orientation distribution function (ODF) based on the formulation of Liu and his coworkers [35] using on orientation order parameters.

The orientation order parameters of $<P_2 (\cos\theta)>$ and $<P_4(\cos\theta)>$ which are, respectively, the average values of $P_2(\cos\theta)$ and $P_4(\cos\theta)$ for the SWNTs bulk product. The Pi(cosθ) is the Legendre polynomial of degree i which is defined as $P_2(\cos\theta) = (3 \cos^2\theta-1)/2$ and $P_4(\cos\theta) = (35 \cos^4\theta-30 \cos^2\theta+3)/8$ for the second and fourth degree , respectively. More specifically the $<P_2(\cos\theta)>$ is known as the Herman's orientation factor which varies between values of 1 and 0 corresponding, respectively, to nanotubes fully oriented in tbrefiber direction and randomly distributed. The orientation order parameters can be determined by solving the simultaneous algebraic equations based on numerical values of $I^{VV}_{G,RBM}(\psi = 0°)/I^{VH}_{G,RBM}(\psi = 0°)$, $I^{VV}_{G,RBM}(\psi = 90°)/I^{VH}_{G,RBM}(\psi = 0°)$ [35]. The numerical values for the orientation order parameters are given in Table 1 for as spun and 5x drabre fibers based on both G band and RBM intensity ratios.

The orientation distribution function (ODF) can be formed as given in equation 1 from information of $P_2(\cos\theta)$ and $P_4(\cos\theta)$: [56].

$$f(\theta) = A \exp\left[-\left(\lambda_2 P_2(\cos\theta)\right) + \lambda_4 P_4(\cos\theta)\right] \qquad (1)$$

Three polynomial equations [35] were solved simultaneously for three unknown parameters of A, λ_2 and λ_4 (orientation distribution coefficients) by knowing the amount of $<P_2(\cos\theta)>$ and $<P_4(\cos\theta)>$. The orientation distribution coefficients for as spun and 5X dribrefibers of PAni-ES/AMPSA/SWNT (0.76%w/w) also listed in Table 2. Based on these coefficients, the ODF of PAni-ES/AMPSA-SWNT composibrefiber before and after stretching were plotted for a range of angles between nanotubes and ibrefiber axis (Figure 8). It can be seen that in the 5X dribrefiber of PAni-ES/AMPSA-SWNT, most of nanotubes were oriented at ± 30° versus ibre fiber axis and only a small portion of SWNTs lies between 30° and 150°. However, in the as sibre fiber, the majority of SWNTs shows a distribution between 60° and 120° and a small portion oriented at ± 60° veribre fiber axis. For simplification, the data has bisednormalized to the numerical amount of ODF at θ = 0°. The ODF of highly stretcibre fiber of PMMA/SWNT reported in the literature [35] was also compared with PAni-ES/AMPSA-Sibrefiber in Figure 8. The previously reported results show that the majority of SWNTs were oriented with an angle of less that ± 30° respect to ibrefiber axis. In addition, the ODF function for PAni-ES/AMPSA-Sibre fiber compared to PMMA-SWNT shows similar nanotubes orientation distribution. Therefore, ibrefiber is symmetric with respect to the plane perpendicular to ibrefiber axis. The similar order of orientation of PAni-ES/AMPSA-SWNT (5X draibrefiber and the highly orienibrefiber of PMMA-SWNT suggests a similar adhesion for PAni matrix to SWNT bundles compared with PMMA-SWNT composite. The strong adhesion may result in simultaneous orientation of SWNT and matrix during the thermal stretching process.

TABLE 1 Depolarized Raman scattering intensity ratio of G BAND and RBM bands in VV configuration (θ = 0°, 90°) versus VH configuration at θ = 0° for as-spun and 5x dribrefibers of PAni-Es/AMPSA-SWNT (0.76 %w/w)

	Ψ	$I_{VV}(\psi)/I_{VH}(0)$		$<P_2(\cos\theta)>^i$		$<P_4(\cos\theta)>^{ii}$	
	Degree	**G BAND**	**RBM**	**G BAND**	**RBM**	**G BAND**	**RBM**
5x drawn	0	10.5	9				
	90	2.5	2	0.42	0.4	0.32	0.27
As spun	0	3.2	ND[iii]				
	90	2.9	ND	0.022	-------	0.006	-------

$<P_2(\cos\theta)> = (3 \cos^2\theta-1)/2$
$<P4(\cos\theta)> = (35 \cos^4\theta-30 \cos^2\theta +3)/8$
Not detected

TABLE 2   Orientation order parameters for 5x drawn and as-spun fibers calculated based on the information listed in Table 1 for construction of orientation distribution function

	A		$\lambda 2$		$\lambda 4$	
5x drawn	G BAND	RBM	G BAND	RBM	G BAND	RBM
	0.0042	0.0044	-1.223	-1.483	-3.496	-2.747
As spun fiber	0.0063	------	-0.206	------	-0.135	-------

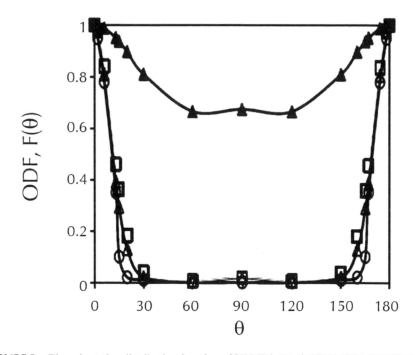

FIGURE 8   The orientation distribution function of SWNT in PAni-ES/AMPSA-SWNT (0.76% w/w) composite fibers constructed with the orientation parameters, $<P_2(\cos\theta)>$ and $<P_4\cos(\theta)>$ listed in Table 6.1.(O) PMMA-SWNT fiber(Ref 35) ($\triangle$)PAni-ES/AMPSA-SWNT(0.76%),G band ($\square$) PAni-ES/AMPSA-SWNT(0.76%), (RBM)($\blacktriangle$) PAni-ES/AMPSA-SWNT(0.76%) (As spun).

The quantitative comparison also shows a good agreement between the SWNT orientation in SWNT fiber produced previously [52] and the PAni-ES/AMPSA-SWNT fibers produced in the present study. Poulin and coworkers [52] have shown that 50% of nanotubes were oriented between 0° and 50° and 75% of SWNT lies between 0° and 70° for SWNT fiber developed by Vigolo [57]. These wet spun fibers contained more than 60% SWNT in a PVA matrix. Moreover, Gommans, et.al [53] developed SWNT fiber in another study and reported that 86% of the nanotubes lies within ± 31° of the fiber axis. Based on the knowledge of the orientation of SWNT in the PAni, it is possible to predict the elastic modulus of the composite fiber based on the fiber composite theory [16]. One approach that has been successfully applied to composites with a high loading of SWNT (>60%) has highlighted the importance of shear modulus of the nanotube bundles [35] and therefore nanotube bundle diameter, as well as nanotube orientation. However, the influence of the matrix has been ignored in these previous studies and the elastic modulus of those composites was assumed to be due only to the nanotubes. The composites described in the present study have less than 1% CNTs, so it is not appropriate to ignore the effect of the matrix on the composite modulus. Krenchel's rule of mixtures can be used to estimate the elastic modulus of composite fibers based on the known elastic modulus and volume fraction of matrix and nanotubes and by assuming perfect orientation of the nanotubes in the fiber direction. From the previous discussion, it is known that the SWNTs are not fully aligned in the fiber direction so it is expected that the rule of mixtures would slightly over estimate the composite modulus. Table 3 shows the predicted elastic modulus for PAni-ES/AMPSA-SWNT composite containing 0.76 wt % SWNTs. In this simple analysis using a value of 640 GPa for the elastic modulus of fully oriented nanotubes, the elastic modulus of the composite was calculated as 8.6 GPa, which is slightly higher than the experimental value of 7.3 GPa [15]. However, as can be seen from Figure 8 the nanotubes ropes were not fully aligned, therefore we expect a lower effective modulus. The rule of mixtures predicts an effective modulus of 520 GPa for the SWNTs based on the measured composite modulus. This value is 81% of the modulus expected for fully aligned SWNTs. These results are in accord with a high degree of alignment of the SWNTs in the PAni matrix.

The CNT alignment is required to maximize the degree of reinforcement. However, this can only occur with efficient load transfer. This property was investigated in-situ micro Raman spectroscopy.

TABLE 3   Comparison of the predicted elastic modulus from rule of mixture and experimental values

	PAni-ES/AMPSA/SWNT (0.76 %w/w)
u_{SWNT} [1]	0.008
E_c[GPa]($Eeff_{SWNT}$=640 Gpa)	8.6 [2]
Eexp[GPa]	7.3

IN-SITU MICRO RAMAN SPECTROSCOPY

When a strain is applied to a material, the interatomic distances change, and thus the vibrational frequencies of some of the normal modes change, causing a Raman peak shift [27,28]. Amongst specific enhanced peaks in SWNTs, the D* band is more sensitive to applied strain. Figure 9 shows a typical Raman peak shift in tension for two extreme strains (i.e. 0 % and 3%).

The frequency peak number was identified by a Gaussian/Lorentzian function. It can be clearly seen that a 5 cm^{-1} shift occurred after applying 3% strain. The reproducibility of the data was examined by several tests carried out with the same experimental conditions. Figure 10 shows the results of Raman peak shift in tension for a PAni fiber containing 0.76%w/w SWNTs as a function of applied strain for three samples. The shift in the Raman peak position with strain in tension is negative by 90-130 wave numbers/applied strain.

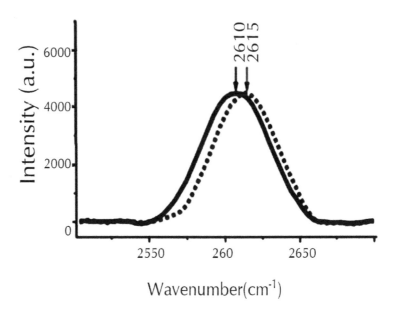

FIGURE 9 Effect of mechanical strain on the frequency shift of D* band of SWNTs incorporated in PAni-Es/AMPSA matrix (dashed line) before applying strain (solid line) after applying 3% strain.

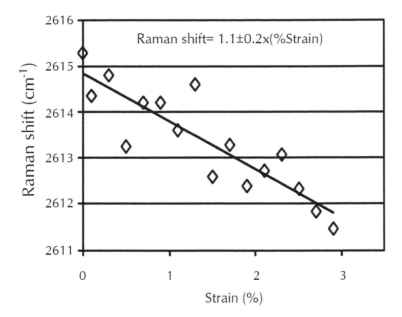

FIGURE 10   The Raman shift of D* peak of nanotube under axial tension for sample of PAni-ES/AMSA containing SWNT (0.76% w/w).

Under increasing tension, the Raman wave number decreases linearly down to about 2610 cm^{-1} (corresponding to a strain of about 3.5%).

The Raman shift of the D band of SWNTs during tensile test suggests a degree of load transfer between PAni matrix and SWNTs which results from a strong nanotube-polymer interface.

9.7.4 CONCLUSION

It has been shown that Raman spectroscopy is a valuable tool that can be used to determine the alignment and load transfer capabilities of CNTs within PAn nanocomposite fibers. The intensities of the G and RBM bands were significantly increased by decreasing the angle between the fiber axis and the polarization plane using the VV configuration of polarizer and analyzer. This behavior was correlated to the orientation of nanotubes through measurement of the average angle between nanotubes and the fiber axis. It has been found that thermal stretching of the as spun fiber orients most of the nanotubes to ± 30° with respect to the fiber axis. The Herman's orientation factor was found to increase from 0.02 to 0.43, for as spun and 5x drawn fibers. Micro Raman spectroscopy has also been used to demonstrate load transfer from the polymer matrix to the nanotubes. A significant shift between 90–130 cm^{-1}/strain in the D* band has been observed.

9.8   CASE STUDY II

In this case study, new battery materials are presented here that consist of either a solid PAni fiber or a similar fiber but containing CNTs. An ionic liquid ethylmethyl imidazolium bis(trifluoromethanesulfonyl) amide (EMI.TFSA) was chosen as electrolyte. The electrochemical properties of PANi or PANi/CNT fibers were investigated using cyclic voltammetry, AC impedance and galvanostatic charge/discharge techniques. The PANi fiber with a CNT content of 0.25% (w/w) exhibited a discharge capacity of 12.1 mAh g^{-1}.

9.8.1   INTRODUCTION

Electronically conducting polymers exhibit a wide range of electrochemical properties, and they have been applied in areas such as sensors [1], electrochromic devices [2], charge-storage devices including supercapacitors and batteries [3,4,5]. The PAni has been extensively studied for use as a battery material. This organic conductor has good redox reversibility and high environmental stability. The PAni has usually been employed as a cathode material in batteries with Zn or Li as the anode [6]. The electrolytes used usually contain inorganic acids such as HCl, $HClO_4$ or H_2SO_4 [7, 8,9]. A battery voltage of 1.2 V and discharge capacities of up to 121 mAh g^{-1} have been reported [8]. There have also been a few reports about the application of PAni in all-polymer batteries. The use of PAni anode and poly-1-naphthol cathode with methyl cyanide containing lithium perchlorate and perchloric acid as electrolyte has been described [10] and a discharge capacity of 150 mAh g^{-1} was reported. In other work, a discharge capacity of 79 mAh g^{-1} was reported for a cell composed of PAni cathode and polyindole anode with sulfuric acid electrolyte [11]. There is, therefore, a need to develop batteries without these corrosive electrolytes as well as in alternative configurations rather than as a conventional cell.

The aim of this work is to develop a wearable power source for wearable diagnostic systems. The integration of electronic components into conventional garments to introduce novel fashion effects, visual displays or audio and computing systems has generated interest in recent years [12,13]. Such systems are also proving useful in the development of wearable diagnostic systems for monitoring of vital data (heart, pulse rate) in medical and military applications. Some promising results for a highly flexible fiber battery have been obtained in our group [14,15]. In those studies, polypyrrole-hexafluorophosphate (PPy/PF_6) was electrodeposited on expensive platinum wire to be used as cathode whereas a stand-alone PAni fiber was employed as an electrode directly in this work. This very thin fiber was prepared by a wet-spinning method with a diameter of about 70–100 μm. This process has the advantage of easy processability and lower cost. The CNTs were also added into the PAni to improve the electrical and mechanical properties of polymer fibers [16].

Moreover, ionic liquid was chosen as electrolyte in this work other than lithium hexafluorophosphate (LiPF$_6$) in 1:1 (ethylene carbonate: dimethyl carbonate). High ionic conductivity, large electrochemical windows, excellent thermal and electrochemical stability and negligible evaporation make ionic liquids an ideal electrolyte for such wearable diagnostics systems [17]. No similar work has been reported so far to our knowledge.

The PAni fibers and its CNT nanocomposite were investigated in ionic liquid EMITFSA in this work, and their electrochemical characteristics were investigated by cyclic voltammetry, AC impedance and galvanostatic charging/discharging techniques.

9.8.2 EXPERIMENTAL

FABRICATION OF PANI FIBERS

The PANi-AMPSA spinning solution containing CNTs was prepared based on the method reported previously [18]. The fibers were then spun according to previously reported procedures [18].

The surface morphologies of the PAni fibers were investigated with a scanning electron microscope (SEM, Leica Model Stereoscan 440) with a secondary electron detector. To obtain clear cross-section images, the fibers were frozen in liquid nitrogen before snapping them.

FABRICATION OF POLYMER ELECTRODE

The PAni fibers or PAni fibers containing CNTs were cut into small segments of 3 cm lengths. Three such segments were bound together with platinum wire ($\phi = 0.125$ mm), and the Pt wire was also used as a lead. The electrodes were soaked in ionic liquid for 2 hours to allow the ionic liquid to penetrate into the inner part of the fibers before electrochemical measurements.

ELECTROCHEMICAL CHARACTERISATION

All the electrochemical characterization of PANi or PANi/CNT fibers were performed in a standard one compartment three-electrode cell with a stainless steel mesh counter electrode, Ag/Ag+ (EMITFSA) reference electrode, and ionic liquid EMITFSA electrolyte. The charge/discharge and cyclic voltammetry (CV) investigations were carried out using an EG and G PAR 363 Potentiostat/Galvanostat, a MacLab 400, and EChem v 1.3.2 software (AD Instruments). The AC impedance spectrum was measured using CH instruments electrochemical workstation 660B (CHI company, USA) in the range of 0.1 to 1×10^5 Hz with 5 mV amplitude.

In charge/discharge tests, the cells were charged galvanostatically at a current density of 0.1 mA cm^{-2} to a cell voltage of 2.0V, and then discharged at the same current density to a cut-off voltage of -1.2V. In the CV test, the scan rate used was 10 mV s^{-1}.

9.8.3   RESULTS AND DISCUSSIONS

SURFACE MORPHOLOGY

The content of CNTs in PANi fiber is given as a weight fraction with respect to the weight of Pani-AMPSA in the solid fiber, and it is 0.25 %w/w in this work. The cross-section images of PANi or PANi/CNT fibers at low or high magnification are shown in Figure 1. Some differences can be clearly observed between these two types of solid fiber. In Figure 1a pure PANi fiber shows an even and featureless cross-section surface, consistent with a fully dense structure. The PANi fiber containing CNTs showed fracture marks on its cross-section, but there was also evidence of some porosity in these fibers (Figure 1b). In any case, the whole volume was electrochemically active.

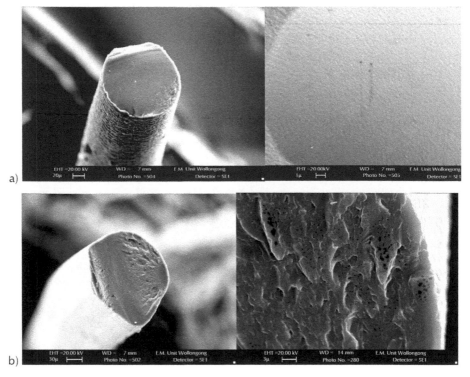

FIGURE 1   SEM images of the cross-section of the PANi (a) and PANi/CNT (b) fibers at low and high magnification.

CYCLIC VOLTAMMETRY

To identify the oxidation and reduction potentials and electrochemical reactions of the fiber electrodes, cyclic voltammetry was performed in EMI.TFSA. Cyclic voltammograms are shown in Figure 2. Two pairs of redox peaks are shown for both PANi and PANi/CNT fiber electrodes, and the potential for redox peaks obtained from the cyclic voltammograms is listed in Table 1. The difference between the oxidation and reduction peaks, $\Delta E_{O,R}$, is calculated and also listed in Table 1. $\Delta E_{O,R}$ is taken as an estimate of the reversibility of the redox reaction [16]. $\Delta E_{O,R}$ values of 0.31 and 0.85 V are obtained for the redox reactions of PANi with CNT incorporation. They are smaller than those $\Delta E_{O,R}$ values of 0.41 and 0.92 V for pure PANi fiber, respectively. These results suggest that the redox reactions appear to occur more reversibly after the addition of CNTs.

The PANi Emeraldine salt (ES) is electrochemically reduced to the PANi leucoemeraldine base (LEB) structure by gaining two electrons and two EMI cations per tetrameric repeat unit at low potential, meanwhile, PANi(LEB) loses two electrons and two EMI cations to form the ES structure when it is cycled to higher positive potentials.

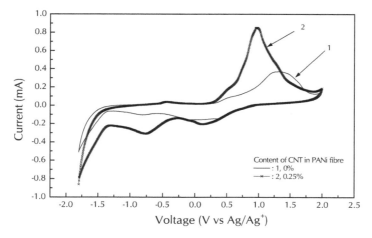

FIGURE 2   Cyclic voltammograms of a PANi fiber and a PANi/CNT fiber in EMI.TFSA.

TABLE 1   Data obtained from the cyclic voltammograms

Potential	E_{OX} (V)		E_{RD} (V)		$DE_{O,R}$ (V)	
Sample	E_{OX1}	E_{OX2}	E_{RD1}	E_{RD2}	$DE_{O,R1}$	$DE_{O,R2}$
PANi/CNT (0.25% w/w)	−0.46	0.98	−0.77	0.13	0.31	0.85
PANi	−0.39	1.30	−0.8	0.38	0.41	0.92

The mechanism of maintaining electron neutrality in a solid fiber during a charging/discharging process has been explained by intercalation/de-intercalation of the cations between the fiber and the electrolyte and it has been claimed that the transfer of protons is not probable to achieve electron neutrality during the redox process. Therefore, the pernigraniline base oxidation state is unlikely to be formed in ionic liquid [12].

It has also been noted that the redox peaks for PANi/CNT fiber became sharper and the voltammogram area became larger as shown in Figure 2. All these results indicate that the oxidation and reduction reactions became more facile after CNTs incorporation, and higher charge and discharge capacity are expected.

AC IMPEDANCE

To investigate the electrochemical behavior of PANi fiber electrodes at the electrode/electrolyte interface, AC impedance measurements were carried out, with the results shown in Figure 3.

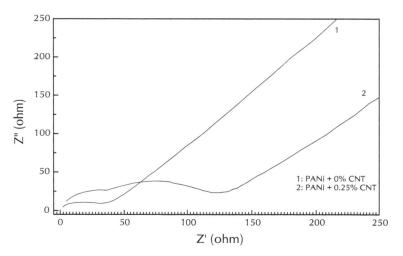

FIGURE 3   Impedance spectroscopy of (1) a PANi fiber and (2) a PANi/CNT fiber in EMI. TFSA.

A semicircle was found at high frequency and a Warburg diffusion (45°line) occurred where the resistance of the circle diminished in the Nyquist plot for PANi without CNT. These features indicate that the electrode process was limited by both charge-transfer kinetics and diffusion processes. However, two semicircles were shown at high frequency after the addition of CNT into PANi fiber, which implies that the cell was kinetically controlled by more complicated processes rather than simple charge transfer. Perhaps the additional features were caused by the interaction between the PAni and CNTs. At low frequency a typical Warburg diffusion process was again observed.

CHARGE/DISCHARGE CHARACTERISTICS

The charge/discharge characteristics of the PANi fibers were evaluated galvanostatically, and the charge/discharge curves are shown in Figure 4. The charge/discharge capacity was calculated based on the weight of the PANi fiber electrodes. It can be seen that the charging potential increases and discharging potential decreases as the depth of the charging/discharging process is increased; which is characteristic of electrode materials used in rechargeable batteries and proves that PAni fibers can be used as electrode materials in ionic liquid. A discharge capacity of 11.2 mAh g^{-1} and charge capacity of 12.4 mAh g^{-1} were obtained for PAni fiber with 0.25% CNT. They are much higher than those obtained for pure PAni fiber. PAni fiber with 0% CNT exhibited a discharge capacity of 4.1 mAh g^{-1} and charge capacity of 4.5 mAh g^{-1}. It has also been noted that the charge voltage became lower and discharge voltage became higher after CNTs were incorporated into PANi fiber; which indicates that energy consumed for (IR) was reduced and effective energy storage was improved.

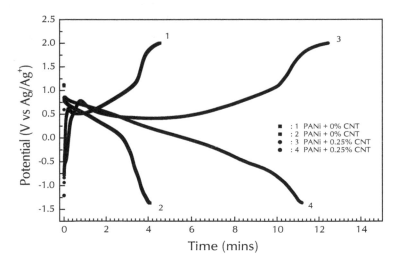

FIGURE 4   Charge/discharge curves of a PANi fiber (1 and 2) and a PANi/CNT fiber (3 and 4) in EMI.TFSA.

This improvement was perhaps because the inner resistance of the electrode was reduced after the CNTs were incorporated. However, the charge and discharge capacity obtained in this work is much lower than that for PANi reported in electrolyte containing inorganic acid such as 121 mAh g^{-1} reported by Mirmohseni and Solhjo [8]. The main reason is perhaps because the fiber prepared in this work was very dense and solid, resulting in the electrochemically accessible surface area becoming smaller and leading to a reduction in the number of electrochemically active sites. This problem can be solved if the electrochemically active area is increased for PANi fiber. The electrolyte, ionic liquid, may also have contributed to the low capacity. The significance of this work is that stand alone PANi fiber can be used as electrode materials in ionic liquid EMI.TFSA and is a promising electrode for wearable diagnostics systems.

CYCLE LIFE

The discharge capacity as a function of the cycle number is shown in Figure 5. The discharge capacity of PANi fiber with 0% CNT increased with the cycle number, and it reached 9.7 mAh g^{-1} after 10 cycles from an initial capacity of 3.9 mAh g^{-1}. This suggests that a slow activation process occurred during the charge/discharge process. For PANi/CNT fiber, however, an initial discharge capacity of 10.9 mAh g^{-1} was obtained that reached 12.1 mAh g^{-1} after 10 cycles. No obvious activation process was shown for PANi/CNT fiber. It can be concluded that the incorporation of CNTs into PANi fiber improves the discharge capability and shortens the activation process.

9.8.4   CONCLUSIONS

Solid PAni fibers were prepared by a wet-spinning method. The neat PANi fibers were observed by SEM to be fully dense, while PAni /CNT composite fibers had some porosity. PAni fiber with 0.25% CNT exhibited a smaller $\Delta E_{O,R}$ than a fiber with 0% CNT; which indicates that the redox processes occur more reversibly. A higher charge/discharge capacity, lower charge voltage and higher discharge voltage were shown for PANi/CNT fiber. All the results show that solid PAni fiber can be used directly as an electrode in ionic liquid EMI.TFSA, and that the electrochemical properties of the fiber as an electrode material have been improved by CNT incorporation. Therefore, PANi/CNT fiber is a promising electrode material for wearable diagnostics systems.

9.9   CASE STUDY III

In this case study, High strength, flexible and conductive PAni - carbon nanotube (SWNT) composite fibers have been produced using wet spinning. The use of DCAA containing 2-acrylamido-2-methyl-1-propane sulfonic acid (AMPSA) has been shown to act as an excellent dispersing medium for CNTs and for dissolution of PAni. The viscosity of DCAA-AMPSA solution undergoes a transition from Newtonian to non-Newtonian viscoelastic behavior upon addition of CNTs. The ultimate tensile strength and elastic modulus of PANi-AMPSA fibers were increased by 50% and 120%, respectively, upon addition of 0.76% (w/w) CNTs. The elongation at break decreased from 11% to 4% upon addition of CNTs, however reasonable flexibility was retained. An electronic conductivity percolation threshold of ~ 0.3% (w/w) CNTs was determined with fibers possessing electronic conductivity up to ~ 750 Scm^{-1}. Raman spectroscopic evidence confirmed the presence of CNTs in the PAni and also the interaction of the quinoid ring with the nanotubes to provide a doping effect.

9.9.1   INTRODUCTION

Polymer fibers are readily produced from the insulating Emeraldine base (EB) (Scheme 1.a) or leucoemeraldine base (LEB) forms of PAni. They are then rendered conductive by doping using an aqueous acid comprised of small anions [1-6]. The main disadvantage of this approach involves the adverse influence of the acid doping process on mechanical properties. Nonhomogenous doping and the possibility of dedoping through diffusion of small anions from the skin of the fibers cause heterogeneities in the fiber structure. The doping of PAni in solution, however, results in more homogenous doping and a more uniform material after casting [7].

The doping of PAni with sulfonic acids results in an extended coil conformation and a high level of crystallinity resulting in high electronic conductivity [8]. In this regard, a number of studies involving the processing of PAni solution blended with sulfonic acids in a suitable solvent have been conducted [9,10]. The production of an electronically conductive Emeraldine salt (ES) (Scheme1.b) form of PAni fiber using 2-acrylamido-2-methyl-1-propane sulfonic acid (AMPSA) (Scheme 1.c) (in dichloracetic acid) as the dopant has recently been reported [11]. Figure 1 shows the Schematic of PANi (EB), PANi (ES) and AMPSA are shown in The ultimate tensile strength, elastic modulus and electrical conductivity reported were 97 MPa, 2 GPa and 600 Scm^{-1}, respectively, after annealing ofefibers that had been drawn to 5x their original length.

The synthesis andncharacterization of PANi-SWNT composites has also been investigated previously [12-15]. Direct dissolution of pristine nanotubes (without chemical fictionalization) in aniline can occur via formation of a donor/acceptor charge complex [16] Using this approach SWNT/PANi composite films have been produced by chemical [17] or electrochemical [18] polymerization of aniline containing dispersed SWNTs. Alternatively, the SWNTs have been blended with preformed PAne in solvents such as N-methylpyrrolidone (NMP) [13]. The enhanced electroactivity and conductivity of SWNT/PANi composite films has been attributed to the strong molecular level interactions that occur between SWNTs and PANi.

In this work the possibility of producing AMPSA-doped PAneefibers containing CNes has been investigated. It has been shown that DCAA containing AMPSA and PAne provides excellent dispersing power for the SWNTs. Theefibers produced from this spinning solution show excellent electronic and mechanical properties suitable for many applications, including artificial muscleefibers.

9.9   EXPERIMENTAL DETAILS

MATERIAL

PAne (PANi, SFST, 280 000 g/mol), DCAd (DCAA, Merck, 9 %) and 2-acrylamido-2-methyl-1-propane sulfonic acid (AMPSA, Aldrich, 9 %) were all used without any further purification. Purified SWCNT (SWNTs, Hipco@CNI) were used as purchased and contained 5 w % iron residue as determined by elemental analysis.

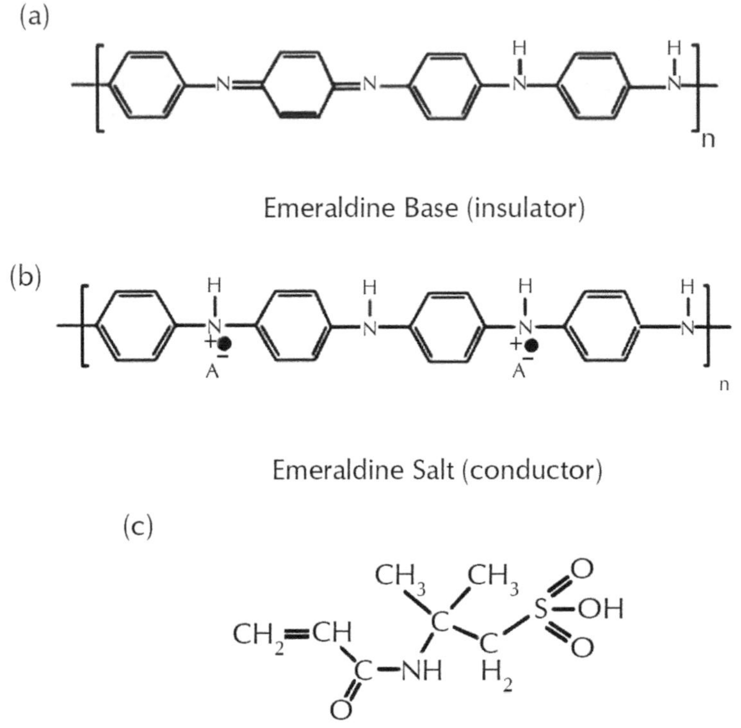

(a)

Emeraldine Base (insulator)

(b)

Emeraldine Salt (conductor)

(c)

SCHEME 1   Structures of (a) PANi (EB), (b) PANi(ES) and (c) AMPSA.

PREPARATION OF SPINNING SOLUTION

The 0.4 g of AMPSA was sonicated for a few seconds in 20 g DCAA to produce a colourless solution. SWNTs (6, 9, 12, or 18 mg) were then added to this solution and sonicated for 30 min. 0.9 g of AMPSA was added to 1.0 g of PANi (EB) powder and this mixture was ground using a mortar and pestle to obtain a grey powder. The resultant powder was added to the previously prepared solution containing SWNTs and AMPSA in DCAA. The solution was continuously stirred over a period of 30 min under a N_2 atmosphere. The spinning solution was stirred for another 30 min at 2000 rpm and °C. Bubbles in the viscous solution were removed using a dynamic vacuum over a period of one hour. The viscous bubble free solution was transferred to a N_2 pressure vessel to drive the spinning solution through a filter (200 μm, Millipore) then through a single hole spinneret with L/D = 4 and D = 250 μm and finally to an acetone coagulation bath at room temperature.

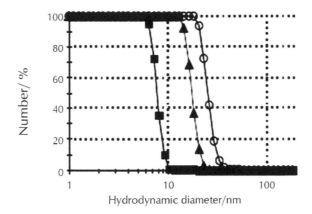

FIGURE 1   Influence of concentration of AMPSA in DCAA on hydrodynamic diameter of SWNTs (0.05 % (w/w)). AMPSA concentration was varied between (□) 0 %, (▲) 1 % and (■) 2 % (w/w). Hydrodynamic diameters were determined 30 min after sonication.

In the coagulation bath the solution solidified and the emergingefiber was taken up on a first bobbin (D = 2.5 cm) using a linear velocity of 3 m/min. Semi-solidefibers were passed through warm air immediately above a hot plate at 100 °C before being collected by a second bobbin (D = 5 cm) using a linear velocity of 6 m/min. Theefibers were then left at room temperature for at least 30 min prior to the hot drawing process. This involved stretching the as-spunefiber 5x across a soldering iron wrapped with Teflon tape and heated to 100 °C. The fibers were then dried at room temperature for 48 hr beforencharacterization.

INSTRUMENTATION

Fiber samples were cut after cooling in liquid N_2 to obtain circular undamaged cross sections. Small pieces ofefiber were fixed vertically on an aluminium stub using conductive glue. A sputter coater (Dynavac) was used for coating of a thin layer of gold on the cross section and side wall of theefibers (35 mA for 12 sec under 200 mbar Ar). A fully digital LEO Cambridge/Leica Stereoscan 440 Scanning Electron Microscope (SEM) with tungsten filament using 20 kV beam energy was used for morphological studies of the compositeefibers.

The hydrodynamic diameter of carbon nanotubes (SWNTs) solutions were measured using dynamic light scattering (ZS, Malvern, UK). A red laser beam of 632 nm (He/Ne) was used. This system uses the NIBS (non-invasive back scatter) technology where the back scatter at 173 °C is detected. The use of NIBS technology reduces multiple scattering effects since scattered light does not have to travel through the entire sample, so that the size distribution at higher concentrations of sample can be measured.

The dynamic mechanical analysis (DMA) was carried out using a Model Q-800 (Thermal Analysis). The strain rate mode can be used to collect stress versus strain data equivalent to that obtained from a universal testing machine. In this mode, a 10 mm gauge length ofefiber sample was stretched at a strain rate of 500 μ /min until the sample broke or yielded at 2 °C.

The viscosity of SWNTs solutions was recorded using a Brookfield viscometer (LV-DV II+) using DIN spindle 85 and 87.

In order to measure the conductivity ofefibers a homemade four point probe conductivity cell at constant humidity and temperature was employed. The electrodes were circular pins with constant separation of 0.33 cm andefibers were connected to the pins using silver paint (SPI). A constant current was applied between two outer electrodes using atpotentiostatGgalvanostat (Princeton Applied Research Model 363). The potential difference between the inner electrodes was recorded using a digital multimeter 34401A (Agilent).

A three electrode electrochemical cylindrical cell (15 mm×50 mm) coupled to a Bioanalytical Systems (Model CV27) potentiostat was used for cyclic voltammetry. A 10 mmefiber was used as the working electrode with an Ag/AgCl reference electrode and a Pt mesh counter electrode.

Raman spectra were obtained with the Jobin Yvon Horiba Raman spectrometer model HR800. The spectra were collected with a spectral resolution of 1.8 cm in the backscattering mode, using the 632.8 nm line of a He/Ne laser. The nominal power of the laser, polarised 500:1, was 20 mW. A Gaussian Lorentzian-fitting function was used to obtain band position and intensity. The incident laser beam was focused onto the specimen surface through a 100 × objective lens, forming a laser spot ~ 5 μm in diameter, using a capture time of 50 s. Raman signals were obtained with the half wave plate rotated at 170°C with a confocal hole set at 1100 μm and the slit set at 300 μm.

9.9   RESULTS AND DISCUSSION

For effective integration of CNes into a conducting polymer host it is first necessary to achieve a homogenous dispersion in the polymer matrix at weight fractions sufficient enough to allow the percolation threshold to be exceeded. In this regard, a solvent system capable ofgstabilizing the CNes and dissolving the polymer matrix is required. The combination of DCAA and AMPSA here resulted in an efficient solvent system for dissolution of PAne in the ES form.

The ability of DCAA to act as a dispersant for SWNTs and the effect of the addition of AMPSA on this ability was investigated using the dynamic light scattering technique. The hydrodynamic diameter distribution for DCAA-SWNT (0.0 % (w/w)) solutions containing %(w/w) AMPSA was determined (Figure 1).

DCAA-SWNT samples containing AMPSA showed lower hydrodynamic diameters compared to neat DCAA-SWNT, indicating that AMPSA assists in de-bundling of SWNTs. The addition of AMPSA also assists ingstabilizing SWNT dispersions in DCAA. The resulting dispersions were stable for more than 1 day; neither sedimentation nor aggregation of nanotube bundles was observed in the samples. Conversely, in samples that did not contain AMPSA aggregation of particles was observed after 1 hour, as confirmed by the dynamic light scattering technique. The increased stability of the SWNT dispersions after addition of AMPSA was presumably due to interaction of SWNTs and AMPSA, resulting in increased electrostatic repulsion and reduced aggregation [19-21].

The ultrasonication is necessary for dispersion of SWNTs, however, Zakri [22] has shown that increased sonication time significantly reduces the length of CNes and the conductivity ofefibers produced from them. It has also been shown that the aspect ratio of nanotubes has a significant impact on the tensile strength of composites containing them [23], reinforcing the importance of maintaining tube length bngminimizing sonication time. The influence of sonication time on the homogeneity of SWNT dispersions was monitored using viscometry. At shorter sonication times significant random fluctuations in viscosity are observed with shear rate, indicating that the dispersion was not fulledhomogenized. It was found that at least 30 minutes of sonication was required to remove these fluctuations and to obtain a homogeneous dispersion of SWNTs.

An understanding of the rheological properties of dispersions can be used to determine the optimum composition for wet spinning and as a useful tool for prediction of interconnectivity of nanotube bundles in the matrix after addition of polymer [24-26]. Figure 2 shows the viscosity vs. shear rate plots obtained as a function of increasing SWNT content. For successful wet spinning, a viscosity of 2000–3000 cP is required [27]. A number of parameters needed to be optimised to achieve this viscosity while also achieving the desired loadings of SWNTs and AMPSA. Using a feed solution containing a mole ratio of 1:0.6 for PANi: AMPSA has been shown to result in PAnrefibers with high conductivity, [11,28] so this ratio was maintained in the present study.

It was also found that a concentration of approximately 10 % (w/w) PANi-ES in DCAA was required to achieve the desired solution viscosity. Addition of Cbes to the spinning solution significantly increased viscosity at low shear rates (Figure 2).

However, the viscosity of these dispersions decreased rapidly with increasing shear rates. At the shear rate estimated for fiber spinning (10–10^3 s^{-1}), the viscosities of all SWNT-PANi-DCAA dispersions were very similar and in the range required. The higher viscosity and shear thinning behavior observed for SWNT dispersions suggested that the SWNTs aided in forming entanglements with the polymer chains. Specific interactions between the SWNTs and the PANi are considered below.

Another relevant issue forefiber spinning of PANi is gelation of the feed solution, which renders it unsuitable forefiber spinning. The viscosity of the feed solution vs. time was determined (Figure 3). It was found that solutions containing either neat PANi-AMPSA or with 0.9 % (w/w) SWNTs added gelled after 10 hrs or 6.5 hrs, respectively. Gelation times for all other SWNT contents were between 7 and 10 hours. All dispersions were sufficiently stable to allow enough time for spinning prior to gelation.

The use of feed solutions containing 0.3–0.9 % (w/w) SWNTs in DCAA resulted in formation of continuourefibers with reasonable mechanical properties (Table 1). SWNT loadings are given as weight fractions with respect to the weight of PANi-AMPSA in solirefibers. Forefibers containing 0.6 % (w/w) SWNT, a 0 % increase in tensile stress (σ_b), a 10 % increase in Young's modulus (E) and a 0 % decrease in elongation at break (ε_b) compared with neat PANrefibers were observed. The improvement in mechanical properties orefibers with the addition of SWNTs is illustrated by the stress-strain curves given in Figure 4.

Ramamurthy and his colleagues [29] investigated the influence of addition of MWNTs to PANi films fabricated by solution processing. The physical characterization of these composites by tensile testing and dynamic thermal mechanical analysis indicated that PANi containing 1% w/w MWNTs is more mechanically and thermally stable than PAni itself. However, only marginal improvements in mechanical properties were reported.

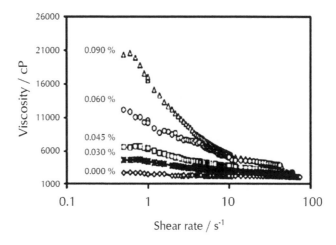

FIGURE 2   The influence of CNT content on the viscosity of solutions containing PANi-AMPSA (11.5% (w/w)). The weight fraction of SWNT is given with respect to the weight of DCAA solvent.

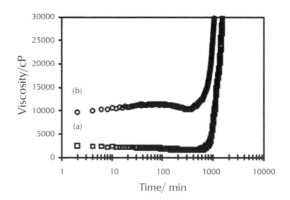

FIGURE 3   Viscosity vs. time up to gelation for solutions containing (a) PANi/AMPSA in DCAA (11.5% (w/w)), (b) as in (a) but with 0.09% (w/w) SWNTs added.

TABLE 1   Influence of SWNTs loading on mechanical properties of PANi-AMPSA fibers

% (w/w) SWNT in PANi-AMPSA	σb (MPa)	E (GPa)	eb (%)
0.00	170 ± 22	3.4 ± 0.4	9 ± 3
0.26	196 ± 17	3.9 ± 0.4	8 ± 3
0.38	199 ± 9	5.6 ± 0.3	7 ± 2
0.52	229 ± 28	6.2 ± 0.3	7 ± 2
0.76	255 ± 32	7.3 ± 0.4	4 ± 1

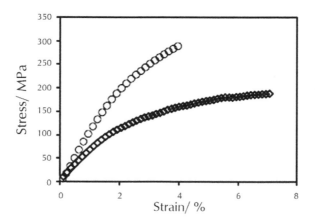

FIGURE 4   Typical stress – strain curves for (◊) neat PANi-AMPSA and (○) PANi-AMPSA-SWNT (0.76% (w/w) SWNTs).

The SEMs of the cross sections of fibers containing no SWNTs and 0.76 % (w/w) SWNTs are illustrated in Figure 5 and Figure 6, respectively. In both cases uniform and non porous fibers were produced.

Although, the strain at break decreases considerably with increasing CNT content, knots formed in PANi-AMPSA fibers with or without SWNTs showed high degrees of flexibility, as demonstrated in the SEMs in Figure 7 a,b. The fiber containing nanotubes also shows high level of twistability (i.e. 36 twist per inch (TPI)) in S (counter clock wise) and Z (clockwise) styles which is essential to enable the fiber to form a strong thread to be as warp or woof in textile matrix (Figure 7 c,d).

Since, the processing parameters including take-up, drawing and thermal stretching ratio were the same for all fibers, the improvement of mechanical properties with the addition of SWNTs could be attributed mainly to reinforcing of the PANi matrix through load transfer to oriented nanotube ropes. It has been shown that the formation of a crystalline coating around CNT bundles can assist load transfer by improving the adhesion between the polymer and nanotubes [23]. However, the XRD studies showed no evidence of crystallinity in the composite fibers produced here [30, 31].

The effect of SWNT addition to PANi fibers on conductivity was also investigated (Figure 8). A percolation threshold of ~ 0.35% (w/w) SWNTs was determined using the basic percolation power law [32]. The value for conductivity percolation threshold obtained here is in good agreement with another polymer-nanocomposite system that showed a 0.5 w/w % percolation level [33]. In addition, a significant increase in electrical conductivity was observed even at low SWNTs loadings, while the neat PANi-AMPSA fibers prepared here had similar conductivities (~ 500 Scm^{-1}) compared to those obtained in a previous study (600 Scm^{-1}) [11].

A number of workers have investigated the addition of CNTs to PANi on the subsequent electrical properties [34-37, 12,14]. For example Yu and coworkers [34] reported that the addition of MWNTs (1% w/w) to PAni increased the conductivity by an order of magnitude. Using a similar approach Deng and coworkers [14] prepared PANi with SWNT loadings using in situ emulsion polymerization. The conductivity of the resultant composite containing 10 % w/w SWNT was 6.6×10^{-2} S cm^{-1}, which was more than 25 times that obtained for the neat PANi (2.6×10^{-3} Scm^{-1}) using the same synthesis conditions.

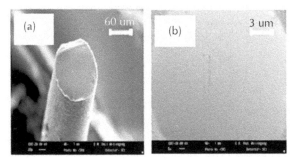

FIGURE 5   Cross-sectional SEM of neat PANi-AMPSA fiber. Scale bar (a) 20 μm (b) 1 μm.

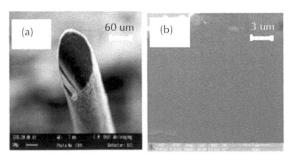

FIGURE 6   Cross-sectional SEM of PANi-AMPSA-SWNT (0.76 % (w/w) SWNTs) fiber. Scale bar (a) 30 μm (b) 1 μm. Inner core was sectioned by ion beam grinder.

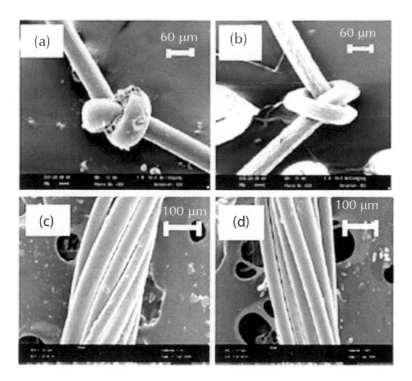

FIGURE 7   SEMs of knotted (a) PANi-AMPSA and (b) PANi-AMPSA-SWNT (0.76 % (w/w) SWNTs). 16 ply twisted fiber (TPI=36) of PANi - AMPSA- SWNT (0.76% w/w) with (c) S shaped twisting(d) Z shaped twisting.

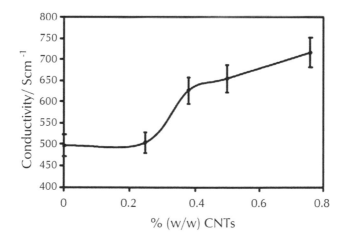

FIGURE 8   Electrical conductivity of PANi-AMPSA-SWNT composite fibers as a function of CNT loading.

The PANi produced in the presence of functionalized (carboxy containing groups) MWNTs also showed a 50–70% higher conductivity than PANi without the nanotubes [36]. Addition of SWNTs to PANi after polymer synthesis has also been shown to significantly enhance conductivity [35,12].

Raman spectroscopy was used to provide evidence of the interaction between CNTs and the PANi in the composite fibers prepared here (Figure 9). The important features of the Raman spectra for singe walled carbon nanotubes (SWNTs) (Figure 9(f)) occur in the range 200–2700 cm⁻¹. The radial breathing mode (RBM) is observed in the range 170–300 cm⁻¹ and gives information on nanotube diameter and chirality.

The tangential G band observed at 1594 cm⁻¹ originates from the graphite like structure, while the disorder (D) band and its second order harmonic known as the D* band appear at 1309 cm⁻¹ and 2604 cm⁻¹, respectively. As shown in Figure 9, Raman absorption peaks due to SWNTs increased in intensity when more nanotubes were incorporated into the fibers. Similar observations have been made previously [13]. A summary of assigned Raman absorption frequencies for PANi-SWNTs composite fibers is provided in Table 2.

Raman spectra can reveal information regarding the interaction between SWNTs and PAni. The dissolution of EB PANi in DCAA–AMPSA resulted in the formation of the ES form of PANi, as described previously [38, 39] (bands at 1337 cm⁻¹, 1496 cm⁻¹, 1592 cm⁻¹). The peak at 1337 cm⁻¹ corresponding to the bipolaron (C-N^{x+}) band of ES PANi was further enhanced by addition of CNTs, suggesting an increase in doping level [37, 40]. With increased SWNT loading a decrease in the ratio of absorption of the intensity of the 1496 cm⁻¹ band (C-N stretching of benzoid ring) compared to the 1169 cm⁻¹ band (C-H stretching of benzoid ring) was observed. This could be attributed to site selective interaction of PANi and SWNTs [37]. Such behavior in PAni fibers has been observed previously in studies involving the introduction of SWNTs after spinning [31]. According to previous studies by Do Nascimento and coworkers [41] the presence of the D band at 2604 cm⁻¹ suggests that some metallic SWNTs do not interact strongly with the PANi backbone.

Cyclic voltammtry for the neat PANi fiber and PANi-SWNT (0.76 %(w/w) SWNTs) composite fiber in 1 M HCl is shown in Figure 10. To avoid the pernigraniline oxidation state and subsequent degradation, the potential range was limited to between –0.2 and 0.5 V.

TABLE 2   Assigned Raman absorption frequencies originating from PANi-AMPSA-SWNT composite fibers. λexc = 632.8 nm.

Raman shift Assignment
2604 D* band of SWNTs
1594 G band of SWNT - C-C stretching of benzoid ring
1496 C-N stretching of benzoid ring
1337 C-N+ stretching of bipolaron structure
1169 C-H stretching of benzoid ring
190-255 (RBM) SWNTs

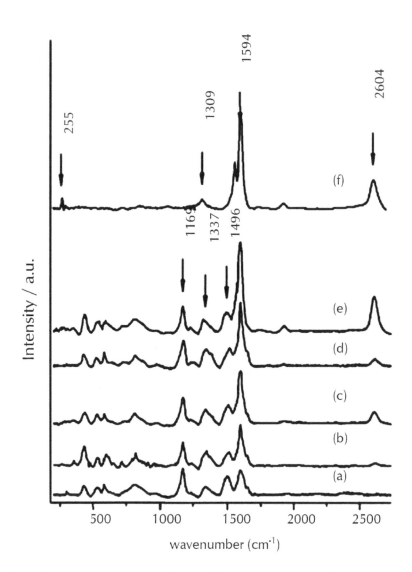

FIGURE 9 Enhanced Raman spectra (λ_{exc} = 632.8 nm) obtained for PANi-AMPSA-SWNT composite fibers containing various SWNT loadings (a) 0% (w/w), (b) 0.26% (w/w), (c) 0.38% (w/w) and (d) 0.52% (w/w). Enhanced Raman spectrum of SWNTs bucky paper (f).

This potential range covers the first redox process of PAni, that is the transition from the leucoemeraldine to the Emeraldine oxidation state. For the fiber containing SWNTs, the oxidation/reduction responses (attributed to the PAni redox chemistry) are better defined when compared with the neat PANi fiber.

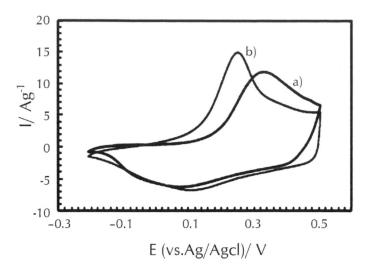

FIGURE 10 Cyclic voltammograms of (a) neat PANi-AMPSA and (b) SWNT-reinforced PANi-AMPSA fibers. Potential was scanned between − 0.2 V and + 0.5 V (vs. Ag/AgCl) in 1 M $HCl_{(aq)}$ at 5 mVs^{-1}.

The oxidation potential (+ 0.3 V) shifted to a lower value (+ 0.2 V) with addition of nanotubes. However, the reduction peak was only slightly changed with the addition of CNTs. The presence of the CNTs obviously facilitates the transition from the less conductive leucoemeraldine state to the Emeraldine state.

9.9.4 CONCLUSION

The addition of CNTs to PANi-AMPSA fibers processed from DCAA resulted in materials with high conductivity, high mechanical strength and modulus. Conductivities as high as 750 Scm^{-1} were achieved in continuously spun fibers up to 50–100 m in length.

Tensile strength of 250–300 MPa and moduli of 7–8 GPa for PANi-SWNT composite fibers were approximately two times higher than for neat PANi fiber. The fibers produced were tough enough to be knotted or twisted. Electroactivity was enhanced by the addition of nanotubes, as shown by cyclic voltammetry.

The AMPSA in DCAA was shown to be a highly effective dispersant for CNTs. Rheological studies and Raman spectroscopy studies indicated strong interactions between the SWNTs and the PANi. These interactions very likely contributed to the effective transfer of load and charge between the PANi matrix and the SWNT fibers.

The unique properties of high strength, robustness, good conductivity and pronounced electroactivity make these fibers potentially useful in many electronic textile applications. PANi-SWNTs composite fibers are currently being considered for application in artificial muscles, sensors, batteries and capacitors.

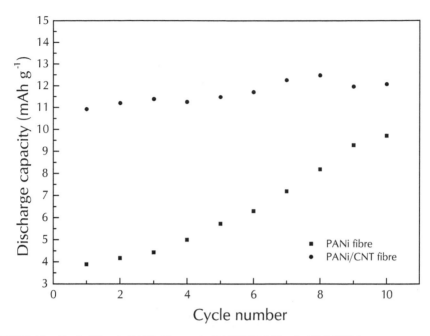

FIGURE 11   Cycle life of a PANi fiber and a PANi/CNT fiber in EMI.TFSA.

9.10   CASE STUDY IV

In this case study, the integration of CNTs into PAni fibers has been achieved using wet spinning techniques. It is shown that addition of small amounts of CNTs significantly improves the mechanical, electronic and electrochemical properties of the PAni fiber. These improved properties result in increased electromechanical actuation when used as artificial muscle fibers.

9.10.1   INTRODUCTION

The PAni may exist in six different forms, ranging from fully reduced (luecoemeraldine) or the fully oxidized (pernigraniline). The most conductive form is Emeraldine salt as highlighted in Scheme 1.

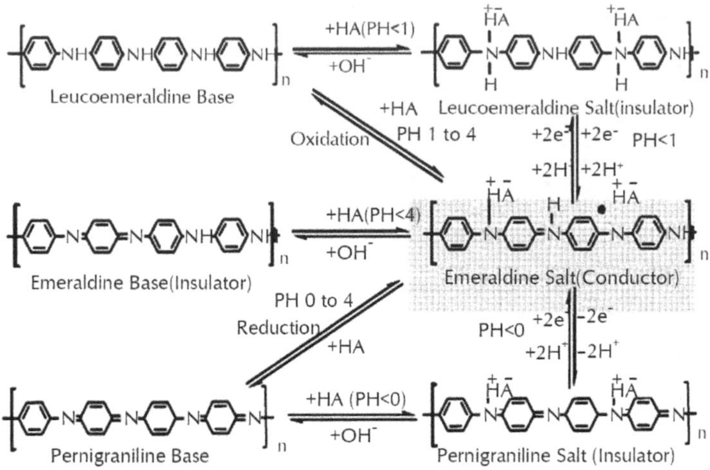

SCHEME 1   The different forms of PAne.

Wet spinning processes have been developed to produce high tenacity, conductive, electroactivrefibers [1,2,3]. The initial requirement for wet spinning is the formation of high concentration solutions of appropriate viscosity and stability. PAne solutions are known to undergo gelation upon standing such that they become unstable for fiber spinning [4]. Recent studies [5,6] have demonstrated that PAne emeraldine base (EB) is more soluble in N,N'-dimethyl propylene urea (DMPU) than N-methyl-2-pyrrolidinone (NMP), and that the onset of gelation is delayed in the DMPU solvent. To delay gelation, further gel inhibitors such as 2-methyl azaridine have been added [4]. Alternatively, the use of a reducing agent such as phenyl hydrazine to produce luecoemeraldine (LEB) can overcome the problem of gelation. Threfibers produced from EB and LEB can be acid doped to produce the conductive emeraldine salt (ES) form. Doping raises conductivity considerably but is detrimental to the mechanical properties [8].

In attempts to improve mechanical properties, blends of PANi with other polymers such as poly(p-phenylene terephthalamide) [9] and poly-omega-amino-undecanole[10] have been produced. Unfortunately, the addition of high tenacity insulator polymers to the PANi matrix significantly decreases the conductivity. There is, therefore, a need to introduce reinforcing fillers that enhances mechanical properties of PANrefiber while having a positive impact on electronic properties. In this regard the use of Cbes as conductive reinforcement for PANi has been investigated and reported here.

The discovery CNs by Iijima in 1991 [1] captured the attention of researchers in a wide range of areas. The CNTs have high tensile strength (100 GPa), high Young's modulus (0.6 TPa) [12] and exceptional electrical conductivity (5000 S/cm) [13]. CNTs have been used as a reinforcement filler in polyvinyl alcohol [14], epoxy [15], poly-(propionylethylenimine-co-ethylenimine) [16] host matrices. The addition of CNTs has also been found to increase the strength and conductivity of polypyrrole (PPy) [17]. The addition of CNTs to polyoctylthiophene (POT) has been shown to enhance the conductivity by 5 orders of magnitude with an 11% (w/w) percolation threshold

[18]. The synthesis and characterization of PANi-CNT composites has also been investigated [1,-22] Direct dissolution of pristine CNTs (without chemical functionalization) in aniline can occur *via* formation of donor acceptor charge complexes [23]. Using this approach PANi-CNT composite films have been fabricated by chemical [24] or electrochemical[22, [25] polymerization of aniline containing dissolved CNTs. Alternatively, CNTs have been blended with PAne in solvents such as NMP, DMPU [26,27]. The enhanced electroactivity and conductivity of the PANi-CNT composite films has been attributed to the strong interaction between CNTs and PANi, which enhances the effective degree of electron delocalization [26]. It was also shown that PANi containing 1% CNT is more mechanically and thermally stable than neat PANi [27].

The present work describes wet spinning of PANi-CNTs composite fibers. Protocols that enable a stable feed solution containing 10% (w/w) PANi and 2% (w/w) CNTs in DMPU have been developed. The resultant composite fibers show significant improvement in conductivity, electroactivity, mechanical strength and performance as electromechanical actuators (artificial muscles).

9.10.2 EXPERIMENTAL DETAILS

Free stand ink bucky paper was obtained from purified SWCes (CNTs) (Hipco@CNI) using a previously reported procedure[28]. The CNTs were dispersed at 0.1 mg/ml in a 0.5 wt % aqueous solution of Triton X-100 using 30 min sonication. The bucky paper was obtained after filtration of dispersed solution on a 0.02 µm Whatman anodisc followed by washing with methanol and water and drying at 130 °C for 30 min. The appropriate amount (10, 20, 40 mg) of the CNT bucky paper was sonicated in 20g DMPU (98%, Sigma Aldrich) for 2 hr at 60 °C to drive the CNTs into solution. 50 mg of EB powder (Santa Fe Science and Technology) (MW = 280000) was then added to the CNT dispersion while sonicating for a few seconds. Then 2 g of PAne (EB) and 0.65 g phenyl hydrazine were added gradually over a 2 hour period to the PANi-CNT–DMPU dispersion at 5 °C. The spinning solution was stirred another 2 hrs (40 rpm) at 0°C. Finally, a further 2 hours stirring under a dynamic vacuum ensured a bubble free solution ready forefiber spinning.

Prior to spinning, the solution was passed through a 400 µm filter then transferred to a nitrogen (N_2) pressure vessel. Nitrogen pressure was used to drive the spinning solution through a second filter (200 µm, Millipore) then through a single hole spinneret with L/D = 4 and D = 250µm and finally to the coagulation bath. The lengths of the first and second bath were 2m and 1m, respectively. Both baths contained 10% (w/w) NMP in water at 20°C. The N_2 pressure was adjusted to between —80 psi to control the injection rate for the spinning dope. In the first coagulation bath the solution solidified and the emerginrefiber was taken up on the first bobbin (D = 2.5 cm) at a linear velocity of 2 m/min. The semi-solirefiber was then passed through the second bath and collected using a linear velocity of 4 m/min on the second bobbin (D = 5 cm). Threfiber was then kept for 2 hr in a water bath to reduce the solvent content, and then for 12 hr in air at room temperature in preparation for the hot drawing process.

The drawing process involved hand stretching of the as-spun fiber across a soldering iron wrapped with Teflon tape and heated to 100 °C. This process was used to stretch the fibers to approximately twice their original length.

The complete removal of residual solvent was carried out firstly by extraction in a water bath for 24 hr followed by drying in a vacuum dryer at 50 °C for 12 hr. The fibers were then doped in methane sulfonic acid (1M) for the required time followed by air drying.

A Shimadzu 1601 and a Cary 5000 were employed for UV-Vis and NIR spectroscopy, respectively. Mechanical tests were carried out using a dynamic mechanical analysis (DMA) instrument Q-800 (TA series) in strain rate mode to record the force -displacement data generated at constant temperature (10 mm gauge length, 0.5 mm/min).

The Particle size distribution of CNT dispersions was determined using a dynamic light scattering technique(Zet sSizer model ZS (Malvern, UK)) with red laser 632 nm (He/Ne). The system uses NIBS (non-invasive back scatter) technology wherein the optics are not in contact with the sample and back scattered light is detected. The use of NIBS technology reduces multiple scattering effects and consequently size distributions in higher concentrations of sample can be measured.

Raman spectra were obtained with a JOBIN Yvon Horiba Raman spectrometer model HR800. The spectra were collected with a spectral resolution of 1.5 cm in the backscattering mode, with 632.8 nm line from a Helium/Neon laser. The nominal power of the laser, polarized 500:1, was 20 mW. A Gaussia/ Lorentzian-fitting function was used to obtain band position and intensity. The incident laser beam was focused on to the specimen surface through a 100 × objective lens, forming a laser spot of approximately 1 μm in diameter, using a capture time of 50 s. The Raman signals were obtained with the half wave plate rotated at 170 °C with a confocal hole set at 1100 μm and the slit set at 300 μm.

The viscosity of CNT dispersion was measured using a Brookfield viscometer (RV-DV II+) using spindle Din 85–87. The viscometer was equipped with software (Rheocalc 2.4) to collect and store data and allow it to be analyzed. Echem software (ADI instruments) coupled with a CV27/Potentiostat (BAS) and a Maclab (ADI instruments) were employed for recording cyclic voltammograms. A three-electrode electrochemical cell was used, which consisted of the PANi fiber covered with Pt as the working electrodea Ag/AgCl reference electrode, and a Pt mesh counter electrode.

The conductivity of th 5five samples of conductivrefibers with similar length (1 cm) and diameter (80 ± 20 μm) was measured using the 4-point probe technique. Since, the samples are quite thin. It can be presumed that sample fiber are uniform enough to obtain reliable results. A homemade four probe electrical conductivity cell operated at constant humidity and temperature has been employed. The electrodes were circular pins with separation distance of 0.33 cm anrefibers were connected to pins by silver paint (SPI). Between the two outer electrodes (1cm distance) a constant DC current was applied b Ppotentiosta/Ggalvanostat model 363 (Princeton Applied Research). The generated potential difference between the inner electrodes along the current flow direction was recorded by digital multimeter 34401A (Agilent). The conductivity of the solirefiber with circular cross section can then be calculated according to the cross-sectional area of threfiber, DC current applied and the potential drop across the two inner electrodes.

The actuation performance of PAN fibers with different CNT loading was tested with a Dual Mode Lever system (Aurora Scientific) in a 1M HNO_3 electrolyte. A two-electrode configuration was used while a pulsed current was applied as electrochemical stimulation. The positive and negative current was applied for 30 sec alternatively from 1mA to the current at which PANi actuators produced the maximum strain. The current density range was about 20mA/cm² to 250mA/cm². Isotonic actuation tests were conducted at steadily increasing loads until samples were broken. The applied current for the isotonic actuation tests was equivalent to the current density where the maximum actuation strain was obtained.

9.10.3 RESULTS AND DISCUSSION

NANOTUBE DISPERSION

To obtain a continuous distribution of CNTs throughout the compositrefiber it was essential that a high concentration, homogenous spinning solution be prepared. High viscosity without gelation is also required for wet spinning. Reproducible mechanical reinforcement provided by CNTs and any enhancement in charge transfer characteristics would be favored by the presence of long and thin bundles that are homogenously exfoliated throughout the matrix [29]. Therefore, the ideal processing conditions would faour the breaking up of nanotube bundles without reducing their lengths.

To achieve these aims, a solvent system capable of effectively dispersing CNTs and also dissolving PANi was required. We have found that DMPU meets both of these requirements. The dissolution of the LEB form of PANi in DMPU is facilitated by hydrogen bonding [5,30]. DMPU has a high dielectric constant and is classified as a polar aprotic, Lewis base solvent as it shows a higher degree of free electron availability (ß) without hydrogen donicity (π*) [31,32]. Since (ß) and (π*) are key properties in determining the solvation of CNTs [28], DMPU has proven to be effective in this regard.

Figure 1 shows the pronounced absorption bands observed in the UV-visible-NIR spectra of CNTs in DMPU. The bands observed at higher wavelength (1.07, 0.95, and 0.86 eV) are attributed to transitions between DOS singularities in semi conducting tubes and bands observed at lower wavelength (2.25, 1.90, and 1.70 eV) can be attributed to the Fermi level transitions observed with metallic SWNTs [4,-35]. The bands observed at other wavelengths are attributed to the heterogeneity of the samples in terms of tube diameters and helicity.

The enhanced features in different spectral regions significantly match with reported spectra from the known band structure of the SWNTs. The UV-Vis- NIR of CNTs dispersed in DMPU shows all the same peaks as for SWNTs dispersed in NMP which has previously been shown to be a good dispersant for SWNTs [28]. It presumably suggests that DMPU as a non-hydrogen-bonding Lewis bases can provide good dispersability for SWNTs.

Using dynamic light scattering the hydrodynamic diameter of the nanotube bundles was found to be between 150 and 400 nm after 30 min of sonication. After 60 min and 120 min sonication, the hydrodynamic diameter was reduced to 10–200 nm and 0–80 nm, respectively (Figure 2).

The viscosity as a function of shear rate was also determined after different sonication times (Figure 3). The fluctuations in viscosity observed after 30 or 60 minute sonication times are indicative of heterogeneity in the dispersion.

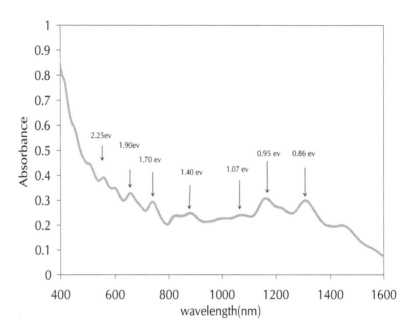

FIGURE 1   UV-Vis NIR spectra of CNT (0.01%(w/w) in DMPU after sonication for 120 min.

The smooth curve observed after 120 minutes indicates homogeneity. Note also this dispersion has the lowest viscosity which can be attributed to the smaller size of nanotube bundles after sonication.

The results presented above indicate that ultrasonication was effective in dispersing CNTs in DMPU. However, the nanotube bundles were found to re-agglomerate upon standing after sonication. As shown in Figure 4, the size distribution of CNT/DMPU dispersion (without PANi) shows a considerable increase in hydrodynamic diameter when measured 60 minutes after sonication, compared with the sample measured immediately following sonication.

Importantly, the addition of PANi to the CNT-DMPU dispersion was found to stabilise the exfoliated nanotubes and limit aggregation. Figure 4 shows that the hydrodynamic diameter of the CNTs after sonication and in the presence of PANi was ~20 nm compared with ~30 nm when PANi was not added. When the size distributions were measured again 60 minutes after sonication, the hydrodynamic diameter with PANi present had increased slightly to ~50nm, but had increased significantly to 100-800 nm when no PANi was present.

This observation suggests that nanotubes may interact physically through either wrapping or covering of nanotube surface by solubilized polymeric chain of PAni. Another possibility offers a π-π chemical interaction of planner surface of nanotube with quinoid ring of PAni chain [20]. More experimental study needs to be carried out to clarify the origin of this phenomenon.

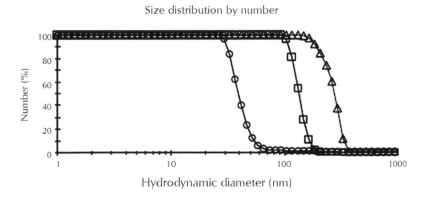

FIGURE 2 The hydrodynamic diameter distribution of nanotube bundles in DMPU for different sonication times: (Δ) 30 min (□) 60 min (○) 120 min.

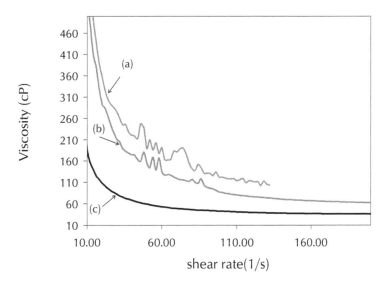

FIGURE 3 Viscosity of CNT dispersions in DMPU (0.2% w/w) obtained at different shear rates after different sonication times (a) 30 min (b) 60 min (c) 120 min.

Disregarding of cause of this effect, the sonication of CNTs in the presence of polymers that are structurally close to the matrix polymer was reported as a desirable approach to ensures compatibility of the functionaliznotubes with the polymer matrix to avoid any potential microscopic phase separation in the nanocomposits6, 37,38].

Interfacial adhesion of polymer matrix to nanotube without any moieties used for functionalizing and/or soublizing also reduce the impurity of final nanocomposite and improve the charge and load transfer efficency.

The maximum concentration of CNTs that could be dispersed in thniline containing DMPU solution was 2.0% (w/w). At 2.5% (w/w) large agglomerates could be seen visually even after 2 hours sonication. The minimum concentration of PANi required to give the viscosity favored for spinning is about 10 (w/w %). The use of a 10% (w/w) solution does produce low porosity fibres if the solution is aged [29].

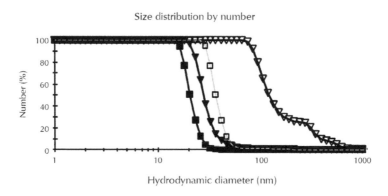

FIGURE 4 Size distribution shown as hydrodynamic diameter of CNTs in DMPU with (square) and without (inverted triangle) PANi. The distribution immediately after sonication is shown by filled symbols while the distribution measured 60 min after sonication was stopped is shown by unfilled symbols.

Here, it was found that using higher concentration of PANi (> 10% w/w) no additional tubes could be dispersed (due to limited capability of solvent for dispersion of CNTs) and so to keep the weight ratio of CNT: PANi as high as possible the 10% PANi solution was used. Different amount of CNTs: 0.5, 1.0 or 2.0% (w/w) were added to this solution and in each case fibers were successfullys

MECHANICAL PROERTIES

The effect of the CNT loading on mechanical properties of the come fibrefiber (2x drawn) was investigated and the tensile stress-strain curves are shown in Figure 5. For undoped PANi, the addition of 2% w/w CNTs increased the yield stress by 100%, the tensile stress (σ_b) by 50%, the Young's Modulus (E) by more than 200% compared to the neaniline fiber. A 30% decrease in elongation at break (ε_b) also occurred (Table 1).

TABLE 1   The influence of CNTS loading on mechanical properties before and after doping

% w /w CNT$_s$	$\sigma_{y=0.2}$ (MPa)		σ_b (MPa)		E (GPa)		$\varepsilon_{b(\%)}$	
	Before doping	After doping	Before doping	After doping	Before doping	After doping	Before doping	After doping
0%	120	35	168	45	7	3.5	3.1	4.6
0.5%	195	110	198	138	11	7	2.1	3.1
1%	230	145	236	145	14	9	2.1	2.7
2%	250	165	260	197	17	11	2.2	2.8

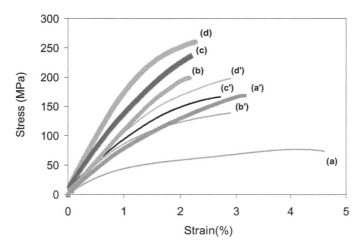

FIGURE 5   The Stress strain curves of PANi- CNT composite fibers with different CNT contents, before doping (thick lines) and after doping for 24 hr in MSA (thin lines): (a, a') Neat PAne, (b,b') PANi-CNT- 0.5 wt% (c,c') PANi-CNT- 1 wt% (d,d') PANi-CNT- 2 wt%, before and after doping.

After acid doping a 40% decrease in the Young's modulus and a 25% decrease in tensile strength were observed for theefiber containing 2% (w/w) CNTs. In comparison a 50% decrease in Young's modulus and a 70% decrease in tensile strength occurred upon doping the neat PANi fiber. Presumably, the reduction in modulus and strength of theefibers after acid doping is related to the introduction of charges on the polymer chain and counterions as well as the swelling that occurs during doping. However, the smaller reduction of tensile modulus and strength for the compositeefibers further illustrates the efficient reinforcement produces by the CNTs. The SEM images were obtained for drawn fibers with and without CNTs (Figure 6) and in spite of significant difference in mechanical properties, no significant difference in porosity was observed.

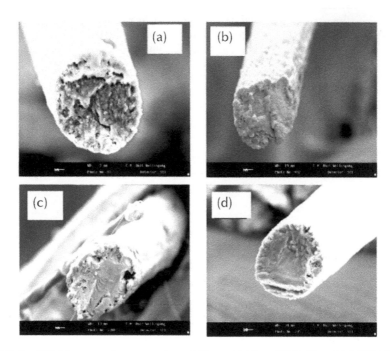

FIGURE 6   SEM images of cross section of fibers (a) Neat PANi fiber (b) PANi-CNT (0.5 w/w)% (c) PANi-CNT (1 w/w)% (d) PANi-CNT (2 w/w)%. Scale bar 10 μm.

The improvement of mechanical properties of polymers with added CNTs has been previously observed in several studies. Cadek [39] found that Young's modulus increases by a factor of 1.8 with addition of 1 %(w/w) SWNTs to PVA and 2.8 and a 50% improvement in the elastic modulus of poly(p-phenylene enzobisoxazole) (PBO) was observed upon addition of Nes [40]. An increase in tensile strength and elastic modulus by a factor of 1.5 and 1.7 respectively was observed upon addition of 5 wt% SWNT to polyacrylonitrile (PAN)[41].

The mechanical properties of the fiber obtained in this study exceed those reported previously for neat PANiefibers. The tensile strength of fibers produced from PANi (EB) [8] or PANi/AMPSA [42] are in the range of 95–157 MPa. For the 2% PANi-CNTefibers, the tensile strengths were 260 MPa (before doping) and 197 MPa (after doping). Moreover, the Young's modulus of PANi-CNT fibers were up to 17 GPa, again much higher than compared with PANi fibers produced in previous studies (2–4 GPa)

RAMAN SPECTROSCOPY

The interaction between PAne and the CNTs in the compositeefiber before doping was investigated using Raman Spectroscopy (Figure 7). PAne in the LEB state (Figure 7-e) is fully reduced and Raman bands attributed to the quinoid ring (refer tosScheme 1) should not be observed. However, LEB is a very unstable oxidation state and tends to switch to the EB form when the spinning solution is exposed to the atmosphere as evidenced by the appearance of the blue color. The Raman bands observed for the PAne fiber after spinning are summarized in Table 2

TABLE 2   Raman bands observed for the PAN-CNT fiber spun from the solution containing CNT (0- 2 w/w %)

Raman shift(cm⁻¹) The origin of enhanced spectra
1592 C-C stretching of benzoid ring
1523 N-H bending of bipolaronic structure
1472 C=N stretching of quinod ring
1334 C-N⁺ stretching of bipolaron structure
1218 C-N stretching of benzoid ring
1162 C-H stretching of quinoid ring
1172 C-H stretching of benzoid ring

The bands observed at 1162 cm⁻¹and 1472 cm⁻¹ are attributed to C-H stretching and C=N stretching of the quinoid ring, respectively. The presence of the benzoid ring results in bands at 1592 cm⁻¹ corresponding to C-C stretching and 1218 cm⁻¹ attributed to C-N stretching [43-45].

With increased amounts of CNTs the Raman spectra changes significantly. A decrease in intensity of the 1472 cm⁻¹ band (C=N stretching of quinoid ring) and an up shift of the 1162 cm⁻¹ band (C‾H stretching of quinoid ring) to 1172 cm⁻¹ band (C‾H stretching of benzoid ring) are obvious. It has been shown that a decrease in intensity of peaks assigned to the quinoid ring provides strong evidence for site selective interaction between the quinoid ring of PANi and CNTs [46]

It is worth noting that the up shift of the 1162 cm⁻¹ band to 1172 cm⁻¹ with increasing amounts of nanotubes indicates a transition towards a salt structure in the polymer[44,45]. The bond at 1334 cm⁻¹ corresponds to C-N stretching of the radical cation and is a measure of the degree of delocalization. In the case of PANi- CNT fibers before doping, CNTs play the role of a doping agent. The nanotubes appear to act as radical anion fragments interacting with positive charges induced on the polymer chain [26].

The emergence of bands at 1397 cm⁻¹ and 1654 cm⁻¹ have not been reported in previous work on PANi-CNT composites, presumably due to differences in the method of sample preparation.

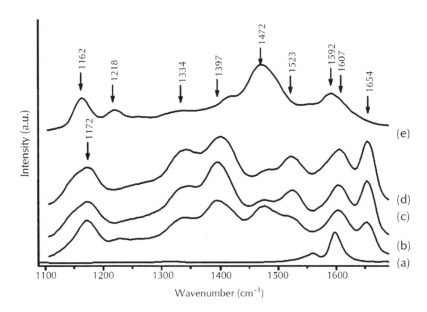

FIGURE 7   The Raman Spectra (λexc=632.8) obtained from PANi-CNT compositeefibers (before doping). (a) CNT bucky paper (b) PANi-0.5%w/w CNT, (c) PANi-1%(w/w) CNT, (d) PANi-2 % (w/w) CNT (e) Neat PANiefiber.

Colomban and coworkers [47] found thatefibers of ES-camphor sulphonic acid (HCSA) doped in m-cresol (secondarily doped) and ES doped in HCl have bands at 1390, 1650 cm⁻¹ upon red light (647.1nm) excitation. These bands were attributed to deprotonated PANi segments and this may be the source of additional band observed here.

CONDUCTIVIT

Table 3 shows the 4-point probe electrical conductivity before and after doping (dried at room temperature and 100 °C) of 2x drawn fibers with different CNT contents. For undoped PANi it was observed that the addition of % (w/w) CNes produces a two orders of magnitude increase in conductivity over the undoped neat PANi fiber.

Presumably two mechanisms, doping and interconnectivity of CNTs, contribute to the formation of a conductive passage way in the composite before doping. The potential interaction between highly delocalized π-electrons of CNTs and the π-electrons of the polymer backbone increase the effective electron delocalization and as a consequence improvs the conductivity of the host polymer. Most specifically, the enhance-ment of 1337 cm^{-1} band before doping in Figure 7 corresponds to C-N^{+} stretching of bipolan structure, revealing the formation of a conductive passage way.

TABLE 3   The influence of CNT loading on four point probe conductivity of 2x drawnefiber after doping with MSA

% w/w CNT	Conductivity (S/cm) Before doping	Conductivity (S/cm) After doping	
		Dried at 100 °C	Dried at 25 °C
0.0	1.3×10^{-6}	1.2 ± 0.4	96.1 ± 18.9
0.5	5.1×10^{-4}	6.2 ± 1.4	120.2 ± 20.8
1.0	6.3×10^{-4}	24.7 ± 6.3	121.6 ± 17.5
2.0	7.1×10^{-4}	32.4 ± 3.7	128.0 ± 25

After acid doping, fiber samples were dried initially at room temperature and at 100°C. After drying at room temperature, the conductivity of PAni-CNT fibers con-taining 2.0 w/w % CNTs showed a 30% increase relative to neat PAni fibers. In sam-ples dried at room temperature the conductivity was probably dominated by the host doped PAni matrix. However, after drying at 100°C, It was likely that the low molecu-lar weight dopant MSA was removed by evaporation and that conductivity was mostly dominated by the presence of SWNTs. After drying at 100°C, the conductivity of neat PAni was quite low (~1 S/cm) while the conductivity was increased 30 times upon addition of 2.0% w/w SWNT. The low conductivity of neat PAni fiber after drying at 100°C can be attributed to deprotonation of the PAni segments through evaporation of the small dopants. The improvement of conductivity in PAni composites containing SWNTs could be attributed to the formation of a percolating network of SWNTs.

CYCLIC VOLTAMMETRY

Cyclic voltammetry (CV) was carried out on PANi-CNTs composite fibers as part of the mechanical actuator testing (Figure 8). The resolved CV ensure ion and charge transfer required for actuation performance. To improve the conductivity of fibers, very thin Pt layer coated around the fibers. Much improved CVs were obtained with the Pt covered fibers. The influence of CNTs on the electroactivity of Pt covered PAni fibers in 1M HCl aqueous solution is illustrated in. It can be clearly seen that the addi-tion of CNTs enhanced the amplitude of the redox peaks, indicating a higher electro-activity compared to neat PANi fibers. Peak potentials also shifted with the addition of CNTs which can be attributed to interaction between PAni and CNTs. Three oxidation peaks at +0.10 V, +0.42 V and 0.70 V were observed for samples containing CNTs. Reduction peaks corresponding to the oxidation peaks at +0.42 V and +0.70 V were incorporated in a broad peak that composed of two unresolved peaks at +0.54 V and +0.22 V. The oxidation peak at + 0.10 V showed no obvious reverse cathodic peak. In contrast, two non resolved oxidation peaks at + 0.06 V and + 0.64 V and one broad reduction peak at 0.46 V were observed in the CV of neat PANi fibre. Upon comparison of the peak potentials, it was revealed that addition of CNTs to PAni results in easier electrochemical switching between leucoemeraldine, emeraldine and pernegraniline oxidation states compared to neat PANi. The more facile switching between oxidation states can be attributed to the interaction of the large π bonded surface of CNTs with the conjugated structure of PAni via π-π stacking [19,22]. This increases the extent of electron delocalization and results in improved charge transfer along the fiber axis.

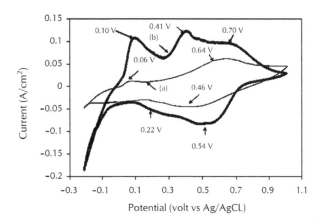

FIGURE 8   Cyclic voltametry of (a) neat PAni fibre (b) PAni/ SWNTs (2 % w/w) composite fibre. Scan rate= 50 mVs-1, electrolyte: 1.0 M HCl (aq)

ACTUATION PERFORMANE

The enhancement of mechanical properties along with the improvement of electrical and electrochemical properties by the inclusion of nanotubes can be used to produce low voltage actuators which can exert significant force and/or movement as compared to neat PANi. The electrochemically induced strain observed for neat PANi and PANi with 2% w/w CNT content is shown in Figure 9. For both materials, the strain increases to a maximum value before declining, as the applied current is steadily increased. The fibre with 2% (w/w) CNTs gave a maximum actuation strain of 2.2%, which is double the actuation obtained from neat PANi. These results are likely due to the higher conductivity and improved electroactivity derived from CNT loading in PANi.

   Figure 10 shows the effect of isotonic stress on the actuation strain of neat PANi fibre and the fibre with 2% w/w CNT loading. In both cases, the actuation strain was determined at higher isotonic stresses until the sample broke. The maximum stress the fibre could withstand was significantly increased from 5.3 to 15.3 MPa by addition of CNT.

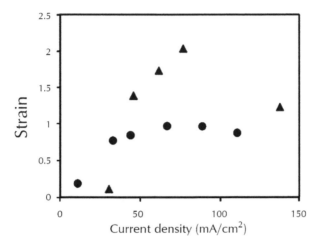

FIGURE 9   The influence of CNT loading on the electromechanical actuation performance of composite fieres. (▲) PANi-CNTs (2% wt) (●)Neat PANi.

Using the neat PANi fibre, the actuation stri / isotonic stress relationship gives a sharp negative slope; the actuation strain decreased quickly with the increasing stress. While for the fiere with 2% w/w CNT, the actuation strain was much more stable with increased stress. The effect of applied stress on actuation strain is known to be related to the change in elastic modulus of the actuator material that occurs during the actuation process [48].

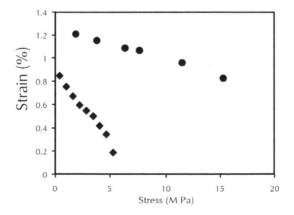

FIGURE 10   The effect of isotonic stress on the actuation strain of (■) neat PANi, (●) PANi-CNT (2% w/w).

The incorporation of CNT into PANi greatly improves the actuation performance both in terms of the higher strain attainable and better performance under load.

9.10.4 CONCLUSION

The PAni/CNT composite fibres have been produced using a wet spinning process. Considerable refinement of the mixing process was necessary to produce a stable dispersion of CNT in a PANi solution that had suitable rheological properties for wet spinning. iberes containing 2% (w/w) CNTs have superior mechanical, electrical and electrochemical properties compared with the neat PANi fibers. These improved properties resulted in significantly improved electromechanical actuation performance of the composite fibers when compared with neatiline iberes. The improvement in properties was attributed to the strong interaction betweeniline and CNTs as indicated using rheological studies and Raman Spectroscopy. Further enhancements in these properties are expected through the optimization of the CNT functionality so as to further control the solution rheology and stress transfer between the CNTs and PNi.

9.11   CASE STUDY V

In this case study, the influence of CNTs on the magnitude of morphological disorder in polymer composites was quantitatively measured using electrical transport measurements over a broad range of temperatures. A heterogeneous model which explains the non-metallic and metallic character of the iline/nanotubes composite is then described. The serial sum of quasi-1-D metallic resistance in crystalline areas and the amorphous resistance based on the fluctuation-induced tunneling between metallic islands including a constant term for metallic disorder shows a highly precise fit on the experimental data. Results revealed that the metallic transport in crystalline areas and metallic disorder in amorphous areas of the polymer is increased respectively by a factor o 1.5--2.5 an 1.2--1.4 by addition of the SWTs.

9.11.1   INTRODUCTION

The transport regime and magnitude of conductivity in doped p-conjugated polymers is determined by the degree of structural disorder. In general, as the disorder increases, ationlocalization is induced and the material becomes more insulating upon decreasing the temperature [1]. It has been widely sh,4,5,-6] that the measurement of macroscopic (DC) electrical conductivity over a wide range of temperatures can provide a comprehensive insight into nanoscale polymer morphology and the extent of disorder.

Several studies have indicated thatlinePAni)-emeraldine salt(S) / 2-acrylami-do-2 mthyl -1-propane sulfonic acid (AMPSA) [7,8,9,10] cast id (DCAA) is in the metallic regime toward the insulator-metal transition. The metallic beaviour was attributed to the criterion of finite conductivity when T⊔0 K, the positive slope of the temperature dependence reduced activation energy (i.e: W (T) = d [log σ(T)]/d [log(T)]),liseddelocalized electronic states at the Fermi level that allow conduction without thermal activation. However, this polymer does not show metallic behavior over the whole range of temperatures. They exhibit semiconducting (dσ/dT) > 0) and metal-like beaviours (dσ/dT < 0) respectively at temperatures lower and higher than room temperature.

The linear metallic and a crossover to nonmetallic beaviours at lower temperature has also been observed in the temperature dependence resistivity of ropes and mats of tubes [11,12]

It has been already shown that either quasi-one-dimensional variable range hopping (quasi-1D VRH) or 3-dimensional VRH models [13][1,] do not apply across the entire temperature range studied, showing both metallic and non metallic regime 5 K --340 K)[10]. This complex temperature dependence in both PAni-ES/AMPSA and SWNTs suggest that the heterogeneous model proposed by Kaiser [2] may be suitable for the quantitative determination of electronic transport parameters in both the metallic islands with highly ordered domain and the disordered amorphous regions-sisedemphasized in the literature the electrical transport occurs through metallic islands of oriented crystalline polymer chains (subscript c) that are interconnected by amorphous regions [14,15,16,] By taking into account that metallic-semiconducting transitions may occur in the amorphous region with both non metallic (subscript n) and metallic disorder (subscript d), the heterogeneous transport model assumes that the metallic and amorphous regions act as resistors in series. A simplified view of the oriented polymer microstructure is shown in Figure 1.

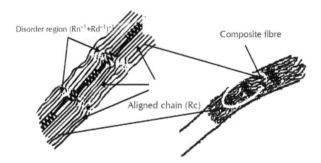

FIGURE 1   Crystalline regions separated by disordered regions constitutes the heterogeneous morphology in compositefibere.

The resistivity of the composite, therefore, is determined by the contribution of metallic (crystalline) and non metallic (amorphous) resistivity. The role of each portion in the determination of DC conductivity is described by temperature functions which identify the contribution of each part in the total electrical transport mechanism. The metallic part is described by the quasi VRH one dimensional exponential model [2,17,18] and an expression describing the amorphous part of the resistivity which is based on the fluctuation-induced tunneling between metallic islands[3].

The measurement of electrical transport over a very broad range of temperatures from 5 K to 340 K has been carried out. This data has been used in an attempt to understand the influence notubes addition on morphological disorder and the metallic properties of a PAni-ES/AMPSA compositefibere through extraction of conductivity data, such as reduced activation energy, and by fitting of the obtained data to a range of functions integrated into the heterogeneous mdel.

9.11.2   EXPERIENTAL

FABRICATION OF PANI/AMPSA-SWNT FIBRE SMLES

The 5X drawn fibres of PAni-ES/AMPSA containing different weight fractions of SWNTs (0.00 %w/w 0.25 % w/w 0.76 %w/w) were prepared based on the method reported previously[19][1]. Continuous mono filaments of PAni-AMPSA/SWNTfibere were then spun and thermally stretched five times versus initial length according to our previously reported procedure[19] [1]. They were then dried at 50 °C for 24 hr in vacuum. High purity silver paint (SPI) for connecting metallic wires tofibere samples, specific paper as an electrically insulating substrate and a specific type of varnish (known as "G" varnish) have been used for preparation of samples for transport measuremnt.

SAMPLE PREPARATION FOR TRANSPOR TEST

The sample holder equipped with 12 ports was designed for 3 samples to measure resistance *via* the 4 point probe method. It was cleaned using acetone to remove all erroneous particles left from previous experiments prior to use. A piece of eclectically insulating paper which can properly conduct thermal energy is attached to the surface using "g" varnish which is thermally conductive and electrically non–conductive. A 1 cm length of 3 fibre samples having different nanotube content was located on the paper and fixed using silver paste. A small needle was used to apply a very small amount of silver paste on each junction. The four point probe apparatus is designed in each port to pass current from the outer connections and to measure the voltage from the inner connections. The distance between inner points should be as far apart as possible (5-7 mm). Four wire contacts from each channel were connected to the four measurement points using silver paste. One hour vacuum drying at 50 °c is required to ensure that the silver paste was fully dried. Finally "g" varnish is used to ensure connection of the surface of the sample holder for efficient heat tranfer.

INSTRUMENTATION

The physationcharacterization equipment of PPMS (Physical Properties Measurement System, Oxford, UK) was used the range of temperatur of 5 °K to 340 °K. The sequence was defined for the experiment based on increasing or decreasing the temperature using the lowest possible scan rate. In this experiment the scan starts at 300 K and then goes up to 340 K and back to 5 K using a scan rate 3 K/min. The data fitting module of MATLAB was used for experimental data fitting on proposed models.

9.11.3 RESULT AND DISCUION

ELECTRICAL TRANSPORT ANLSIS

Conductivity (σ) versus temperature (T) plots for neat PAni-ES/AMPSAfibere andfiberes containing 0.25 and 0.76 % w/w SWNT (5X thermally stretched) are illustrated in Figure 1. The conductivity parameters for these different samples were also listed in Table 1.

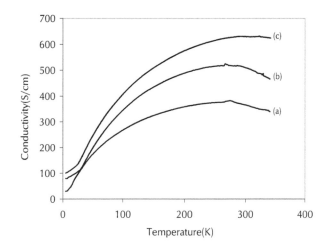

FIGURE 2 The temperature dependence conductivity for PAni-ES/AMPSA fibre and their composite containing 0.25% and 0.76 %w/w SWNT.

TABLE 1 The conductivity parameters for PAni-ES/AMPSA fibre and its SWNT composite fibres

% w/w SWNT in PAni-ES/AMPSA matrix	$\sigma_{(298\ K)}$ (S/cm)	σ_{max}(S/cm)	$T_{\sigma\,max}$(K)	$T_{metallic}$
0.00	360	378	260	<10 K
0.25	480	483	290	40
0.76	620	620	320	50

For all samples a strong dependence of conductivity and temperature can be seen over the whole temperature range investigated. At higher temperatures, a turnover temperature (260 K<$T_{\sigma max}$<320 K) corresponding to the temperature of maximum conductivity and a change in slope of the curve from nonmetallic ($d\sigma/dT>0$) to metallic ($d\sigma/dT<0$) was observed for all samples.

This transition occurs when sufficient thermal motion of the AMPSA counter ion and the polymer chain [7] eliminates the disorder induced in the amorphous regions between metallic islands [16, 20]. The higher $T_{\sigma max\ for}$ composite fibers contain nanotube compared to neat PAni fiber suggests that the SWNTs restrict the degree of motion of polymer chains so that higher energy must be absorbed to cross the transition temperature. The lower mobility of the polymer chains in the PAni-ES/AMPSA-SWNT composite fiber compared with neat PAni-ES/AMPSA has been shown elsewhere by differential scanning calorimetery (DSC) and dynamic mechanical analysis (DMA) data[21]. In addition, the extraction of temperature dependent reduced activation energy data from conductivity data had shown in Table 1 or Figure 1 indicates the location of the metal- insulator phase boundary. Figure 3 shows a log-log plot of reduced activation energy versus temperature, W (T), for the different samples.

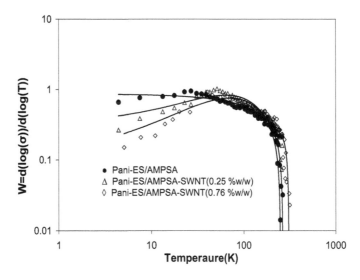

FIGURE 3   Log-Log plot of reduced activation energy (W) vs. temperature (●) PAni-Es-AMPSA (Δ) PAni-Es-AMPSA-SWNT (0.25 % w/w) (◊) PAni-Es-AMPSA-SWNT (0.76 % w/w)

As a general rule the plot of W(T) for the metallic, critical and insulator regimes show positive, zero and negative slope respectively. It can be observed that at low temperatures the PAni-ES/AMPSA fibre exhibits a metallic profile (i.e. the temperature below which metallic transport is present) less than 10 K. Addition of Cbes shifts the metallic-insulator boundary towards higher temperatures (~50 K). In addition, the neat PAni-ES/ AMPSA fibre compared with those fibres containing SWNT sustain a smaller positive slope region in proximity of the critical region or the insulator –metal boundary (zero slope at M-I boundary) at low temperature. The low metallic transport regime for neat PAni-ES/AMPSA is in good agreement with literature [9, 22] for PAni film and fieres doped with AMPSA. This phenomena was explained by simultaneous improvement of the metallic properties of the highly ordered domains (HOD) and increasing of disorder in the amorphous regions which separates the metallic islands in thermally stretched fieres [21]. The PAni-ES/AMPSA-SWNT composite fieres, however, show a well defined metallic behavior (positive slope of W vs T). Presumably, the nanotubes act by interlinking the metallic regions of the fiere ensuring a smaller coulomb gap in the density of states near the Fermi level [23,24]. Table 1 lists conductivity parameters for different fiere samples extracted from the experimental temperature dependent conductivity data.

HETEROGENEOUS MOEL

The qualitative explanation of temperature dependence on conductivity can be verified by fitting of the conductivity data to the heterogeneous model described above. The resistance in a heterogeneous model can be formulated as the sum of the resistances of the crystalline parts (subscript c) and barrier parts of the conduction path along the fiere (Figure 4). It was considered that the M-I transition may happen in the barrier region, with both non-metallic (subscript n) and disordered metallic (subscript d) portions present in parallel resistance form [2, 25, 26,27,28,29. The behaiour of this system can be represented by Equation 1.

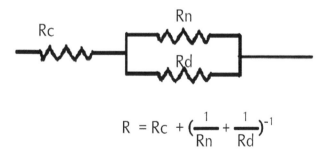

$$R = Rc + \left(\frac{1}{Rn} + \frac{1}{Rd} \right)^{-1}$$

FIGURE 4   Schematic of the resistance arrangement in the proposed model of charge transport described in equation 7.1.

$$\sigma^{-1} = f_c \rho_c + \left[\left(f_n \rho_n \right)^{-1} + \left(f_d \rho_d \right)^{-1} \right]^{-1} \tag{1}$$

Where fc, fn, fd are the ratio of effective length and cross sectional area of each region with resistivity ρc, ρn, ρd respectively for metallic, nonmetallic and metallic disorder regions .

Based on earlier work [30], Kaiser [2] proposed that the model of Sheng [3] for fluctuation-induced tunneling between extended metallic regions can be employed to describe the mechanism of non-metallic temperature dependence for the barrier resistance including a constant parameter for the disordered metal resistance denoted by ρ_d (the second term in Equatn . 1.

Using this term, the influence of electron scattering on impurities and lattice perfection is considered leading to temperature independent residual resistivity ($f_d\rho_d$). In addition, a quasi-one dimensional exponential model has been inserted for describing the crystalline conductivity in the metallic regions (first term of Equation .1). Based on these proposed terms for metallic and non metallic resistivity, the following comprehensive equation can be fitted on the temperature dependence conductivity data:

$$
\sigma^{-1} = f_c \rho_m \exp\left(\frac{-T_m}{T}\right) + \left[\left[f_n \rho_t \exp\left[\left(\frac{F_t}{T+Ts}\right)\right]\right]^{-1} + \left(f_d \rho_d\right)^{-1}\right]^{-1} \qquad (2)
$$

MATLAB software equipped with curve fitting tools based on least square algorithms was employed for data fitting on Equatiof 2. The equation comprises six parameters (i.e. T_m, T_t, T_s, $f_c\rho_m$, $f_d\rho_d$, $f_n\rho_t$) that can be fitted on imported data based on initial starting value and domain of variation for each parameter. The sensitivity of best fit with respect to variation of starting valu (versus constant parameters Tm, Tt, etc) are reported in Table 2. For clarity table is divided into six parts where each part highlights the influence of varying each individual parameters.

For example, in the first column of part 1, different values of Twasere selected between 1 and 500 K while the value of the other parameters are equal to 1. In each measurement based on these initial values, fitting parameters were calculated and also the amount of R-square was presented in the last column. This should be obvious, as R-squared approaches unity as a regression approaches a perfect fit. The comparison among the values of fitting parameter in all parts reveals their sensitivity versus the initial values.

Even in part six it can be seen that for some initial values (20, 40, 50) the calculation does not converge to any consistent result. However, just in this part, for the other initial values shown convergence is observed but it was not the best fit compared to those data shown as bold.

TABLE 2 The sensitivity of the model to the initial starting values for the fitting Mmn

Variable starting value Tm		T_m	T_s	T_t	$f_c\rho_m$	$f_d\rho_d$	$f_n\rho_t$	R^2
Part one	1	2077	9.759	67.5	0.2071	0.04197	0.002155	0.9972
	5	2342	5.287	55.96	0.3176	0.0363	0.00237	0.9984
	10	2335	4.627	54.11	0.2428	0.035	0.0024	0.9984
	20	1813	23.61	97.82	0.1799	0.107	0.001708	0.9916

Variable starting value Tm		T_m	T_s	T_t	$f_c\rho_m$	$f_d\rho_d$	$f_n\rho_t$	R^2
	40	2298	10.98	69.82	0.458	0.044	0.00211	0.9968
	50	2371	10.23	67.19	0.489	0.0426	0.00218	0.9969
	100	2310	25.76	92.67	0.6461	30.21	0.00178	0.9906
	500	2338	26.03	93.88	0.6901	142.5	0.00176	0.9906

Variable starting value Ts		T_m	T_s	T_t	$f_c\rho_m$	$f_d\rho_d$	$f_n\rho_t$	R^2
Part Two	1	2077	9.759	67.5	0.2071	0.04197	0.002155	0.9972
	5	2186	4.87	54.77	0.1653	0.036	0.0024	0.9984
	10	2199	4.39	53.41	0.1387	0.035	0.0024	0.9984
	20	2377	5.85	57.45	0.3285	0.036	0.0023	0.9983
	40	2377	4.24	53.1	0.2512	0.035	0.0024	0.9984
	50	2380	4.3	53.25	0.2597	0.035	0.0024	0.9984
	100	2148	9.03	63.51	0.2274	0.04	0.00219	0.9975
	500	2324	3.59	51.22	0.1296	0.034	0.0025	0.9982

Variable starting value Tt		T_m	T_s	T_t	$f_c\rho_m$	$f_d\rho_d$	$f_n\rho_t$	R^2
Part Three	1	2077	9.759	67.5	0.2071	0.04197	0.002155	0.9972
	5	2391	3.887	52.16	0.249	0.035	0.0024	0.9983
	10	2342	3.716	51.77	0.2516	0.035	0.0024	0.9983
	20	2254	4.92	54.91	0.2066	0.036	0.0024	0.9984
	40	2038	25.3	91.39	0.2344	2.713	0.0018	0.9906
	50	2379	0.02	48.38	0.128	0.0317	0.0026	0.9946
	100	2398	4.693	54.28	0.305	0.0358	0.0024	0.9984
	500	2324	3.599	51.22	0.1296	0.034	0.0025	0.9982

TABLE 2   *(Continued)*

Variable starting value Tm		T_m	T_s	T_t	$f_c\rho_m$	$f_d\rho_d$	$f_n\rho_t$	R^2
Part Four	1	2077	9.759	67.5	0.2071	0.04197	0.002155	0.9972
	5	2371	6.898	60.1	0.3914	0.0376	0.0023	0.9981
	10	2397	1.293	44.53	0.00635	0.03359	0.0026	0.9968
	20	2336	7.046	60.31	0.3695	0.03829	0.00228	0.9980
	40	2293	10.29	68.13	0.4291	0.0426	0.00214	0.9750
	50	2320	17.23	74.11	0.3148	0.08867	0.002136	0.9916
	100	2214	8.934	65.06	0.3058	0.0407	0.00219	0.9976
	500	2324	3.599	51.22	0.1296	0.03483	0.0025	0.9982

Variable starting value fdpd		T_m	T_s	T_t	$f_c\rho_m$	$f_d\rho_d$	$f_n\rho_t$	R^2
Part Five	1	2077	9.759	67.5	0.2071	0.04197	0.002155	0.9972
	5	1864	26.14	94.3	0.1546	9.26	0.001756	0.9907
	10	989.7	26.57	96.57	0.01268	73.07	0.001714	0.991
	20	1979	22.85	81.95	0.1544	0.8071	0.001992	0.9893
	40	2347	25.88	92.94	0.8636	40.51	0.00178	0.9905
	50	2246	24.36	87.47	0.4929	142.5	0.001863	0.9903
	100	1800	25.96	93.79	0.1323	12.7	0.001764	0.9907
	500	2324	3.599	51.29	0.1296	0.03483	0.0025	0.9982

Variable starting point fnpt		T_m	T_s	T_t	$f_c\rho_m$	$f_d\rho_d$	$f_n\rho_t$	R^2
Part Six	1	2077	9.759	67.5	0.2071	0.04197	0.002155	0.9972
	5	2284	4.806	54.69	0.2261	0.03958	0.002405	0.9984
	10	2394	8.353	63.65	0.4874	0.03942	0.002232	0.9977
	20							
	40	No convergence						
	50							
	100							
	500	2324	3.599	51.22	0.1296	0.0328	0.0025	0.9982

It is worth noting that fitting parameters in same value of R-square show very low discrepancy and also very low sensitivity versus the initial values. For example in part one in same R^2 value of 0.9984, the value of fitting parameters shows very small difference. Good fitting of the prescribed model according to fitting parameter on experimental data is shown for neat PAni-ES/ AMPSA and PAni –ES/AMPSA-SWNT (0.25 -0.76 %w/w) fibers in Figure 5.

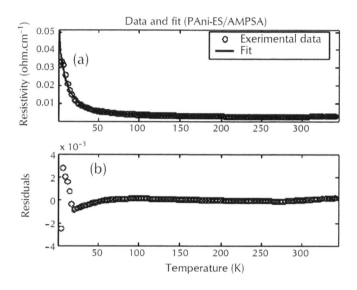

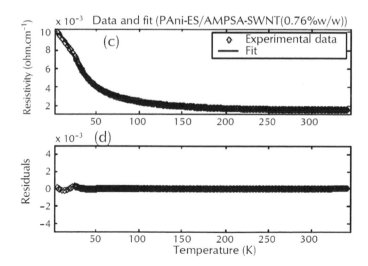

FIGURE 5 The experimental data and fit line based on the proposed model (a)PAni-ES/ AMPSA (c) PAni-ES/AMPSA –SWNT (0.75%w/w) and residual amount of least square of fitting calculation. (b)PAni-ES/AMPSA (d) PAni-ES/AMPSA –SWNT (0.76%w/w).

The precision of the data in Figure 5 shows a residual amount of nearly zero in the temperature range between 50-300 K, however the scattered residual data beyond this temperature range even is in the order of 1×10^{-3}. The data given in Table 3 shows the fitting parameter of conductivity data for PAni–ES/AMPSA fiber and its composites with added CNT.

T_m sets the energy of phonons that can backscatter charge carriers (It is different from T metallic in Table 1) , and T_t and T_s are the tunneling parameters which shows the magnitude of electrical barriers [3]. Highly conductive samples show lower Tt. Also Lower T_s values are obtained for high conductive samples. It can be seen that the numerical amount of Tt and Ts significantly decrease by addition of SWNTs that suggests a lower morphological barrier in the amorphous region for electrical transport. The data given in Table 3 can be employed to deduce the contribution of metallic resistivity in bulk resistivity at room temperature (ω_1) and the contribution of metallic disorder (fdρd) in amorphous resistivity at room temperature in the amorphous term (ω_2).

TABLE 3   The fitting parameters presented in Equation 2 for different samples.

% w /w SWNT/ PAni-ES/AMPSA	$f_c\rho_m$	T_m	$f_n\rho_t$	T_t	T_s	$f_d\rho_d$	R^2
0.00	0.2071	2077	0.0021	127.05	9.75	0.0095	0.9984
0.25	0.3134	2244	0.0015	97.74	4.82	0.012	0.9982
0.76	0.0540	2276	0.0009	87.84	3.92	0.14	0.9978

$$\omega_1 = \frac{f_c\rho_m \exp\left(\dfrac{-T_m}{T}\right)}{\sigma^{-1}} \tag{3}$$

$$\omega_2 = \frac{f_d\rho_d}{\left(\left[f_n\rho_t \exp\left(\dfrac{T_t}{T+Ts}\right)\right]^{-1} + \left(f_d\rho_d\right)^{-1}\right)^{-1}} \tag{4}$$

Table 4 shows the numerical amounts of ω_1 and ω_2 based on Equation 3 and 4. Both of these terms shows that, the fraction of the quasi-1D metallic transport in crystalline region of ω_1 (at room temperature) and the fraction of metallic disorder in amorphous area of ω_2 is increased by addition of SWNTs to PAni-ES/AMPSA.

TABLE 4   The contribution of metallic resistivity in bulk resistivity and the metallic disorder in amorphous resistivity

% w /w SWNT in PAni-ES/AMPSA	$\omega_1(\%)$	$\omega_2(\%)$
0%	5.4	0.83
0.25 %	12	0.84
0.76 wt %	14	1.01

The highly ordered domain of metallic islands in the sample without SWNTs is separated by amorphous region (after thermal stretching) that shows a higher barrier magnitude (lower Tt) compared to samples with SWNTs.

The increasing of ω_1, ω_2, and decreasing of T_t and T_s after addition of SWNT, suggests the formation of a interconnected network of metallic SWNTs in barriers of the amorphous area between crystalline PAni islands. Another possibility is related to iron content of nanotubes that should not be ignored as elemental analysis results listed in Table 4 showed higher iron content by increasing of nanotubes.

In addition, the high orientation of SWNTs in PAni-ES/AMPSA matrix [20] after thermal stretching may result in a extended structure and higher metallic transport. Also, the doping effect of nanotubes while reacted with PAni is further supported by the elemental analysis results (Table 4). For example using the N content as reference, the addition of CNT material to PAni-ES/AMPSA should not affect the S/N ratio if the nanotubes have no doping effect. However, if CNT competes with S, then the S/N ratio should decrease in the presence of CNTs. This assumption is definitely observed in the obtained data. The S/N ratio is 0.61 in neat PAni-AMPSA and 0.28 in PAni-ES/AMPSA-SWNT (0.76% w/w).

TABLE 5   The elemental analysis data of PAni-ES/AMPSA fiber with different SWNT content

% w/w SWNT in PAni-ES/AMPSA	% w/w C	% w/w H	% w/w N	% w/w Fe	% w/w S	S/N
0.00	54.02	5.89	9.72	0.00	5.99	0.61
0.25	49.99	5.37	8.45	0.07	4.19	0.49
0.76	57.86	5.34	10.36	0.18	2.94	0.28

9.11.4   CONCLUSION

In summary, the temperature dependent conductivity of PAni-ES/AMPSA fibre with and without SWNTs was evaluated. The results of the experimental data were fitted on a model that proposes a heterogeneous structure comprising crystalline and amorphous resistivities in series for the conducting polymer and SWNTs which are used to identify the conductivity parameters and/ or transport regime.

Metallic behavior accompanied with a peak in the conductivity and a turn over to non-metallic behavior was observed for highly conducting PAni and its SWNT composites. The temperature corresponding to maximum conductivity is increased by addition of SWNTs suggesting a change in morphological disorder. In addition, the reduced activation energy indicates greater metallic behavior in PAni-ES-AMPSA containing SWNT compared with neat PAni-ES/AMPSA.

The qualitative analysis of the metallic behavior of composite fibers was further investigated using a quantitative approach. It was shown that the higher metallic transport of PAni-ES/AMPSA-SWNT composite fiber compared to neat PAni-ES/AMPSA fiber is based on the numerical amount of the fraction of metallic resistivity to the total resistivity and the contribution of metallic disorder in resistivity of amorphous regions (i.e. $\omega 1$, $\omega 2$). These parameters indicated that the addition of SWNTs improves the metallic property in the crystalline areas and boosts the metallic disorder contribution in amorphous areas.

In the PAni fiber without SWNTs, the induced separation in highly ordered domains (HOD) of metallic islands during the stretching process is filled by nonmetallic disorder of the amorphous regions. Meanwhile, the addition of SWNTs may produce a network including a mixture of metallic and nonmetallic ropes interacting with PAni in the crystalline and amorphous areas. It is therefore expected that the gap between metallic regions is filled by metallic ropes which are well distributed and aligned in the composite.

REFERENCES

1. Iijima, S. Helical microtubules of graphitic carbon. *Nature*, 354(6348), 56 (1991).
2. Dresselhaus, M. S., Dresselhaus, G., and Eklund, P. C. *Science of fullerenes and carbon nanotubes*. New York: Academic Press; (1996).
3. Saito, R., Dresselhaus, G., and Dresselhaus, M. S. *Physical properties of carbon nanotubes*. Imperial College Press; (1998).
4. Harris, P. J. F. *Carbon nanotubes and related structures: New materials for the twenty-first century*. Cambridge, United Kingdom: Cambridge University Press; 279 (1999).
5. Wagner, H. D. and Vaia, R. A. Nanocomposites: Issues at the interface. *Mater Today*, **7(11)**, 38 (2004).
6. Zeng, Q. H., Yu, A. B., Lu, G. Q., Multiscale modeling and simulation of polymer nanocomposites, *Prog. Polym. Sci.* **33**, 191–269 (2008).
7. Iijima, S. Helical microtubules of graphitic carbon. *Nature*, **354**, 56–8 (1991).
8. Bethune, D. S., Klang, C. H., De Vries, M. S. and et al. Cobalt-catalysed growth of carbon nanotubes with single-atomic-layer walls. *Nature*, **363**, 605–7 (1993).
9. Dresselhaus, M. S., Dresselhaus, G., and Saito, R. Physics of carbon nanotubes. *Carbon*, **33**, 883–91 (1995).
10. Thostenson, E. T., Ren, Z. F., and Chou, T. W. Advances in the science and technology of CNTs and their composites: a review. *Compos Sci Technol.*, **61**, 1899–912 (2001).
11. Yakobson, B. I. and Avouris, P. Mechanical properties of carbon nanotubes. *Top Appl Phys.*, **80**, 287–327 (2001).
12. Ajayan, P. M., Schadler, L. S., and Braun, P. V. *Nanocomposite science and technology*. Weinheim: Wiley-VCH, 77–80 (2003).

13. Li, J., Ma, P. C., Chow, W. S., To, C. K., Tang, B. Z., and Kim, J. K. ,Correlations between percolation threshold, dispersion state and aspect ratio of carbon nanotube, *Adv Funct Mater.,* **17**, 3207–15 (2007).

14. Ajayan, P. M., Stephan, O., Colliex, C., and Trauth, D. Aligned carbon nanotube arrays formed by cutting a polymer resin-nanotube composite. *Science,* **265**, 1212–4 (1994).

15. Summary of Searching Results. <http://www.scopus.com>.

16. Du, J. H., Bai, J., and Cheng, H. M. The present status and key problems of carbon nanotube based polymer composites. *Express Polym Lett.,* **1**, 253–73 (2007).

17. Lee, J. Y, Baljon, A. R. C., Loring, R. F., and Panagiotopoulos, A. Z. Simulation of polymer melt intercalation in layered nanocomposites. *J Chem Phys.,* **109**, 10321–30 (1998).

18. Smith, G. D., Bedrov, D., Li, L. W., and Byutner, O. A molecular dynamics simulation study of the viscoelastic properties of polymer nanocomposites. *J Chem Phys.,;* **117**, 9478–89 (2002).

19. Smith, J. S., Bedrov, D., and Smith, G. D. A molecular dynamics simulation study of nanoparticle interactions in a model polymer–nanoparticle composite. *Compos Sci Technol.,* **63**, 1599–605 (2003).

20. Zeng, Q. H., Yu, A. B., Lu, G. Q., and Standish, R. K. Molecular dynamics simulation of organic-inorganic nanocomposites: Layering behavior and interlayer structure of organoclays. *Chem Mater.,* 15, 4732–8 (2003).

21. Vacatello, M. Predicting the molecular arrangements in polymer-based nanocomposites. *Macromol Theory Simul.,* **12**, 86–91 (2003).

22. Zeng, Q. H., Yu, A. B., and Lu, G. Q. Interfacial interactions and structure of polyurethane intercalated nanocomposite. *Nanotechnology,* 16, 2757–63 (2005).

23. Allen, M. P. and Tildesley, D. J. *Computer simulation of liquids.* Oxford: Clarendon Press; (1989).

24. Frenkel, D. and Smit, B. *Understanding molecular simulation: From algorithms to applications.* 2nd ed. San Diego: Academic Press, (2002).

25. Metropolis, N., Rosenbluth, A. W., Marshall, N., Rosenbluth, M. N., and Teller, A. T. Equation of state calculations by fast computing machines. *J Chem Phys.,* **21**, 1087–92 (1953).

26. Carmesin, I. and Kremer, K. The bond fluctuation method: a new effective algorithm for the dynamics of polymers in all spatial dimensions. *Macromolecules,* **21**, 2819–23 (1988).

27. Hoogerbrugge, P. J. and Koelman, J. M. V. A. Simulating microscopic hydrodynamic phenomena with dissipative particle dynamics. *Europhys Lett.,* **19**, 155–60 (1992).

28. Gibson, J. B., Chen, K., and Chynoweth, S. Simulation of particle adsorption onto a polymer-coated surface using the dissipative particle dynamics method. *J Colloid Interface Sci.,* **206**, 464–74 (1998).

29. Dzwinel, V. and Yuen, D. A. A two-level, discrete particleapproach for large-scale simulation of colloidal aggregates. *Int J Mod Phys C.,* **11**, 1037–61 (2000).

30. Dzwinel, W. and Yuen, D. A. A two-level, discrete-particle approach for simulating ordered colloidal structures. *J Colloid Interface Sci.,* 225, 179–90 (2000).

31. Chen, S. and Doolen, G. D. Lattice Boltzmann method for fluid flows. *Annu Rev Fluid Mech.,* **30**, 329 64 (1998).

32. Cahn, J. W. On spinodal decomposition. *Acta Metall.,* **9**, 795–801 (1961).

33. Cahn, J. W. and Hilliard, J. E. Spinodal decomposition: a reprise. *Acta Metall.,* **19**, 151–61 (1971).

34. Cahn, J. W. Free energy of a nonuniform system. II. Thermodynamic basis. *J Chem Phys.,* **30**, 1121–4 (1959).

35. Lee, B. P., Douglas, J. F., and Glotzer, S. C. Filler-induced composition waves in phase-separating polymer blends. *Phys Rev E.*, 60, 5812–22 (1999).

36. Ginzburg, V. V., Qiu, F., Paniconi, M., Peng, G. W., Jasnow, D., and Balazs, A. C. Simulation of hard particles in a phaseseparating binary mixture. *Phys Rev Lett.*, **82**, 4026–9 (1999).

37. Qiu, F., Ginzburg, V. V., Paniconi, M., Peng, G. W., Jasnow, D., and Balazs, A. C. Phase separation under shear of binary mixtures containing hard particles. *Langmuir*, 15, 4952–6 (1999).

38. Ginzburg, V. V., Gibbons, C., Qiu, F., Peng, G. W., and Balazs, A. C. Modeling the dynamic behavior of diblock copolymer/ particle composites. *Macromolecules*, **33**, 6140–7 (2000).

39. Ginzburg, V. V., Qiu, F., and Balazs, A. C. Three-dimensional simulations of diblock copolymer/particle composites. *Polymer*, 43, 461–6 (2002).

40. He, G. and Balazs, A. C. Modeling the dynamic behavior of mixtures of diblock copolymers and dipolar nanoparticles. *J Comput Theor Nanosci.*, 2, 99–107 (2005).

41. Altevogt, P., Ever, O. A., Fraaije, J. G. E. M., Maurits, N. M., and van Vlimmeren, B. A. C. The MesoDyn project: software for mesoscale chemical engineering. *J Mol Struct.*, 463, 139–43 (1999).

42. Kawakatsu, T., Doi, M., and Hasegawa, A. Dynamic density functional approach to phase separation dynamics of polymer systems. *Int J Mod Phys C*, **10**, 1531–40 (1999).

43. Morita, H., Kawakatsu, T., and Doi, M. Dynamic density functional study on the structure of thin polymer blend films with a free surface. *Macromolecules*, **34**, 8777–83 (2001).

44. Doi, M. OCTA-a free and open platform and softwares of multiscale simulation for soft materials /http://octa.jp/S. (2002).

45. Tandon, G. P. and Weng, G. J. The effect of aspect ratio of inclusions on the elastic properties of unidirectionally aligned composites. *Polym Compos.*, **5**, 327–33 (1984).

46. Fornes, T. D. and Paul, D. R. Modeling properties of nylon 6/clay nanocomposites using composite theories. *Polymer*, 44, 4993–5013 (2003).

47. Odegard, G. M., Pipes, R. B., and Hubert, P. Comparison of two models of SWCN polymer composites. *Compos Sci Technol.*, **64**, 1011–20 (2004).

48. Odegard, G. M., Gates, T. S., Wise, K. E., Park, C., and Siochi, E. J. Constitutive modeling of nanotube-reinforced polymer composites. *Compos Sci Technol.*, **63**, 1671–87 (2003).

49. Pipes, R. B. and Hubert, P. Helical carbon nanotube arrays: mechanical properties. *Compos Sci Technol.*, **62**, 419–28 (2002).

50. Rudd, R. E. and Broughton, J. Q. Concurrent coupling of length scales in solid state systems. *Phys Stat Sol B.*, **217**, 251–91 (2000).

51. Starr, F. W. and Glotzer, S. C. Simulations of filled polymers on multiple length scales, Ed. A. I. Nakatani, R. P. Hjelm, M. Gerspacher, R. Krishnamoorti. In *Filled and nanocomposite polymer materials*, Materials research symposium proceedings. Warrendale: Materials Research Society, pp. KK4.1.1–KK4.1.13 (2001).

52. Glotzer, S. C. and Starr, F. W. Towards multiscale simulations of filled and nanofilled polymers, In: Foundations of molecular modeling and simulation. Ed. P. T. Cummings, P. R. Westmoreland, B. Carnahan: Proceedings of the 1st international conference on molecular modeling and simulation. Keystone: American Institute of Chemical Engineers, pp. 44–53 (2001).

53. Mori, T. and Tanaka, K. Average stress in matrix and average elastic energy of materials with misfitting inclusions. *Acta Metallurgica.*, **21**, 571-575 (1973).

54. Fisher, F. T., Bradshaw, R. D., and Brinson, L. C. Fiber waviness in nanotube-reinforced polymer composites—I: modulus predictions using effective nanotube properties. *Comp Sci and Tech.*, **63**, 1689–1703 (2003).

55. Fisher, F. T., Bradshaw, R. D., and Brinson, L. C. Fiber waviness in nanotube-reinforced polymer composites—II: modeling via numerical approximation of the dilute strain concentration tensor. *Comp Sci and Tech.*, **63**, 1705–1722 (2003).

56. Liu, Y. J. and Chen, X. L. Evaluations of the effective material properties of carbon nanotube-based composites using a nanoscale representative volume element. *Mech of Mat.*, **35**, 69–81 (2003).

57. Schadler, L. S., Giannaris, S. C., and Ajayan, P. M., Load transfer in carbon nanotube epoxy composites. *Applied Physics Letters*, **73(26)**, 3842–3844 (1998).

58. Wagner, H. D., Lourie, O., Feldman, Y., and Tenne, R., Stress-induced fragmentation of multiwall carbon nanotubes in a polymer matrix. *Applied Physics Letters*, **72(2)**, 188–190 (1998).

59. Qian, D., Dickey, E. C., Andrews, R., and Rantell, T., Load transfer and deformation mechanisms in carbon nanotube polystyrene composites. *Applied Physics Letters*, **76(20)**, 2868–2870 (2000).

60. Liu, Y. J. and Chen, X. L., *Modeling and analysis of carbon nanotube-based composites using the FEM and BEM.* Submitted to CMES: Computer Modeling in Engineering and Science. (2002).

61. Liu, Y. J., Xu, N., and Luo, J. F., Modeling of inter phases in fiber-reinforced composites under transverse loading using the boundary element method. *Journal of Applied Mechanics*, **67(1)**, 41–49 (2000).

62. Fu, Y., Klimkowski, K. J., Rodin, G. J., Berger, E., and et al., A fast solution method for three-dimensional many-particle problems of linear elasticity. *International Journal for Numerical Methods in Engineering*, **42**, 1215–1229 (1998).

63. Nishimura, N., Yoshida, K.-i., and Kobayashi, S., A fast multipole boundary integral equation method for crack problems in 3D. *Engineering Analysis with Boundary Elements*, **23**, 97–105 (1999).

64. Qian, D., Liu, W. K., and Ruoff, R. S., Mechanics of C60 in nanotubes. *The Journal of Physical Chemistry B*, **105(44)**, 10753–10758 (2001).

66. Chen, X. L. and Liu, Y. J. Square representative volume elements for evaluating the effective material properties of carbon nanotube-based composites. *Comput Mater Sci.*, 29, 1–11 (2004).

67. Bower, C., Rosen, R., Jin, L., Han, J., and Zhou, O. Deformation of carbon nanotubes in nanotube–polymer composites *Applied Physics Letters*, **74**, 3317–3319 (1999).

68. Gibson, R. F. *Principles of composite material mechanics*, CRC Press, 2nd edition. 97-134 (2007).

69. Wan, H., Delale, F., and Shen, L. Effect of CNT length and CNT-matrix interphase in carbon nanotube (CNT) reinforced composites. *Mech Res Commun.*, 32, 481– 489 (2005).

70 Shi, D., Feng, X., Jiang, H., Huang, Y. Y., and Hwang, K. Multiscale analysis of fracture of carbon nanotubes embedded in composites. *Int J of Fract.*, 134, 369-386 (2005).

71. Buryachenko, V. A. and Roy, A. Effective elastic moduli of nanocomposites with prescribed random orientation of nanofibers. *Comp: Part B*, **36(5)**, 405-416 (2005).

72. Buryachenko, V. A., Roy, A., Lafdi, K., Andeson, K. L., and Chellapilla, S. Multi-scale mechanics of nanocomposites including interface: experimental and numerical investigation. *Comp Sci and Tech.*, 65, 2435–246 (2005).

73. Li, C. and Chou, T. W. Multiscale modeling of carbon nanotube reinforced polymer composites. *J of Nanosci Nanotechnol.*, **3**, 423-430 (2003).

74. Li, C. and Chou, T.W. Multiscale modeling of compressive behavior of carbon nanotube/polymer composites. *Comp Sci and Tech.*, **66**, 2409–2414 (2006).

75. Anumandla, V. and Gibson, R. F. A comprehensive closed form micromechanics model for estimating the elastic modulus of nanotube-reinforced composites. *Com: Part A*, **37**, 2178–2185 (2006).

76. Seidel, G. D. and Lagoudas, D. C. Micromechanical analysis of the effective elastic properties of carbon nanotube reinforced composites. *Mech of Mater.*, **38**, 884–907 (2006).

77. Hashin, Z. and Rosen, B. The elastic moduli of fiber-reinforced materials. *Journal of Applied Mechanics*, **31**, 223–232 (1964).

78. Christensen, R. and Lo, K. Solutions for effective shear properties in three phase sphere and cylinder models. *Journal of the Mechanics and Physics of Solids*, **27**, 315–330 (1979).

79. Selmi, A., Friebel, C., Doghri, I., and Hassis, H. Prediction of the elastic properties of single walled carbon nanotube reinforced polymers: A comparative study of several micromechanical models. *Comp Sci and Tech.*, **67**, 2071–2084 (2007).

80. Luo, D., Wang, W. X., and Takao, Y. Effects of the distribution and geometry of carbon nanotubes on the macroscopic stiffness and microscopic stresses of nanocomposites. *Comp Sci and Tech.*, **67**, 2947–2958 (2007).

81. Fu, S. Y., Yue, C. Y., Hu, X., and Mai, Y. W. On the elastic transfer and longitudinal modulus of unidirectional multi-short-fiber composites. *Compos Sci Technol.*, **60**, 3001–3013 (2000).

82. Lauke, B. Theoretical considerations on deformation and toughness of short-fiber reinforced polymers. *J Polym Eng.*, **11**, 103–154 (1992).

83. Tserpes, K. I., Panikos, P., Labeas, G., and Panterlakis, S. G. Multi-scale modeling of tensile behavior of carbon nanotube-reinforced composites. *Theoret and Appl Fract Mech.*, **49**, 51-60 (2008).

84. Spanos, P. D. and Kontsos, A. A multiscale monte carlo finite element method for determining mechanical properties of polymer nanocomposites. *Prob Eng Mech.*, doi:10.1016/j.probengmech.2007.09.002. (2008).

85. Paiva, M. C., Zhou, B., Fernando, K. A. S., Lin, Y., Kennedy, J. M., and Sun, Y-P. Mechanical and morphological characterization of polymer-carbon nanocomposites from functionalized carbon nanotubes. *Carbon*, **42**, 2849-54 (2004).

86. Zhu, J., Peng, H., Rodriguez-Macias, F., Margrave, J., Khabashesku, V. Imam, A., Lozano, K., and Barrera, E. Reinforcing epoxy polymer composites through covalent integration of functionalized nanotubes. *Advanced Functional Materials*, **14(7)**, 643–648 (2004).

87. Bhuiyan, Md. A., Pucha, R. V., Worthy, J., Karevan, M., and Kalaitzidou, K. Defining the lower and upper limit of the effective modulus of CNT/polypropylene composites through integration of modeling and experiments. *Composite Structures*, **95**, 80–87 (2013).

88. Papanikos, P., Nikolopoulos, D. D., and Tserpes, K. I. Equivalent beams for carbon nanotubes. *Comput Mater Sci.*, **43(2)**, 345–52 (2008).

89. Affdl, J. C. H. and Kardos, J. L. The Halpin–Tsai equations: a review. *Polym Eng Sci.*, **16(5)**, 344–52 (1976).

90. Tandon, G. P. and Weng, G. J. The effect of aspect ratio of inclusions on the elastic properties of unidirectionally aligned composites. *Polym Composite*, **5(4)**, 327–33 (1984).

91. Nielsen, L. E. Mechanical properties of polymers and composites, New York: Marcel Dekker, **2** (1974).

92. Tucker, C. L. and Liang, E. Stiffness predictions for unidirectional short-fiber composites: review and evaluation. *Compos Sci Technol.*, **59(5)**, 655–71 (1999).

93. Shokrieh, M. M. and Rafiee, R., Investigation of nanotube length effect on the reinforcement efficiency in carbon nanotube based composites. *Composite Structures*, **92**, 2415–2420 (2010).

94. Shokrieh, M. M. and Rafiee, R. On the tensile behavior of an embedded carbon nanotube in polymer matrix with non-bonded interphase region. *J Compos Struct.*,22, 23–5 (2009).

95. Press Release. US Consulate. World-record-length carbon nanotube grown at US Laboratory. Mumbai-India; September 15, (2004).

96. Evans, J. *Length matters for carbon nanotubes: long carbon nanotubes hold promise for new composite materials.* Chemistry World News 2004. <http:// www/rsc.org/chemistryworld/news>.

97. Pan, Z., Xie, S. S., Chang, B., and Wang, C. Very long carbon nanotubes. *Nature*, 394, 631–2 (1998).

98. http://www.carbonsolution.com.

99. http://www.fibermax.eu/shop/.

100. http://www.nanoamor.com.

101. www.thomas-swan.co.uk.

102. Rice University's chemical 'Scissors' yield short carbon nanotubes. New process yields nanotubes small enough to migrate through cells. *Science Daily* [July 2003]. <http://www.sciencedaily.com/releases/2003/07/030723083644.htm>.

103. Chowdhury, S. C., Haque, B. Z., Okabe, T., and Gillespie Jr. J. W., Modeling the effect of statistical variations in length and diameter of randomly oriented CNTs on the properties of CNT reinforced nanocomposites. *Composites: Part B*, **43**, 1756–1762 (2012).

104. Tserpes, K. I. and Chanteli, A. Parametric numerical evaluation of the effective elastic properties of carbon nanotube-reinforced polymers. *Composite Structures*, **99**, 366–374 (2013).

105. Gou, J., Minaie, B., Wang, B., Liang, Z., and Zhang, C. Computational and experimental study of interfacial bonding of single-walled nanotube reinforced composites. *Comput Mater Sci.*, **31**, 225–36 (2004).

106. Frankland, S. J. V. and Harik, V. M. Analysis of carbon nanotube pull-out from a polymer matrix. *Surf Sci.*, 525, 103–8 (2003).

107. Natarajan, U., Misra, S., and Mattice, W. L. Atomistic simulation of a polymer-polymer interface: interfacial energy and work of adhesion. *Comput Theor Polym Sci.*, **8**, 323–9 (1998).

108. Lordi, V. and Yao, N. Molecular mechanics of binding in carbon-nanotube polymer composites. *J Mater Res.*, **15**, 2770–9 (2000).

109. Wong, M., Paramsothy, M., Xu, X. J., Ren, Y., Li, S., and Liao, K. Physical interactions at carbon nanotube-polymer interface. *Polymer*, **44**, 7757–64 (2003).

110. Qian, D., Liu, W. K., and Ruoff, R. S. Load transfer mechanism in carbon nanotube ropes. *Compos Sci Technol.*, **63**, 1561–9 (2003).

111. Liu, Y. J. and Chen, X. L. Continuum models of carbon nanotube-based composites using the boundary element method. *J Boundary Elem.*, **1**, 316–35 (2003).

112. Chen, X. L. and Liu, Y. J. Square representative volume elements for evaluating the effective material properties of carbon nanotube-based composites. *Comput Mater Sci.*, **29**, 1–11 (2004).

113. Chen, X. L. and Liu, Y. J. Evaluations of the effective material properties of carbon nanotube-based composites using a nanoscale representative volume element. *Mech Mater.*, **35**, 69–81 (2003).

114. Qian, D., Dickey, E. C., Andrews, R., and Rantell, T. Load transfer and deformation mechanisms in carbon nanotubepolystyrene composites. *Appl Phys Lett.*, **76**, 2868 (2000).

115. Thostenson, E. T. and Chou, T-W. Aligned multi-walled carbon nanotube-reinforced composites: processing and mechanical characterization. *J Phys D Appl Phys.*, 35, 77–80 (2002).
116. Bower, C., Rosen, R., Jin, L., Han, J., and Zhou, O. Deformation of carbon nanotubes in nanotube-polymer composites. *Appl Phys Lett.*, 74, 3317–9 (1999).
117. Cooper, C. A., Cohen, S. R., Barber, A. H., and Wagner, H. D. Detachment of nanotubes from a polymer matrix. *Appl Phys Lett.*, 81, 3873–5 (2002).
118. Qian, D. and Dickey, E. C. In-situ transmission electron microscopy studies of polymer-carbon nanotube composite deformation. *J Microsc.*, 204, 39–45 (2001).
119. Schadler, L. S., Giannaris, S. C., and Ajayan, P. M. Load transfer in carbon nanotube epoxy composites. *Appl Phys Lett.*, 73, 3842 (1998).
120. Wagner, H. D. Nanotube-polymer adhesion: A mechanics approach. *Chem Phys Lett.*, 361, 57–61 (2002).
121. Ajayan, P. M., Schadler, L. S., Giannaris, C., and Rubio, A. Single-walled carbon nanotube-polymer composites: strength and weakness. *Adv Mater.*, 12, 750–3 (2000).
122. Cooper, C. A., Young, R. J., and Halsall, M. Investigation into the deformation of carbon nanotubes and their composites through the use of Raman spectroscopy. *Compos Part A Appl Sci Manuf.*, 32, 401–11 (2001).
123. Hadjiev, V. G., Iliev, M. N., Arepalli, S., Nikolaev, P., and Files, B. S. Raman scattering test of single-wall carbon nanotube composites. *Appl Phys Lett.*, 78, 3193 (2001).
124. Paipetis, A., Galiotis, C., Liu, Y. C., and Nairn, J. A. Stress transfer from the matrix to the fibre in a fragmentation test: Raman experiments and analytical modeling. *J Compos Mater.*, 33, 377–99 (1999).
125. Valentini, L., Biagiotti, J., Kenny, J. M., Lopez Manchado, M. A. Physical and mechanical behavior of singlewalled carbon nanotube/polypropylene/ethylene-propylene-diene rubber nanocomposites. *J Appl Polym Sci.*, 89, 2657–63 (2003).
126. Qian, D., Wagner, G. J., Liu, W. K., Yu, M-F., and Ruoff, R. S. Mechanics of carbon nanotubes. *Appl Mech Rev.*, 55, 495–532 (2002).
127. Yu, M-F., Yakobson, B. I., and Ruo, R. S. Controlled sliding and pulout of nested shells in individual multiwalled nanotubes. *J Phys Chem B.*, 104, 8764–7 (2000).
128. Qian, D., Liu, W. K., and Ruoff, R. S. Load transfer mechanism in carbon nanotube ropes. *Compos Sci Technol.*, 63, 1561–9 (2003).
129. Liao, K. and Li, S. Interfacial characteristics of a carbon nanotubepolystyrene composite system. *Appl Phys Lett.*, 79, 4225–7 (2001).
130. Andrews, R. and Weisenberger, M. C. Carbon nanotube polymer composites. *Curr Opin Solid State Mater Sci.*, 8, 31–7 (2004).
131. Lordi, V. and Yao, N. Molecular mechanics of binding in carbonnanotube- polystyrene composite system. *J Mater Res.*, 5, 2770–9 (2000).
132. Wagner, H. D. and Vaia, R. A. Nanocomposites: Issue at the interface. *Mater Today*, 7, 38–42 (2004).
133. Wernik, J. M., Cornwell-Mott, B. J., and Meguid, S. A. Determination of the interfacial properties of carbon nanotube reinforced polymer composites using atomistic-based continuum model. *Inte. Jour. of Sol. and Struct.*, 49, 1852–1863 (2012).
134. Wernik, J. M. and Meguid, S. A., Multiscale modeling of the nonlinear response of nano-reinforced polymers. *Acta Mech.*, 217, 1–16 (2011).
135. Yang, S., Yu, S., Kyoung, W., Han, D. S., and Cho, M. Multiscale modeling of size-dependent elastic properties of carbon nanotube/polymer nanocomposites with interfacial imperfections. *Polymer*, 53, 623-633 (2012).

136. Ayatollahi, M. R., Shadlou, S., and Shokrieh, M. M., Multiscale modeling for mechanical properties of carbon nanotube reinforced nanocomposites subjected to different types of loading. *Composite Structures*, **93**, 2250–2259 (2011).

137. Jiang, L. Y., Huang, Y., Jiang, H., Ravichandran, G., Gao, H., Hwang, K. C., and et al. A cohesive law for carbon nanotube/polymer interfaces based on the van der Waals force. *J Mech Phys Solids.*, **54**, 2436–52 (2006).

138. Tan, H., Jiang, L. Y., Huang, Y., Liu, B., and Hwang, K. C. The effect of van der Waals-based interface cohesive law on carbon nanotube-reinforced composite materials. *Composites Science and Technology*, **67**, 2941–2946 (2007).

139. Zalamea, L., Kim, H., and Pipes, R. B. Stress transfer in multi-walled carbon nanotubes. *Compos Sci Technol.*, **67(15–16)**, 3425–33 (2007).

140. Shen, G. A., Namilae, S., and Chandra, N. Load transfer issues in the tensile and compressive behavior of multiwall carbon nanotubes. *Mater Sci Eng A.*, **429(1–2)**, 66–73 (2006).

141. Gao, X. L. and Li, K. A shear-lag model for carbon nanotube-reinforced polymer composites. *Int J Solids Struct.*, **42(5–6)**, 1649–67 (2005).

142. Li, C. and Chou, T. W. A structural mechanics approach for the analysis of carbon nanotubes. *Int J Solids Struct.*, **40(10)**, 2487–99 (2003).

143. Tsai, J. L. and Lu, T. C. Investigating the load transfer efficiency in carbon nanotubes reinforced nanocomposites. *Compos Struct.*, **90(2)**, 172–9 (2009).

144. Lu, T. C. and Tsai, J. L. Characterizing load transfer efficiency in double-walled carbon nanotubes using multiscale finite element modeling. *Composites: Part B*, **44**, 394–402 (2013).

CHAPTER 10

ORIENTATION CONTROLLED IMMOBILIZATION STRATEGY FOR β-GALACTOSIDASE ON ALGINATE BEADS

M. S. MOHY ELDIN, M. R. EL-AASSAR, and E. A. HASSAN

CONTENTS

10.1   INTRODUCTION

In recent years, enzyme immobilization has gained importance for design of artificial organs, drug delivery systems, and several biosensors. Polysaccharide based natural biopolymers used in enzyme or cell immobilization represent a major class of biomaterials which includes agarose, alginate, dextran, and chitosan. Especially, alginates are commercially available as water-soluble sodium alginates and they have been used for more than 65 years in the food and pharmaceutical industries as thickening, emulsifying, and film forming agent. Entrapment within insoluble calcium alginate gel is recognized as a rapid, nontoxic, inexpensive, and versatile method for immobilization of enzymes as well as cells. The formulation conditions of the alginate beads entrapment immobilized with the enzyme have been optimized and effect of some selected conditions on the kinetic parameter, Km, have been presented. The β-galactosidase enzymes entrapped into alginate beads are used in the study of the effect of both substrate diffusion limitation and the misorientation of the enzyme on its activity, the orientation of an immobilized protein is important for its function. Physico-chemical characteristics and kinetic parameters, the protection of the activity site using galactose as protecting agent has been presented as a solution for the misorientation problem. This technique has been successful in reduction, and orientation-controlled immobilization of enzyme. Other technique has been presented to reduce the effect of substrate diffusion limitation through covalent immobilization of the enzyme onto the surface of alginate beads after activation of its OH-groups.

Recently an increasing trend has been observed in the use of immobilized enzymes as catalysts in several industrial chemical processes. Immobilization is important to maintain constant environmental conditions order to protect the enzyme against changes in pH, temperature, or ionic strength; this is generally reflected in enhanced stability [1]. Moreover, immobilized enzymes can be more easily separated from substrates and reaction products and used repeatedly. Many different procedures have been developed for enzyme immobilization; these include adsorption to insoluble materials, entrapment in polymeric gels, encapsulation in membranes, crosslinking with a bifunctional reagent, or covalent linking to an insoluble carrier [2]. Among these, entrapment in calcium alginate gel is one of the simplest methods of immobilization. The success of the calcium alginate gel entrapment technique is due mainly to the gentle environment that provides for the entrapped material. However, there are some limitation such as low stability and high porosity of the gel [1]. These characteristics could lead to leakage of large molecules like proteins, thus generally limiting its use to whole cells or cell organelles [3].

Enzymes are biological catalysts with very good prospects for application in chemical industries due to their high activity under mild conditions and high selectivity and specificity [4-7]. However, enzymes do not fulfill all of the requirements of an industrial biocatalyst or biosensor [8, 9]. They have been selected throughout natural evolution to perform their physiological functions under stress conditions and quite strict regulation. However, in industry, these biocatalysts should be heterogeneous, reasonably stable under conditions that may be quite far from their physiological environment [10] and retain their good activity and selectivity when acting with substrates that are in some instances quite different from their physiological ones. β-galactosidase, commonly known as lactase, catalyzes the hydrolysis of β-galactosidic linkages such as those between galactose and glucose moieties in the lactose molecule. While the enzyme has many analytical uses, being a favorite label in various affinity recognition techniques such as ELISAs, or enzyme-linked immunosorbent assays, its main use is the large scale processing of dairy products, whey, and whey permeates.

Immobilization of the enzymes to solid surface induces structural changes which may affect the entire molecule. The study of conformational behavior of enzymes on solid surface is necessary for better understanding of the immobilization mechanism. However, the immobilization of enzymes on alginate beads is generally rapid, and depends on hydrophobic and electrostatic interactions as well as on external conditions such as pH, temperature, ionic strength, and nature of buffer [11, 12]. Enzymes denaturation may occur under the influence of hydrophobic interactions, physico-chemical properties of the alginate beads or due to the intrinsic properties of the enzyme.

One of the main concerns regarding immobilized proteins has been the reduction in the biological activity due to immobilization. The loss in activity could be due to the immobilization procedure employed, changes in the protein conformation after immobilization, structural modification of the protein during immobilization, or changes in the protein microenvironment resulting from the interaction between the support and the protein. In order to retain a maximum level of biological activity for the immobilized protein, the origins of these effects need to be understood, especially as they relate to the structural organization of the protein on the immobilization surface. The immobilization of proteins on surfaces can be accomplished both by physical and chemical methods. Physical methods of immobilization include the attachment of protein to surfaces by various inter actions such as electrostatic, hydrophobic/hydrophilic, and van der waals forces. Though the method is simple and cost-effective, it suffers from protein leaching from the immobilization support. Physical adsorption generally leads to dramatic changes in the protein microenvironment, and typically involves multipoint protein adsorption between a single protein molecule and a number of binding sites on the immobilization surface. In addition, even if the surface has a uniform distribution of binding sites, physical adsorption could lead to heterogeneously populated immobilized proteins. This has been ascribed to unfavorable lateral interactions among bound protein molecules [13]. The effect of ionic strength on protein adsorption has also been studied, and a striking dependence on the concentration of electrolyte was noticed. For example, in the case of adsorption of apotransferrin on a silicon titanium dioxide surface, it was found that an increase in ionic strength resulted in a decrease in adsorption. The increase in ionic strength decreases the negative surface potential and increases the surface pH. This leads to an increase in the net protein charge, creating an increasingly repulsive energy barrier, which results in reduced effective diffusivity of molecules to the surface [14]. Several research groups have attempted to develop a model that can qualitatively and quantitatively predict physical adsorption, especially as it relates to ion-exchange chromatography. The models have been developed to predict average interaction energies and preferred protein orientations for adsorption. Adsorption studies on egg white lysozyme and o~- lactalbumin adsorbed on anion-exchanger polymeric surfaces have shown that it is possible to determine the residues involved in the interaction with the support [15, 16]. It has been also possible to calculate the effective net charge of the proteins. Entrapment and microencapsulation are other popular physical methods of immobilization and have been discussed elsewhere [17]. The immobilization of enzymes through metal chelation has been reviewed recently [18].

Attachment of proteins by chemical means involves the formation of strong covalent or coordination bonds between the protein and the immobilization support. The chemical attachment involves more drastic (non mild) conditions for the immobilization reaction than the attachment through adsorption. This can lead to a significant loss in enzyme activity or in binding ability (in the case of immobilization of binding proteins and antibodies). In addition, the covalent and coordinate bonds formed between the protein and the support can lead to a change in the structural configuration of the immobilization protein. Such a change in the enzymatic structure may lead to reduced activity, unavailability of the active site of an enzyme for the substrates, altered reaction pathways or a shift in optimum pH [19, 20]. In the case of immobilized antibodies and binding proteins this structural change can lead to reduced binding ability.

Oriented immobilization is one such approach [21]. Adsorption, bioaffinity immobilization and entrapment generally give high retention of activity as compared to covalent coupling method [22-26]. The use of macroporous matrices also helps by reducing mass transfer constraints. Calcium alginate beads, used here, in that respect, are an attractive choice. Calcium alginate beads are not used for enzyme immobilization as enzymes slowly diffuse out [31]. In this case, binding of β-galactosidase and galactos to alginate before formation of calcium alginate beads ensured the entrapment of these individual enzymes and undoubtedly, also contributed to enhanced thermostabilization. Such a role for alginate in fact, has already been shown for another alginate binding enzyme.

In this work, β-galactosidase enzyme entrapped into alginate beads are used in the study of the effect of both substrate diffusion limitation and the misorientation of the enzyme on its activity, physico-chemical characteristics and kinetic parameters. The protection of the activity site using galactose as protecting agent has been presented as a solution for the misorientation problem. This technique has been successful in reduction, but not in elimination of the effect of misorientation on the activity as well as the kinetic parameters; Km and Vm. Other technique has been presented to reduce the effect of substrate diffusion limitation through covalent immobilization of the enzyme onto the surface of alginate beads after activation of its OH-groups. The impact of different factors controlling the activation process of the alginate hydroxyl groups using p-benzoquinone (PBQ) in addition to the immobilization conditions on the activity of immobilized enzyme have been studied. The immobilized enzyme has been characterized from the bio-chemical point of view as compared with the free enzyme.

10.2 MATERIALS AND METHODS

10.2.1 MATERIALS

Sodium alginate (low viscosity 200 cP), β-Galactosidase (from Aspergillus oryzae) p-Benzoquinone (purity 99+ %): Sigma-Aldrich chemicals Ltd. (Germany); calcium chloride (anhydrous Fine GRG 90%): Fisher Scientific (Fairlawn, NJ, USA); ethyl alcohol absolute, Lactose (pure Lab. Chemicals, MW 360.31), Galactose: El-Nasr Pharmaceutical Chemicals Co. (Egypt); glucose kit (Enzymatic colorimetric method): Diamond Diagnostics Co. for Modern Laboratory Chemicals (Egypt); Tris-Hydrochloride (Ultra Pure Grade 99.5 %, MW 157.64): amresco (Germany); sodium chloride (Purity 99.5 %): BDH Laboratory Supplies Pool (England).

10.2.2 METHODS

PREPARATION OF CATALYTIC CA-ALGINATE GEL BEADS BY ENTRAPMENT TECHNIQUE

The Ca-alginate gel beads prepared by dissolving certain amount of sodium alginate (low viscosity) 2% w/v in distilled water with continuous heating the alginate solution until it becomes completely clear solution, mixing the alginate solution with enzyme (β-galactosidase enzyme as a model of immobilized enzyme (0.005g)). The sodium alginate containing β- galactosidase is dropped (drop wise) by 10 cm³ plastic syringe in (50ml) calcium chloride solution (3% w/v) as a safety crosslinker to form beads to give a known measurable diameter of beads. Different aging times of beads in calcium chloride solution are considered followed by washing the beads by (50ml) buffer solution, and determination of the activity of immobilized enzyme. In case of oriented immobilization, galactose with different concentrations was mixed first with the enzyme-buffer solution before mixing with the alginate.

ALGINATE BEADS SURFACE MODIFICATION

The Ca-alginate gel beads prepared by dissolving sodium alginate (low viscosity) in distilled water with continuous heating the solution until become completely clear to acquire finally 4% (w/v) concentration. The alginate solution was mixed with PBQ solution (0.02M) and kept for four hours at room temperature to have final concentration 2% (w/v) alginate and 0.01 M (PBQ). The mixture was added dropwise, using by 10 cm³ plastic syringe to calcium chloride solution (3% w/v) and left to harden for 30 min at room temperature to reach 2 mm diameter beads. The beads were washed using buffer-ethanol solution (20% ethanol) and distilled water, to remove the excess (PBQ), before transferring to the enzyme solution (0.005g of β-galactosidase in 20ml of Tris-HCl buffer solution of pH = 4.8) and stirring for 1 hr at room temperature then the mixture was kept at 4°C for 16 hr to complete the immobilization process. The mechanism of the activation process and enzyme immobilization is presented in Scheme 1.

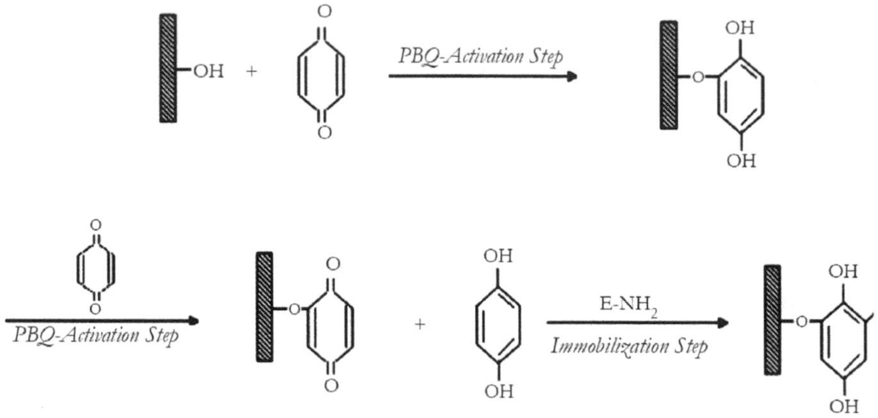

SCHEME 1   Mechanism of activation and immobilization process.

DETERMINATION OF IMMOBILIZED ENZYME ACTIVITY

The catalytic beads were mixed with 0.1 M Lactose-Tris-HCl buffer solution of (pH 4.8 with stirring, 250 rpm, at room temperature for 30 min. Samples were taken every 5 min to assess the glucose production using glucose kit. Beads activity is given by the angular coefficient of the linear plot of the glucose production as a function of time.

10.3   METHODS AND RESULT

10.3.1   ENTRAPMENT IMMOBILIZATION

The impact of different factors affecting the process of enzyme entrapment and its reflection on the activity of the immobilized enzyme, its physico-chemical characters and its kinetic parameters have been studied and the obtained results are given.

EFFECT OF ALGINATE CONCENTRATION

It is clear from Figure 1 that increasing the concentration of alginate has a linear positive effect on the activity. Increasing the activity with alginate concentration can be explained in the light of increasing the amount of entrapped enzyme as a result of the formation of as more densely cross-linked gel structure [28]. This explanation has been confirmed by the data obtained in case of immobilized and/or entrapped amount of enzyme.

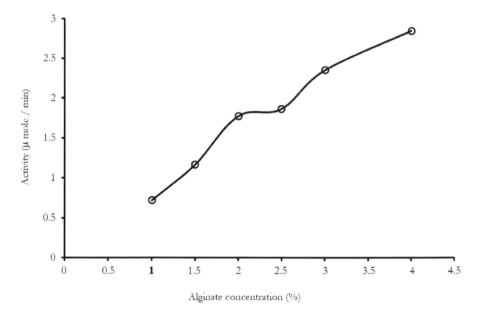

FIGURE 1   Effect of alginate concentrations on beads catalytic activity.

EFFECT OF CROSS-LINKING TEMPERATURE

The opposite behavior of the activity has been obtained with increasing the temperature of cross-linking process in $CaCl_2$ solution (Figure 2). Such results could be explained based on increasing cross-linking degree along with the temperature of calcium chloride solution which leads finally to decrease the amount of diffusive lactose substrate to the entrapped enzyme and hence reduce the activity of the immobilized enzyme.

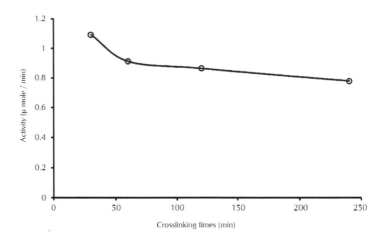

FIGURE 2   Effect of cross-linking temperature on beads catalytic activity.

EFFECT OF CROSS-LINKING TIME

The Figure 3 shows the effect of increasing the cross-linking time in $CaCl_2$ solution on the activity of entrapped enzyme. It is clear that a reduction of about 20% of the activity has been detected with increasing the time from 30 to 60 min. Further increase has no noticeable effect on the activity.

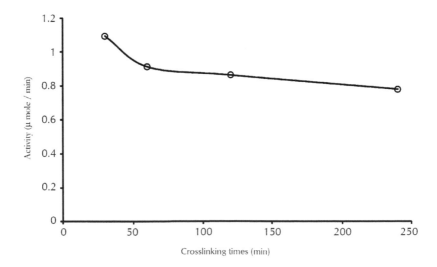

FIGURE 3   Effect of cross-linking time on beads catalytic activity.

EFFECT OF CACL₂ SOLUTION CONCENTRATION

The effect as given in Figure 4 of $CaCl_2$ concentration on the activity of entrapped enzyme shows that increase in $CaCl_2$ concentration decreases the activity linearly. Such results could be interpreted in the light of increasing the gel cross-linking density and hence reducing the amount of diffusive lactose.

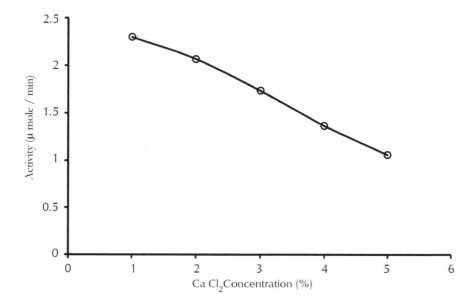

FIGURE 4 Effect of $CaCl_2$ concentration (%) on beads catalytic activity.

EFFECT OF ENZYME CONCENTRATION

The dependence of the catalytic activity of the beads on enzyme concentration is illustrated in Figure 5. It is clear that increasing the enzyme concentration increases the activity linearly within the studied range. Increasing the amount of entrapped enzyme is the logic explanation of such results.

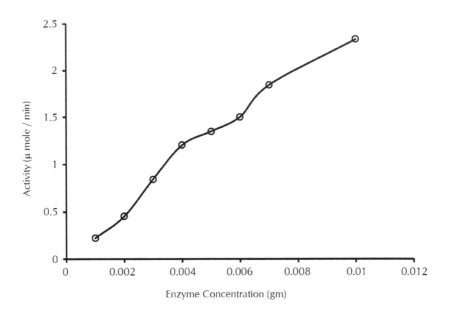

FIGURE 5   Effect of enzyme concentration (gm) on beads catalytic activity.

EFFECT OF SUBSTRATE'S TEMPERATURE

Shift of the optimum temperature towards lower side has been detected upon entrapping of the enzyme (Figure 6). A suitable temperature found to be 45°C for the entrapped enzyme compared with 55°C for the free form. Such behavior may be explained in the light of acidic environment due to the presence of free unbinding carboxylic groups. This explanation is confirmed by the higher rate of enzyme denaturation at higher temperatures, 50–70°C, in comparison with the free form.

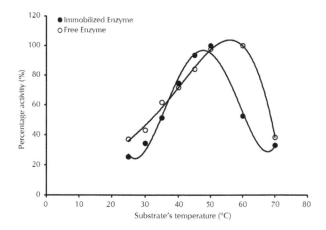

FIGURE 6   Effect of substrate's temperature (°C) on beads catalytic activity.

EFFECT OF SUBSTRATE'S PH

The similar behavior of the immobilized enzyme to the free one has been observed in respond to the changes in the substrate's pH (Figure 7). The optimum pH has not shifted, but the immobilized enzyme shows higher activity within acidic region, pH 2.5–3.0. This resistance to pH could be explained due to the presence of negatively charged free carboxylic groups unbinding by calcium ions. This result is in accordance with those obtained by other authors [29].

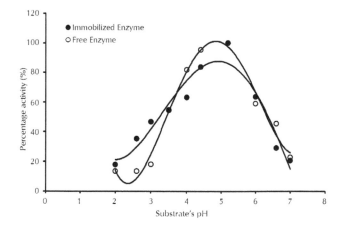

FIGURE 7   Effect of substrate's pH on beads catalytic activity.

KINETIC STUDIES

Figure 13 presents the effect of the immobilization process on the kinetic parameters especially *Km* which can be considered as the reflection of presence or absence of substrate diffusion limitation.

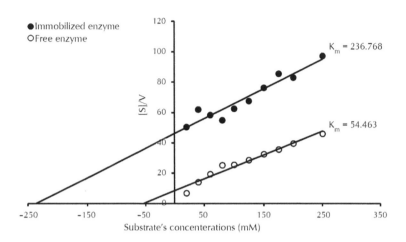

FIGURE 8   Han's plot curve for free and immobilized enzyme.

A Han's plot curve indicates how much the immobilization process can be affected as medicated from the kinetic parameters of immobilized enzyme. That the *Km* value has become five times higher than that of the free one is a reflection of powerful diffusion limitation of the substrate.

In addition, misorientation of the active sites as a result of the immobilization process could be another factor affecting the accessibility of the active sites to the substrate. To study this effect, galactose has been added during the immobilization process in different concentrations to protect the active site (figure 9).

The *Km* values have been reduced with increasing the galactose concentration reaching minimum value at 30 μ mole (figure 16). Beyond this concentration, the *Km* value starts to increase again.

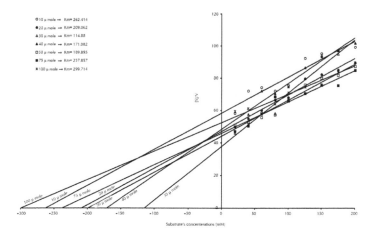

FIGURE 9 Han's plot curve for immobilized enzyme with addition of galactose.

The protection of the active sites has a positive refection on the activity of immobilized enzyme (figure 10).The reduction in the *Km* value and activity enhancement proved the role of misorientation but still the fact that diffusion limitation of the substrate has the main role.

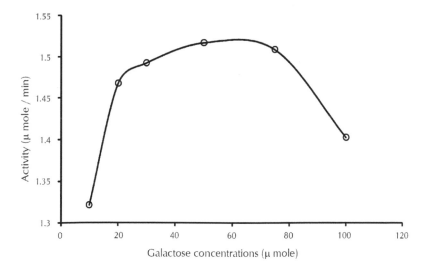

FIGURE 10 Effect of galactose concentration on beads catalytic activity.

To solve the problem of diffusion limitation, the enzyme was used to be covalently immobilized on the surface of the alginate beads after activation with *p*-benzoquinone. The impact of different factors controlling the activation process of the alginate hydroxyl groups using PBQ in addition to the immobilization conditions on the activity of immobilized enzyme have been studied. The immobilized enzyme has been characterized from the bio-chemical point of view as compared with the free enzyme.

10.3.2   ACTIVITY OF THE BEADS [30]

THE EFFECT OF ALGINATE CONCENTRATION

From Figure 11, its clear that increasing the concentration of alginate has negative effect on the activity. A reasonable explanation could be obtained by following the amount of immobilized enzyme which decreases in the same manner. This could be explained as a result of increasing the cross-linking density and a direct reduction of the available pores surface area for enzyme immobilization. Indeed, the retention of activity has not affected so much since the decrease rate of both activity and amount of immobilized enzyme is almost the same.

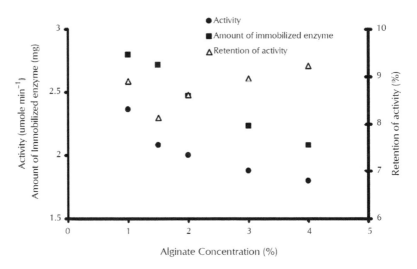

FIGURE 11   Effect of alginate concentrations on beads catalytic activity.

EFFECT OF CACL$_2$ CONCENTRATION

As shown in Figure 12, the activities of the beads have not been affected so much by changing the concentration of CaCl$_2$ solution. Unexpectedly, the amount of immobilized enzyme has been increased gradually with concentration of CaCl$_2$ solution. This behavior could be explained according to the impact of concentration of CaCl$_2$ solution on the number of the formed pores inside the beads which offering additional surface area for enzyme immobilization. Since the activity was found almost constant and the amount of immobilized enzyme increased under the same conditions, so it is logic to have a reduction of the retention of activity.

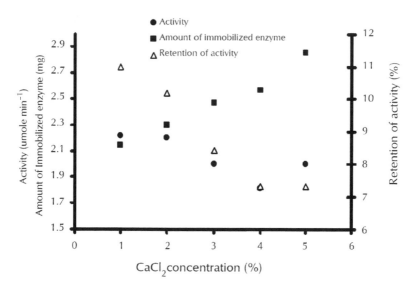

FIGURE 12   Effect of $CaCl_2$ concentrations on beads catalytic activity.

EFFECT OF AGING TIME

It has been found that changing the aging time from 30 to 240 min has a clear effect on the catalytic activity of the beads. This may be due to the cross-linking effect of $CaCl_2$ which affected directly on the pores size producing pores smaller than the enzyme size and hence reducing the available pores surface area for immobilizing enzyme. Since the amount of immobilized enzyme reduced with higher rate than the activity did, so the retention of activity affected positively with aging time increase (Figure 13). It is clear from the results here that not only the pores surface area but also the pores size affects the amount of immobilized enzyme and hence the activity.

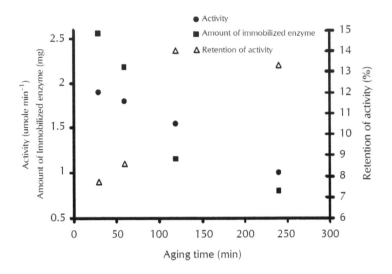

FIGURE 13   Effect of aging time on beads catalytic activity.

EFFECT OF AGING TEMPERATURE

The dependence of the catalytic activity of the beads on aging temperature is illustrated in Figure 14. From the figure, it is clear that increasing the aging temperature resulted in increasing the activity in gradual way regardless the cross bonding decrease of the amount of immobilized enzyme. This is takes us again to the combination between the pores size distribution and pores surface area. The best retention of activity has been obtained with the highest aging temperature.

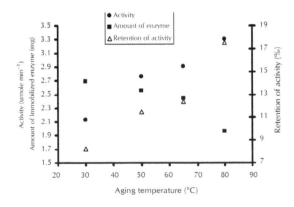

FIGURE 14   Effect of aging temperatures on beads catalytic activity.

KINETIC STUDIES

The kinetic parameters, K_m and V_m, of the immobilized enzyme under selected conditions, aging time and temperature, have been calculated from the Hanes plot curves and tabulated in the following tables, table 1and 2 respectively. In addition, the diffusion coefficient (De) of lactose substrate has been calculated [31] and related to the K_m and V_m values.

TABLE 1 Kinetic parameters and diffusion coefficient of immobilized enzyme prepared under different aging time

Parameters	Aging time (minutes)			
	30	60	120	240
Km (mmole)	61.00	64.66	133.86	221.80
Vm (umole.min^{-1})	2.94	2.46	4.00	3.36
De (cm^2.s^{-1})	4.9×10^{-11}	4.6×10^{-11}	9.73×10^{-11}	0.184×10^{-11}

TABLE 2 Kinetic parameters and diffusion coefficient of immobilized enzyme prepared under different aging temperature

Parameters	Aging temperature (°C)			
	30	50	65	80
Km (mmole)	61.32	73.44	79.17	265.70
Vm (umole.min^{-1})	2.99	3.24	2.27	5.62
De (cm^2.s^{-1})	5.49×10^{-11}	13.98×10^{-11}	7.3×10^{-11}	10.8×10^{-11}

From the obtained results, it is clear that increasing the cross-linking degree of the beads through increasing either the aging time or aging temperature in CaCl$_2$ solution has affected directly the K_m values which tended to increase. This could be explained based on the diffusion limitation of the lactose into the pores of the beads. This conclusion could be supported by taking into account the obtained data of diffusion coefficient which in general is higher than the value of immobilized enzyme on beads free of diffusion limitation of the substrate; 1.66×10^{-8} cm^{2}s^{-1}. On the other hand V_m values are in general less than that of the free enzyme regardless that some of the obtained K_m values are equal to that of the free enzyme.

10.3.3 BIOCHEMICAL CHARACTERIZATION OF THE CATALYTIC ALGINATE BEADS

EFFECT OF SUBSTRATE'S TEMPERATURE

Similar temperature profile to the free enzyme has been obtained for our immobilized form (Figure 15). An optimum temperature has been obtained at 55°C. This similarity indicates that the microenvironment of the immobilized enzyme is identical to that of the free one, so absence of internal pore diffusion, illustrates the main drawback of the immobilization process, and the success of surface immobilization to overcome this problem. The confirmation of this conclusion has come from determination of the activation energy for both the free and immobilized form [29], which was found very close in value, 6.05 Kcal/mole for the free and 6.12 Kcal/mole for the immobilized forms respectively.

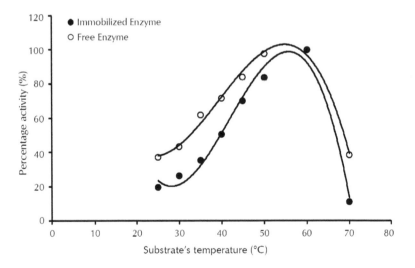

FIGURE 15   Effect of substrate's temperature on beads catalytic activity.

EFFECT OF SUBSTRATE'S PH

The different behaviors of the immobilized enzyme compared with those of the free one have been observed in respond to the change in the catalytic reaction's pH (Figure 16). Both slight and broaden of the optimum pH for the immobilized form has been obtained since the "optimum activity range", activity from 90–100%, was found from pH value of 4.0 to 5.5. This shift to the acidic range could be explained by the presence of negative charges of unbinding carboxylic groups [29]. This behavior is advantage for use this immobilized form in the degradation of lactose from whey waste which is normally has a low pH. It is enough to mention here that at (pH 3.0) the immobilized form beads retained 65%, of its maximum activity compared with 20% of the free enzyme.

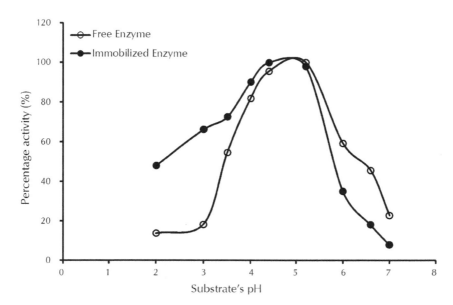

FIGURE 16 Effect of substrate pH on beads catalytic activity.

OPERATIONAL STABILITY

The operational stability of the catalytic beads was investigated (Figure 17). The beads were repeatedly used for 21 cycles, one hour each, without washing the beads between the cycles. The activity was found almost constant for the first five cycles. After that, gradual decrease of the activity has been noticed. The beads have lost 40% of their original activity after 21 hr of networking time. No enzyme was detected in substrate solution, so we cannot claim the enzyme leakage as the cause for activity decline. Incomplete removal of the products from the pores of the beads which can cause masking of the enzyme active sites could be a proper explanation. Washing the beads with suitable buffer solution between the cycles and using higher stirring rate may help in solving this problem.

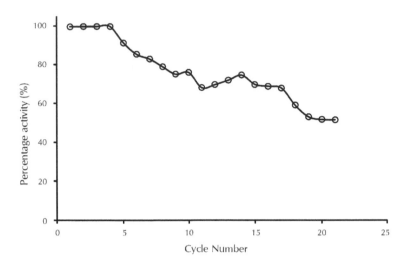

FIGURE 17   Operational stability of the catalytic beads.

10.4   CONCLUSION

From the obtained results, it can conclude that the activation of alginate's OH groups by using PBQ offers a new matrix for covalent immobilization of β-galactosidase enzyme. Both the temperature and the pH of the activation process were found to be determining factors in obtaining high catalytic activity beads. The pH of the enzyme solution is found to have a pronounced effect on the catalytic activity of the beads. Three hours of immobilization time was found optimum using 5 mg of enzyme-buffer solution (pH range 4.0–4.5). The wide "optimum pH range" of the activity of the immobilized β-galactosidase makes it is to be recommended to be used in the degradation of lactose in whey waste to reduce its BOD and the production of de-lactose milk to serve the peoples suffering from lactose tolerance. Finally, the immobilization of β-galactosidase on the surface of PBQ activated alginate beads has succeeded in avoiding the diffusion limitation of lactose.

KEYWORDS

- **p-Benzoquinone**
- **Enzyme concentration**
- **β-Galactosidase**
- **Han's plot curve**
- **Immobilization**

REFERENCE

1. Cheetham, P. S. J. Principles of industrial enzymology: Basis of utilization of soluble and immobilized enzymes in industrial processes, Handbook of Enzyme Biotechnology (Wiseman, A., ed.) Ellis Horwood Limited, Chichester, 1985, pp. 74-86

2. Klibanov, A. Immobilized enzymes and cells as practical catalysts. *Science*, **219**, 722–727 (1983).

3. Smidsrod, O. and Skjak-Braek, G. Alginate as immobilization matrix for cells. *Trends Biotech.*, **8**, 71–78 (1990).

4. Pollard, D. J. and Woodley, J. M. Biocatalysis for pharmaceutical intermediates: the future is now. *Trends Biotechnol.*,**25**, 66–73 (2007).

5. Woodley, J. M. New opportunities for biocatalysis: making pharmaceutical processes greener. *Trends Biotechnol.*, **26**, 321–7 (2008).

6. Schmid, A, Dordick, J. S., Hauer, B., Kiener, A., Wubbolts, M., and Witholt, B. Industrial biocatalysis today and tomorrow. *Nature.*,**409**, 258–68 (2001).

7. Straathof, A. J. J., Panke, S., and Schmid, A. The production of fine chemicals by biotrans-formations. *Curr Opin Biotechnol*, **13**, 548–56 (2002).

8. Schoemaker, H. E., Mink, D., and Wubbolts M. G. Dispelling the myths–biocatalysis in industrial synthesis. *Science*, **299**, 1694–7 (2003).

9. Meyer H. P. Chemocatalysis and biocatalysis (biotransformation): some thoughts of a chemist and of a biotechnologist. *Org Proc Res Dev*, **10**, 572–80 (2006).

10. Durand, J., Teuma, E., and Gómez, M. Ionic liquids as a medium for enantioselective catalysis. *C R Chim*, **10**, 152–77 (2007).

11. Trivedi, A. H., Spiess, A. C., Daussmann, T., and Buchs, J. Effect of additives on gas-phase catalysis with immobilized Thermoanaerobacter species alcohol dehydrogenase ADHT, *Appl. Microbiol. Biotechnol.* **71**, 407–414 (2006).

12. Solis, S., Paniagua, J., Martinez, J. C., and Asmoza, M. Immobilization of papain on meso-porous silica: pH effect, *J. Sol-Gel Sci. Technol.*, **37**, 125–127 (2006).

13. Johnson, R. D., Wang, A. G., and Arnold, E. H. *J. Phys. Chem.*, **100**, 5134 (1996).

14. J.J. Ramsden, J. E. Prenosil, *J. Phys. Chem.*, **98**, 5376 (1994).

15. Noinville, V., Calire, V. M., and Bernard, S. *J. Phys. Chem.*, **99**, 1516 (1995).

16. Sadana, A. *Bioseparation*, **3**, 297 (1993).

17. Marek, P. K. J. and Coughlan, M. E. in: Protein Immobilization (R. E Taylor, ed.), Marcel Dekker, New York, pp. 13–71 (1990).

18. Kennedy, J. E. and Cabral, J. M. S. *Artif Cells, Blood Substitutes Immobilization Biotech-nol.*, **23**, 231 (1995).

19. Mosbach, K. and Mattiasson, B., *Methods Enzymol.*, **44**, 453 (1976).

20. Trevan, M. D. *Immobilized Enzymes: An Introduction and Applications in Biotechnology*, Wiley, Chichester, 1980.

21. Wilchek, M. and Miron T. Oriented versus random protein immobilization. *J Biochem Biophys Methods*, **55**, 67–70 (2003).

22. Gupta MN, Mattiasson B. Unique applications of immobilized proteins in bioanalytical systems. In: Suelter CH, Kricka L, editors. Bioanalytical applications of enzymes, vol. 36. New York: John Wiley & Sons Inc.; 1992. p. 1–34.

23. Mohy Eldin, M. S., Serour, E., Nasr, M., and Teama, H. *J Appl Biochem Biotechnol*, **164**, 10 (2011).

24. Mohy Eldin, M. S., Serour, E., Nasr, M., and Teama, H. *J Appl Biochem Biotechnol*, **164**, 45 (2011).

25. Mohy Eldin, M. S. *Deutsch lebensmittel-Rundschau*, **101**, 193 (2005).

26. Mohy Eldin, M. S., Hassan, E. A., and Elaassar, M. R. *Deutsch lebensmittel-Rundschau*, **101** , 255 (2005).
27. Smirsod, O and Skjak-Braek, G. Alginate as immobilization matrix for cells. *Trends Biotechnol*, **8**, 71–8 (1990).
28. Maysinger, D., Jalsenjak, I., Cuello A. C. *Neurosci. Lett.*, **140**, 71–74, (1992).
29. Bergamasco, R., Bassetti, F. J., De Moraes, F., and Zanin, G. M. *Braz. J. Chem. Eng.* **17**, 4 (2000).
30. M. S. Mohy Eldin, E. A. Hassan, M. R. Elaassar. β-Galactosidase Covalent Immobilization on the Surface of P-Benzoquinone-Activated Alginate Beads as a Strategy For Overcoming Diffusion Limitation. III.Effect of Beads Formulation Conditions on the Kinetic Parameters, The 1st International Conference of Chemical Industries Research Division 6-8 December, 2004.
31. White, C. A. and Kennedy, J. F. *Enzyme microb. Technol.*, **2**, 82–90 (1980).

CHAPTER 11

PREPARATION, CHARACTERIZATION, AND EVALUATION OF WATER-SWELLABLE HYDROGEL VIA GRAFTING CROSS-LINKED POLYACRYLAMIDE CHAINS ONTO GELATIN BACKBONE BY FREE RADICAL POLYMERIZATION

M. S. MOHY ELDIN, A. M. OMER, E. A. SOLIMAN, and E. A.HASSAN

CONTENTS

11.1   INTRODUCTION

This study concerns the preparation, characterization and evaluation of water-swellable hydrogel *via* grafting cross-linked polyacrylamide (PAM) chains onto gelatin backbone by free radical polymerization for agricultural applications. The characterizations by Fourier transform infrared spectroscopy (FTIR), thermogravimetric analysis (TGA), and scanning electron microscope (SEM) provide proofs of grafting process on the backbone of gelatin. The water-holding capacity of grafted hydrogel was found depend on the concentrations of the feed compositions of the hydrogel. Moreover, the hydrogel particle size was found also of determining effect where maximum swelling was observed with particles size ranged from 500 μm to 1mm. The prepared cross-linked PAM-g-gelatin hydrogel shows good thermal stability and moderate sensitivity towards media pH changes in the range of 1–4. Finally, the retention of water by the hydrogel graft copolymer in sandy soil has been used as monitor to evaluate its applicability as soil conditioners. The obtained results recommended our preparation for biotechnological and agricultural applications.

Highly swelling polymeric hydrogels are hydrophilic three-dimensional networks that can absorb water in the amount from 10% up to thousands of times their dry weight [1]. They are widely used in many technological and bio-technological fields, such as disposable diapers, feminine napkins, pharmaceuticals medical applications, and agricultural and horticultural [2-5]. The productivity of sandy soils is mostly limited by their low water-holding capacity and excessive deep percolation losses. Thus the management of these soils must aim at increasing their water-holding capacity and reducing losses due to deep percolation. These polymers were developed to improve the physical properties of soil in view of: increasing their water-holding capacity, increasing water use efficiency, enhancing soil permeability, and infiltration rates. Hydrogels have been commonly utilized in agricultural field mainly as water storage granules [6]. The need for improving the physical properties of soil to increase productivity in the agricultural sector was visualized in 1950s [7]. This led to the development of water-soluble polymers such as carboxymethyl cellulose (CMC) and PAM to function as soil conditioners [8]. The addition of hydrogels to a sandy soil changed the water-holding capacity to be comparable to silty clay or loam [9]. The swellable hydrogel delivery systems are also commonly utilized for controlled release of agrochemicals and nutrients of importance in agricultural applications to enhance plant growth with reduced environmental pollution.

The PAM is one of the most widely employed soil conditioner [10]. Linear PAM dissolves in water, cross-linked PAM is a granular crystal that absorbs hundreds of times its weight in water. Absorption of deionised water by PAMs under laboratory conditions can vary between products in the range from 20–1000 per g [11]. Several attempts have been made in the past to combine the best properties of both by grafting synthetic polymers onto natural polymers [12, 13]. Recently, a new class of flocculating agents based on graft copolymers of natural polysaccharides and synthetic polymers has been reported [14-16]. The PAM has also been used in combination with natural polysaccharides for soil conditioning purposes. For example, Wallace et al.

(1986) showed that a mixture of a galactomannan, extracted from guar bean, and PAM resulted in an additive response when applied to certain soils [17].

This work reports a study on the preparation, characterization, and evaluation of water-swellable hydrogel *via* grafting cross-linked PAM chains onto gelatin backbone by free radical polymerization. We aim to increase the water-holding capacity of gelatin to wide its applications as soil conditioners.

11.2 EXPERIMENTAL

11.2.1 MATERIALS

Gelatin (research grade), (purity 99%) obtained from El-Nasr Pharmaceutical Co. for chemicals. (Egypt). Acrylamide (AM) (M wt. 71.08), purity 97%, was purchased from Sigma–Aldrich Chemie. N,N′-methylene bis-acrylamide(MBA) (M wt. 154.17, MP = 300) purchased from Sigma–Aldrich Chemie was used as cross-linking agent without any pre-treatment. Ammonium per sulfate (APS) (purity 99%, M wt. 228.2) was purchased from Sigma–Aldrich Chemicals Ltd (Germany). Ammonium ferrous sulfate (extra pure AR), M.wt. 392.13, assay min. 99%, obtained from Sisco Research Laboratories Pvt. Ltd., (India). Tetramethylethylenediamine (TEMED) (purity 99%, M. wt.116.21) was obtained from Merck Schuchardt (Germany). Sand soil, was obtained from Alexandria desert (Egypt). Other chemicals were of analytical grade used throughout the experiments.

11.2.2 PREPARATION OF PAM-G-GELATIN HYDROGEL

Different concentrations of gelatin (2–10%) (w/v) were dissolved in a beaker 500 ml using hot distilled water (pH 7.2). A known concentration of monomer AM (1–10%) (w/v) and cross-linker MBA (0.03–0.2%) (w/v) were added and finally a known concentration of initiator redox system.

The APS/FAS (0.025–0.125%)/(0.015:0.075%) (w/v) were added. Temed (25 µM) was then injected in the mixture, the solution was stirred well quickly for 30min. to avoid lumping and then left at (25–60 °C) for 1 hr. The mixture was set aside undisturbed (overnight). The gels are then washed extensively with water to remove soluble moieties and unreacted monomers. Acetone was used as a squeezer of water. The gel so formed was dried overnight at 60°C. The dry gel was crushed and separated into different particle sizes (4 mm–250 µm) using a sieve shakers. Scheme 1 describes the proposed mechanistic pathway for synthesis of gelatin-g-PAM hydrogel.

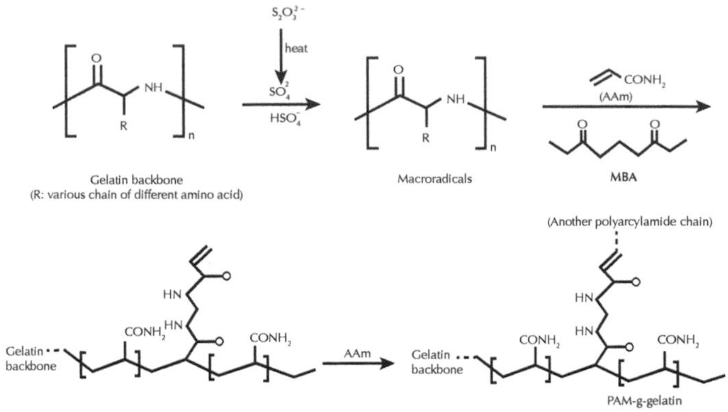

SCHEME 1 Proposed mechanistic pathway for synthesis of PAM g-gelatin hydrogel.

11.2.3 PREPARATION OF PAM (BLANK)

The same preparation method of PAM-g-gelatin hydrogel was used in the preparation of cross-linked PAM hydrogel in absence of gelatin under the same conditions.

Grafting percent (Gp %) and grafting efficiency (GE %) were calculated as follows [17] :

$$Gp\% = [(W1–W0)/W0] \times 100 \qquad (1)$$

$$GE\% = ((Wt. \text{ of grafted PAM/Total Wt. of PAM})) \times 100 \qquad (2)$$

where W1 is the weight of grafted copolymer hydrogel (PAM g-gelatin) and W0 is the weight of native polymer (Gelatin).

Percent weight conversion (WC%) of monomers (AM) into polymeric hydrogel (PAM) was determined from mass measurements [18]. Using the following expression:

$$WC\% = (Total \text{ Wt. of PAM / Total mass of AM in the feed mixture}) \times 100 \quad (3)$$

11.2.4 WATER UPTAKE (SWELLING) EXPERIMENTS

The progress of the swelling process was monitored gravimetrically as described by other workers [19, 20]. In a typical swelling experiment, a pre-weighed piece of sample (0.1 g) was immersed in an aqueous reservoir using distilled water (pH 7.2) and allowed to swell for a definite time period. At least three swelling measurements were performed for each sample and the mean values are reported. The swollen piece was taken out at predetermined time pressed in between two filter papers to remove excess water and weighed. Another sample (1gm) was taken and allowed to swell for a definite time (5 hr) then weighted, the swollen sample was dried at 60°C for 12 hr., and this method was repeated (reswelling) about 10 times of drying. The variation of the swelling degree and weight loss of hydrogels during all drying times were noticed.

The percentage degree of swelling SD (%) of hydrogel [21], can be determined as a function of time as following:

$$SD\ (\%) = ((Mt-M0)/M0) \times 100 \qquad (4)$$

where Mt weight of the swollen hydrogel sample at time t and M0 is the weight of the xerogel sample.

11.2.5 SWELLING IN SANDY SOIL EXPERIMENTS

DETERMINATION OF FLOW RATE OF WATER AND SWELLING DEGREE OF HYDROGEL IN SANDY SOIL

Two plastic measuring cylinders (1000 ml, height 40 cm, area 39.57 cm²) with small holes at the bottom were obtained, a filter paper was placed up the bottom for each cylinder [22], 500 g of sand (particle size 500 μm -1 mm) (height 15 cm) were placed in the first cylinder. Five gram of dry sample was mixed with 495 g of sand (total height 15 cm) with the same particle size were placed in the second cylinder, tap water with height (10 cm) was added up the mixture at the same time, the flow rate of water from up to down in the two cylinders was compared. Also the swelling degree of hydrogel was determined after saturation time (6 hr) [23, 24]. Water releasing rate from sand and sand mixed with hydrogel during 10 days also determined. This method was also repeated with adding 15 cm from sand layers to the cylinders (total height 30 cm). Volume of Flow Water (cm³):

$$Flow\ Rate = Area\ of\ Cylinder\ (cm^2) \times Time\ of\ Flowing\ (min) \qquad (5)$$

DETERMINATION OF WATER LOSS OF SWOLLEN HYDROGEL UNDER EFFECT OF TEMPERATURE

Thirty five gram of swollen hydrogel were mixed with sandy soil (100 g) (particle size 500 μm–1mm) in a plastic measuring cylinder and then covered by 5, 10, 20, and 30 cm of sand layers and weight the contents, then dried at 50°C for (1:5) days. The weight loss of swollen hydrogel with time will be determined as follow:

$$Water\ loss\ \% = ((W1-W2)/W1) \times 100 \qquad (6)$$

where W1 and W2 are weight of mixture before and after drying respectively.

11.2.6 GRAFTING VERIFICATION

Physicochemical characteristics of synthesized PAM, gelatin and PAM-g-gelatin co-polymers were studied using FTIR Spectrophotometer (Shimadzu FTIR-8400 S, Japan) and TGA (Shimadzu TGA -50, Japan). Morphological characteristics were followed using analytical SEM (Joel JSM 6360LA, Japan).

11.3 DISCUSSION AND RESULTS

11.3.1 GRAFTING PROCESS

EFFECT OF POLYMER (GELATIN) CONCENTRATION

Table 1 shows the effect of variation of gelatin in the range of (2–10%) on the grafting percent, grafting efficiency, and weight conversion. It was clear from results that the grafting process was influenced by the amount of gelatin, where with increasing of gelatin concentration up to 6% the grafting percent and grafting efficiency were increased, and then tends to decreases slightly with further increase of gelatin concentration, while weight conversion remains constant. These results may be attributed to that with increasing of gelatin concentration the substrate chains will increases, where the initiated sites on the gelatin backbone will consequently increases and a large number of monomer units will be attached to it, then the grafting percent and grafting efficiency will increases. However, at gelatin concentration higher than 6%, the viscosity of mixture will increase and the free radicals will be hindered and the monomer cannot attached to the active sites on gelatin backbone, and then the grafting percent and grafting efficiency will decrease.

TABLE 1   The effect of gelatin concentration on the percentage of grafting, grafting efficiency at constant conditions (5% AM, 0.05% MBA, and 0.05% APS/0.03FAS at 35°C, overnight)

Gelatin%	GP%	GE%	WC%
2	65	28.57	90.29
4	67.5	61.4	90.29
6	71.66	94.2	90.29
8	53.12	93.2	90.29
10	38	83.33	90.29

The effect of gelatin concentration on the swelling degree of hydrogel was studied as shown in Figure 1. It is clear from results that the swelling degree was increased with increasing of gelatin concentration up to 6% and then tends to decrease with increasing gelatin concentration beyond 6%. Increasing of swelling degree can be explained by the fact that gelatin is a natural water soluble polymer, where its contain a lot of hydrophilic groups which impart hydrophilicity to the molecule, thus increase affinity of water molecules to penetrate in to the gel and swells the macromolecular chains, thus resulting in a greater swelling of hydrogel. However, at much higher concentration of gelatin (beyond 6%) the density of network chains increases so much that both the diffusion rate of water molecules and relaxation rate of macromolecular chains are reduced resulting decreasing the swelling degree of hydrogel.

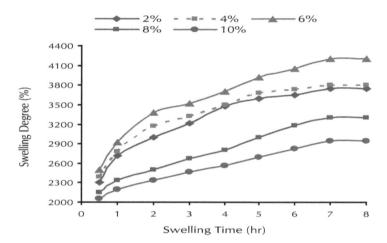

FIGURE 1   The effect of variation in gelatin concentration on the percentage of swelling degree (SD%). [AM] 5%, [MBA] 0.05%, [APS] 0.05%/[FAS] 0.03%, at 35°C, overnight.

EFFECT OF MONOMER CONCENTRATION (AM)

The effect of acrylamide concentration on the grafting percent, grafting efficiency, and total weight conversion was investigated. Table 2 shows that increasing the monomer concentration (AM) clearly increased the percentage of grafting and grafting efficiency. Maximum grafting percent and grafting efficiency were obtained at 5% PAM, also maximum weight conversion of monomer into PAM at the same conditions was observed at concentration 5%. These observations may be attributed to that increasing monomer concentration up to 5% facilates the diffusibility of monomer towards the initiated sites on the gelatin chains, which consequently increases the grafting yield. However, at monomer concentrations higher than 5%, phase separation was observed and the mixture was not homogenous, thus grafting process cannot determine.

TABLE 2   The effect of monomer (AM) concentration on the percentage of grafting, grafting efficiency at constant conditions (6% Gelatin, 0.05% MBA, 0.05% APS/0.03% FAS at 35°C , overnight)

AM%	GP%	GE%	WC%
2	16.66	80	60.97
3	36.66	89.79	80.32
4	55	93.2	87.4
5	71.66	94.2	90.29

Figures 2 (a and b) shows the effect of acrylamide concentration on behavior of the swelling degree gelatin-g-PAM and PAM hydrogels. It was clear from results that the swelling degree of hydrogel increased with increasing of AM concentration up to 5%, this can be   attributed to that acrylamide being a hydrophilic monomer is expected to enhance the hydrophilicity of the hydrogel network when used in increasing concentrations in the reaction mixture. However, beyond 5% phase separation occurred and this may be due to increasing density of AM monomers in the reaction mixture leading to separation from gelatin backbone. Also it was observed from results that the swelling degree of gelatin-g-PAM hydrogel was slightly higher than PAM hydrogel, this indicate that the grafting process was enhanced the swelling process of hydrogel due to increasing the hydrophilic groups in the hydrogel matrix.

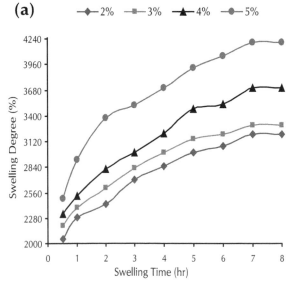

FIGURE 2   *(Continued)*

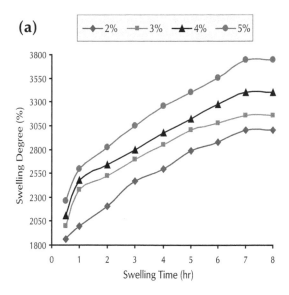

FIGURE 2   The effect of variation in acrylamide concentration (AM) on the percentage of swelling degree (SD%). [Gelatin] 6%, [MBA] 0.05%, [APS] 0.05%/[FAS] 0.03%, at 35°C, overnight, in case of (a) gelatin-g-PAM and (b) PAM hydrogels.

EFFECT OF CROSS-LINKER CONCENTRATION (MBA)

The effect of methylene bis-acrylamide on the grafting yield, grafting efficiency and weight conversion was studied with concentration range (0.03:0.2%) as shown in Table 3. It was clear from results that with increasing the concentration of MBA up to 0.05% the grafting percent and grafting efficiency were increased then tends to decrease slightly with further increase of MBA concentration, but further increasing of MBA concentration beyond 0.2% lead to phase separation. These results may be attributed to that with increasing MBA concentration up to 0.05%, the weight conversion of monomer into PAM will increase, and hence the grafting percent and grafting efficiency depends on the amount of PAM, then they will increases as MBA concentration increases up to 0.05%. While concentrations higher than 0.05% lead to decreasing weight conversion then the grafting yield and grafting efficiency decreased consequently. Also further increase of MBA concentration beyond 0.05% may be leads to formation of more dense three dimensional hydrogel structure which limiting the diffusion of AM monomer and acting in favor of PAM homopolymers formation. This reflects consequently on the drop of both grafting percentage and efficiency.

TABLE 3   The effect of methylene bis-acrylamide (MBA) concentration on the percentage of grafting, grafting efficiency and weight conversion at constant conditions (6% Gelatin, 5% AM, 0.05% APS/0.03% at 35°C, overnight

MBA%	GP%	GE%	WC%
0.03	66	92.59	90.29
0.05	71.66	94.2	90.29
0.1	64.5	92.5	76.6
0.15	62.5	91.13	74.42
0.2	61.66	90.43	73.52

Cross-link's have to be present in a hydrogel in order to prevent dissolution of the hydrophilic polymer chains in an aqueous environment. The cross-linked nature of hydrogels makes them insoluble in water. Figures 3 (a and b) shows the influence of the cross-linking agent in the range (0.03–0.2%) on the swelling degree of PAM g-gelatin and PAM hydrogels. It was clear that maximum swelling degree observed at concentration (0.05%) then tends to decreases, this may be attributed to that on increasing the concentration of MBA beyond (0.05%) in the reaction mixture, the number of cross-link's increases in the gel network. This obviously leads to a slow diffusion of water molecules into the network and restricted relaxation of network chains in the hydrogel thus resulting in a fall in the swelling degree of the network. In fact with concentrations less than 0.03% no gel is prepared and slimy gel is formed due to that the amount of cross-linker was not enough to form the network of hydrogel. Also from figures it was shown that the grafted hydrogel have high swelling degree than PAM hydrogel which indicate that the grafted process increased the swelling degree of hydrogel due to increasing of hydrophilic groups in the hydrogel.

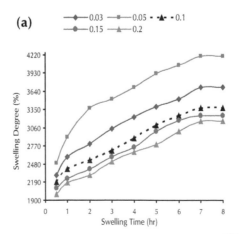

FIGURE 3   (Continued)

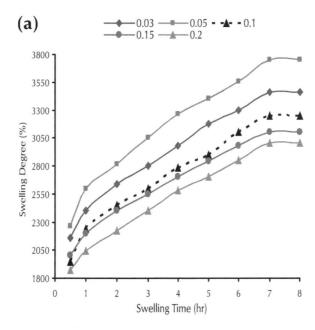

FIGURE 3 The effect of variation in cross-linker concentration (MBA) on the percentage of swelling degree (SD%). [Gelatin] 6%, [AM] 5%, and [APS] 0.05%/[FAS] 0.03%, at 35°C, overnight, in case of (a) gelatin-g-PAM and (b) PAM hydrogels.

EFFECT OF INITIATOR CONCENTRATIONS (APS/FAS)

Table 4 shows the effect of variation of ammonium persulfate (APS) concentration in the range of (0.025–0.1%) and ferrous ammonium sulfate concentration in the range (0.015–0.06%) on the studied grafting parameters. It was clear that the maximum grafting percent and grafting efficiency observed at concentration of (0.05% APS/ FAS 0.03%), and decreased with higher concentrations. These results may be attributed to that with increasing initiator concentration up to (0.05% APS/0.03% FAS) will increasing number of free radicals, also active sites on the substrate will increase consequently, then a great number of monomer units will participate in the grafting process which lead to increasing of grafting yield and grafting efficiency. At higher concentrations (beyond 0.05% APS/0.03% FAS) weight conversion nearly constant. With further increasing of initiator concentrations the number of produced radicals will be more increased and this lead to terminating step *via* bimolecular collision, which leading to decrease molecular weight of the resulting cross-linked PAM, thus shortening the macromolecular chains which leading to decreasing grafting yield and grafting efficiency.

TABLE 4   The effect of redox initiator (APS/FAS) concentrations on the percentage of grafting, grafting efficiency and weight conversion at constant conditions (6% Gelatin, 5% AM, and 0.05% MBA at 35°C overnight)

APS%/FAS%	GP%	GE%	WC%
0.025/0.030	70.8	94	89.5
0.050/0.030	71.66	94.2	90.29
0.075/0.030	70	93.33	89.1
0.100/0.030	66.6	90.9	87.1
0.050/0.0150	63.33	88.37	85.14
0.050/0.045	68.33	93.18	87.12
0.050/0.060	66.6	91.95	86.13

Figures 4 (a and b) shows the effect initiator (APS/FAS) on the swelling degree of hydrogels. The initiator has been varied in the feed mixtures in the concentration range (0.025–0.100)/(0.015–0.06%). It was clear from results that the swelling degree increase with decreasing of initiator concentration in the feed mixture, where higher swelling degree was obtained at concentration (0.05/0.03%) then tends to decrease with further increasing of initiator concentrations these results can be attributed to that the number of hydrophilic groups produced from grafting process was increased and reached to maximum where higher grafting yield (71.66%) observed at concentrations (0.05/0.03%). At concentrations more than (0.05/0.03%) the number of produced radicals will increase and this lead to terminating step *via* bimolecular collision resulting in enhanced crosslink density resulting decrease of the swelling degree.

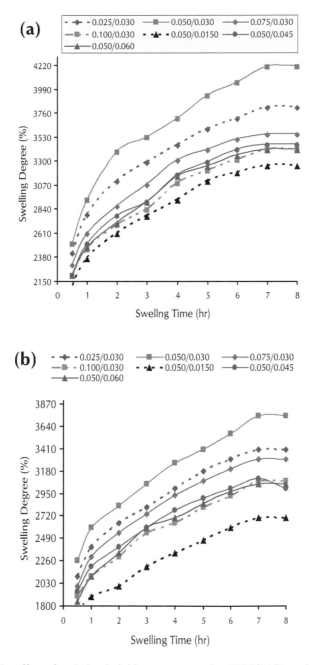

FIGURE 4   The effect of variation in initiator concentration (APS/FAS) on the percentage of swelling degree (SD%). [Gelatin] 6%, [AM] 5%, and [MBA] 0.05%, at 35°C , overnight, in case of (a) gelatin-g-PAM, and (b) PAM hydrogels.

EFFECT OF REACTION TEMPERATURE

The effect of the variation of the reaction temperature (25–45°C) on the grafting parameters was illustrated in Table 5. It was clear from results that the grafting percentage, grafting efficiency also weight conversion increased with increasing of temperature up to 35°C then tends to decreases with further increasing temperature. These results can be attributed to that increasing temperature may be lead to increase diffusion of AM from the solution phase to the swellable gelatin phase, increasing rate of thermal dissociation of initiator, resulting in increasing the solubility of the monomer also formation and propagation of grafted chains. The net effect of all such factors leads to high grafting with increasing the polymerization temperature. While higher temperatures beyond 35°C may be lead to termination step occurring rapidly, then the grafting percentage, grafting efficiency also weight conversion will be decreased.

TABLE 5   The effect of grafting temperature on the percentage of grafting, grafting efficiency, and weight conversion at constant conditions (6% Gelatin, 5% AM, 0.05% MBA, and 0.05% APS /0.03% FAS, overnight)

Grafting Temperature (°C)	GP%	GE%	WC%
25	52.25	86.53	72
30	65.16	90.3	85.74
35	71.66	94.2	90.29
40	61.83	95.087	83.16
45	53	85.48	73.66

The swelling degree of the (Gelatin-g-PAM and PAM) hydrogels prepared with various reaction temperatures is shown in Figures 5 (a and b). From results, it is clear that the swelling degree increase with increasing of reaction temperature up to 35°C then tends to decrease with further increasing temperature, these results may be attributed to that higher temperatures favor the rate of diffusion of the monomers to the gelatin macroradicals as well as increase the kinetic energy of radical centers. The temperatures higher than the optimum value (35°C), however, lead to low-swelling superabsorbents. This swelling loss may be attributed to oxidative degradation of gelatin chains by sulfate radical-anions, resulting in decreased molecular weight and decreased the swelling degree [25].

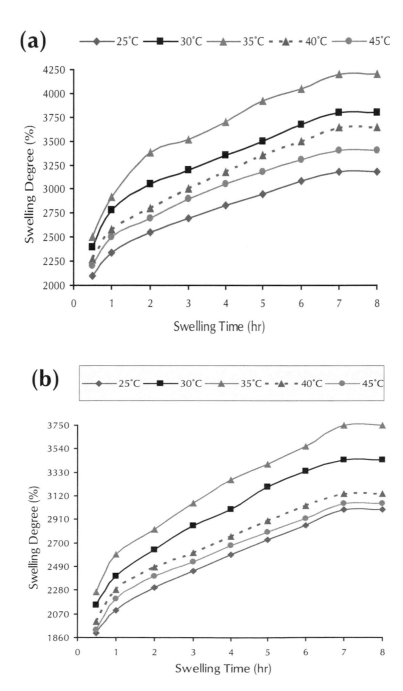

FIGURE 5 The effect of variation in reaction temperature on the percentage of swelling degree (SD%). [Gelatin] 6%, [AM] 5%, [MBA] 0.05%, and [APS] 0.05%/[FAS]0.03%, overnight, in case of (a) gelatin-g-PAM, and (b) PAM hydrogels.

11.3.2 MATERIALS CHARACTERIZATION

INFRARED SPECTROPHOTOMETRIC ANALYSIS (FTIR)

The IR spectra of gelatin, PAM, and PAM g-gelatin respectively are shown in Figure 6(a–c). From the IR spectra of gelatin, it was evident that it shows a broad absorption band at 3431 cm⁻¹, due to the stretching frequency of the –OH and NH_2 groups. The band at 2925 cm⁻¹ is due to C–H stretching vibration. The presence of a strong absorption band at 1649 cm⁻¹confirms the presence of C=O group. The bands around 1446 cm⁻¹ and 1332 cm⁻¹ are assigned to CH_2 scissoring and –OH bending vibration, respectively. In the case of PAM a broad absorption band at 3433 cm⁻¹ is for the N–H stretching frequency of the NH_2 group. Two strong bands around 1693 cm⁻¹ and 1639 cm⁻¹ are due to amide-I (C=O stretching) and amide-II (NH bending). The bands around 1400 cm⁻¹ and 2927 cm⁻¹ are for the C–N and C–H stretching vibrations. Other bands at 1454 cm⁻¹ and 1323 cm⁻¹ are attributed to CH_2 scissoring and CH_2 twisting. The IR spectra of PAM g-gelatin, the presence of a broad absorption band ranged at 3438 cm⁻¹ is due to the overlap of –OH and NH_2 stretching bands of gelatin and PAM. A band at 1641 cm⁻¹ is due to amide-I (C=O stretching) and amide-II band of PAM overlap with each other and lead to a broad band at 1554 cm⁻¹. The bands around 1382 cm⁻¹ and 2923 cm⁻¹ are for the C–N and C–H stretching vibrations. Other band at 1456 cm⁻¹ can be attributed to CH_2 scissoring.

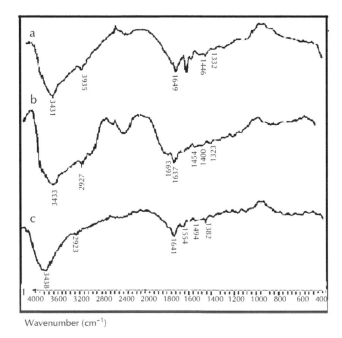

Wavenumber (cm⁻¹)

FIGURE 6   The FTIR of (a) Gelatin, (b) PAM, and (c) PAM-g-Gelatin.

THERMOGRAVIMETRIC ANALYSIS (TGA)

Figure 7 represent the thermal degradation of gelatin, PAM, and PAM g- gelatin hydrogel respectively. In case of gelatin, the initial weight loss can be attributed to presence of small amount of moisture in the sample. The rate of weight loss was increased with increasing temperature. In case of PAM, the initial humidity weight loss followed by a continuous weight loss with increasing temperature, the degradation after that is due to the loss of NH_2 group in the form of ammonia. In case of degradation of PAM g-gelatin, it takes place more closely to gelatin, which indicated to low grafting yield. The data from the TGA was summarized in Table 6. It is clear that the grafting of PAM chains onto gelatin backbone enhanced the thermal stability of the gelatin which was reflected on the increasing of half weight temperature (T_{50}).

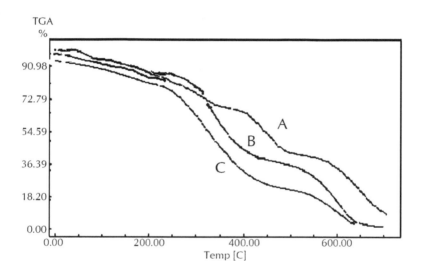

FIGURE 7   The TGA of (A) PAM, (B) Gelatin, and (C) PAM-g-Gelatin.

TABLE 6   Data showing TGA of gelatin, PAM, and PAM g-gelatin hydrogels

		Weight loss %	
Sample	T_{50} °C	0–248 °C	248–430°C
Gelatin	377	13.44	53.16
Gelatin-g-PAM	393	12.551	53
PAM	432	16.67	39.38

SCANNING ELECTRON MICROSCOPE (SEM)

The SEM of gelatin, PAM, and graft copolymer (Gelatin-g-PAM, GP% 71.66) was shown in Figure (8). It was clear that the morphological structure of gelatin was differed than gelatin-g-PAM, where surface morphology of gelatin before grafting shows a smooth surface, which has been changed to surface curly as layers after grafting. The PAM morphology was also changed drastically when grafted onto gelatin backbone.

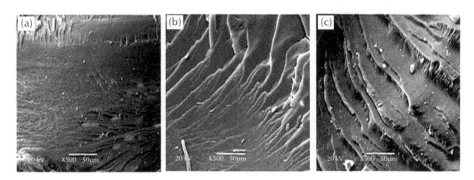

FIGURE 8   The SEM of (a) Gelatin, (b) PAM and (c) PAM.-g-Gelatin.

11.3.3   THE PAM-G-GELATIN HYDROGEL EVALUATION

The effect of variation different operational conditions such as swelling temperature, reswelling ability and pH on the swelling process has been investigated. In addition, effect of hydrogels particle size on the swelling behavior has also been investigated.

EFFECT OF THE SWELLING MEDIUM TEMPERATURE

The swelling has a significant effect on the swelling and deswelling kinetics of hydrogels. The effect of temperature of the swelling medium on the swelling degree of PAM-g-Gelatin hydrogel was studied by variation in temperature of swelling medium in the range of (25–40°C) as shown in Figure 9. It is clear from results that the swelling degree was increased with increasing of temperature up to 30°C then, tends to decrease with further increase of temperature. These results can be attributed to that with increasing temperature up to 30°C, the expansion and flexibility of gel network will increase and this lead to increasing penetration of water molecules into gel network, resulting an increasing of swelling degree. While beyond 30°C the swelling degree decreased and this may be due to that the gel network was malformed and loss its mechanical properties with increasing of swelling time at higher temperatures.

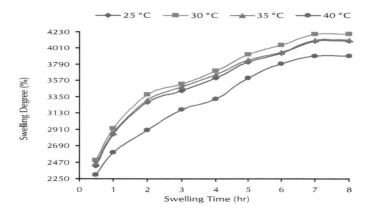

FIGURE 9   The effect of variation in the swelling temperature on the percentage of swelling degree (SD%) of hydrogel [Gelatin] 6% , [AM] 5%, [MBA] 0.05%, and [APS] 0.05%/[FAS] 0.03%,at 35°C, overnight.

EFFECT OF RESWELLING ABILITY ON THE SWELLING DEGREE

The effect of reswelling ability (using distilled water pH 7.2) of hydrogel after drying many times (10 times) was studied as shown in Figure 10. It was clear from results that the swelling degree and weight of sample decrease versus increasing reswelling and drying times (up to 10 times), in which the weight loss of dry sample (1 g) after reswelling and drying 10 time (0.64 g) was 36% from its started weight.

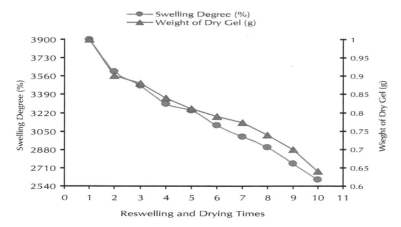

FIGURE 10   The effect of reswelling ability of dry gel and the swelling degree of hydrogel, [Gelatin] 6% , [AM] 5%, [MBA] 0.05%, and [APS] 0.05%/[FAS]0.03%,at 35°C overnight.

EFFECT OF PH MEDIUM ON THE SWELLING DEGREE OF HYDROGEL

As an environmental controlling factor, pH has a significant effect on the swelling degree of PAM-g-Gelatin hydrogel .The swelling degree of hydrogel is affected by the heterogeneous nature of functional groups of amino acids of gelatin that are proton-ated/hydrolyzed on interaction with the swelling medium [26]. Figure 11 show the effect of variation in pH of the swelling medium in the range of 1 to 11. It was clear from results that the prepared hydrogel showed a linear increase in the swelling degree in pH4 and this may be attributed to partial hydrolysis of the amide group, thus, an in-creased volume of network gives rise to greater voids within the gel and, therefore, the water sorption increases leading to an increase of the swelling degree of the hydrogel. On the other hand, cross-linked gelatin has an isoelectric pH (PI) in the range of 4.7 to 5.1, below the PI value the gelatin chains remain protonated. As a result, the chains contain NH_3^+ ions, and the cationic repulsion between them could be responsible for the high swelling degree [27].

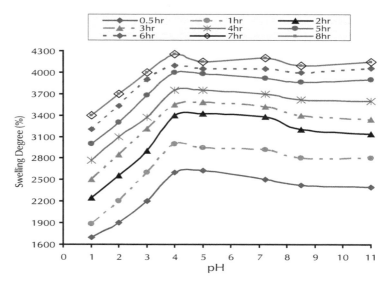

FIGURE 11 The effect of variation in pH of the swelling medium on the percentage of swelling degree (SD%) of hydrogel [Gelatin] 6%, [AM] 5%, [MBA] 0.05%, and [APS] 0.05%/ [FAS]0.03%, at 35°C overnight.

EFFECT OF PARTICLE SIZE OF HYDROGEL

The effect of variation in particle size on the swelling degree of (Gelatin-g-PAM) hydrogel was observed as shown in Figure 12 in the range of 4 mm–250 μm. With a lower the particle size at 1 mm–500 µm, a higher swelling degree was observed. The increasing of the swelling degree would be expected from the increase in surface area with decreasing particle size of hydrogel [28]. Additionally, the ultimate degree of swelling increased as the particle size becomes smaller. This is attributed to more water molecules being held in the volume between the particles. However the swelling degree was decreased at particle size smaller than 1 mm–500 µm may be due to that the network of hydrogel began to destroyed and malformed by crushing and its swelling behavior was decreased.

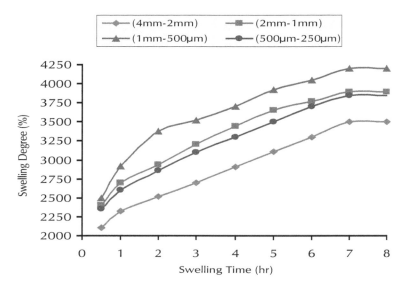

FIGURE 12   The effect of variation in particle size on the percentage of swelling degree (SD%) of hydrogel [Gelatin] 6%, [AM] 5%, [MBA] 0.05%, and [APS] 0.05%/[FAS] 0.03%, at 35°C overnight.

11.3.4 SWELLING OF (GELATIN-G-PAM) HYDROGEL IN SANDY SOIL

Table 7 show that the water flow in sand was decreased when hydrogel mixed with sand soil, these results may be attributed to that the hydrogel retained amount of water and then reduced the flow rate of water. Also it was clear from results that the swelling degree of hydrogel and the flow rate of water increased with decreasing the height of sand soil after saturation of sand soil pores with water (after 7 hr) as shown in Table 8, these results can be attributed to that increasing sand layers will increase the pressure occurring on the hydrogel particles resulting release of water from hydrogel, also hinder the dry gel to swell leading to decreasing the water uptake and the swelling degree of hydrogel.

TABLE 7   Data showing variation of water flow in case of sand soil and sand soil mixed with hydrogel [Gelatin] 6%, [AM] 5%, [MBA] 0.05%, and [APS] 0.05/[FAS] 0.03% at 35°C, overnight. (Using tap water pH 7.2)

Time (hr)	Water Flow in Sand Soil (ml)		Water Flow in Sand Soil mixed with Hydrogel (ml)	
	Height 15 cm Sand	Height 30 cm Sand	Height 15 cm Sand	Height 30 cm Sand
1	3600	2000	3390	2150
2	7000	4100	6680	2790
3	8800	5100	7530	3500
4	9900	5900	8400	4150
5	10850	6500	9200	4750
6	11800	7100	10000	5380
7	12800	7700	10800	5940
8	13800	8300	11600	6500
9	14800	8900	12400	7060
10	15800	8500	13200	7620

TABLE 8   Data showing variation of flow rate of water and swelling degree of hydrogel mixed with hydrogel [Gelatin] 6%, [AM] 5%, [MBA] 0.05%, and [APS] 0.05/[FAS] 0.03% at 35°C , overnight. (Using tap water pH 7.2)

Flow Rate of Water in Sand Soil (cm/min)		Flow Rate of Water in Sand Soil mixed with Hydrogel (cm/min)		Swelling Degree (SD%) of Hydrogel in Sand Soil	
Sand Height (15 cm)	Sand Height (30 cm)	Sand Height (15 cm)	Sand Height (30 cm)	Sand Height (15 cm)	Sand Height (30 cm)
0.421	0.252	0.336	0.235	2300	1700

Table 9 show that the water loss from hydrogel which covered with different sand layers heights increase with increasing of drying time, these results can be attributed to that with increasing drying times soil-water content rapidly changes as water drains under gravitational forces and temperature. Also from results it was found that the prepared hydrogel from gelatin had lost about 50% from its water contents after 5 days from drying at 50°C which indicate that it have good mechanical properties.

TABLE 9 Data showing variation of water loss of the swollen hydrogel mixed with sand soil under effect of temperature 50°C, hydrogel compositions [Gelatin] 6%, [AM] 5%, [MBA] 0.05%, and [APS] 0.05/[FAS] 0.03% at 35°C, overnight. (Using tap water pH7.2)

Time	Water loss %	
(days)	Sand Height (30cm)	
1	12.2	Sand Height (15cm)
2	20.6	16.5
3	28.9	23.3
4	36.15	33.3
5	41.67	41.66

11.4 CONCLUSION

From this study, it was found that the grafting of naturally protein such as gelatin by synthetic polymers as PAM will increase of the water uptake capacity of this natural polymer to have water swellable hydrogel. The water-holding capacity of grafted hydrogel was found depend on the concentrations of the feed compositions of the hydrogel. The characterizations by FTIR, TGA, and SEM provide further proof of grafting process on the backbone of gelatin. The hydrogel particle size was found also of determining effect where maximum swelling was observed with particles size ranged from 500 µm to 1mm. The prepared cross-linked PAM-g-gelatin hydrogel shows good thermal stability, high water absorbency, and high pH, which recommended our preparation for biotechnological and agricultural applications. Finally, grafted copolymer hydrogel have been evaluated using swelling and flow rate of water in sandy soil mixed with hydrogels experiments at optimum conditions. The retention of water by the hydrogel graft copolymer in sandy soil has been used as monitor to evaluate its applicability as soil conditioners.

KEYWORDS

- **Acrylamide**
- **Gelatin**
- **Methylene bis-acrylamide (MBA)**
- **Polyacrylamide (PAM)**
- **Scanning electron microscope (SEM)**

ACKNOWLEDGMENT

The authors acknowledge the help of Prof. M. R. Abdel Fatah, Professor of soil chemistry-arid land institute-Scientific research and technological applications city-Alexandria (Egypt), in performing the soil filtration experiments.

REFERENCES

1. Buchholz, F. L. and Graham, A. T. *Modern Superabsorbent Polymer Technology*, Wiley, New York, NY, 1997.
2. Peppas, L. B. and Harland, R. S. In: *Absorbent Polymer Technology*, Elsevier: Amsterdam, 1990.
3. Po, P. *J Macromol Sci Rev Macromol Chem Phys*, **34**, 607 (1994).
4. HoffmanIn, A. S. and Salamone, J. C. *Polymeric Materials Encyclopedia*, ed., C. R. C. Press, Boca Raton, FL, **5** (1996).
5. Mohy Eldin, M. S., El-Sherif, H. M., Soliman, E. A., Elzatahry, A. A., and Omer, A. M. *Journal of Applied Polymer Science*, **122**, 469–479 (2011).
6. Burillo, G. and Ogawa, T. *Radiation Phys. Chem.*, **18**, 1143–1147 (1981).
7. Hedrick, R. M. and Mowry, D. T. *Soil Sci.*, **73**, 427–441 (1952).
8. Azzam, R. *Commun. Soil Sci. Plant Analysis*, **11**, 767–834 (1980).
9. Huttermann, A., Zommorodi, M., and Reise, K. *Soil and Tillage Research.*, **50**, 295–304 (1990).
10. Omer, A. M. Preparation and Characterization of Graft Copolymer Hydrogels to be used as Soil Conditioners, MS.C. Al-Azhar University, Egypt, (2008).
11. Johnson, M. S. and Veltkamp, C. J. *Journal of the Science of Food and Agriculture*, **36**, 789–793 (1985).
12. Singh, R. P. Plenum Press, New York, NY, p. 227 (1995).
13. Swarson, C. L., Shogren, R. L., and Fanta, G. F. *J. Environ. Polym. Degrad.*, **1**, 155–166 (1993).
14. Nayak, B. R. and Singh, R. P. *J. Appl. Polym. Sci.* **81**, 1776–1785 (2001).
15. Rath, S. K. and Singh, R. P. *J. Appl. Polym. Sci.* **66**, 1721–1729 (1997).
16. Mohy Eldin, M. S., Omer, A. M., Soliman, E. A., and Hassan, E. A. *J. Desalination and Water Treatment*, **51**, 3196–3206 (2013).
17. Mohy Eldin, M. S. Preparation and Characterization of Some Cellulosic Grafted Membranes, MSC, (1995).
18. Guven, O. and Sen, M. *Polymer*, **32**, 2491–2495 (1991).
19. B. D. Ratner and D. F. William (Eds.), CRS Press, Inc., Florida, **2**, pp. 145–175 (1981).

20. Gudeman, L. F. and Peppas, N. A. *J. Appl. Polym. Sci.*, **55,** 919–928 (1995).
21. Aleksandar, K., Borivoj, A., Aleksandar, B., and Jelena., J. *J. Serb. Chem. Soc.*, 72(11), 1139–1153 (2007).
22. Akhter, J. *Plant soil environ.*, **50**(10), 463–469 (2004).
23. Johnson, M. S. *J. Sci. Food Agric.*, **35**, 1196–1200 (1984b).
24. Bowman, D. C. and Evans, R. Y. *HortScience.*, **26**(8), 1063–1065 (1991).
24. Hebeish, A. and Cuthrie, J. *"Chemistry and Technology of Cellulosic Copolymers"*, p. 46, (1981).
26. Ghanshyam, S. *Journal Applied Polymer Science*, **90**, 3856–3871(2003).
27. Krishna, B. *Journal Applied Polymer Science*, **82**, 217–227 (2001).
28. Ali, P. *journal of e-polymers*, **57** (2006).

CHAPTER 12

APPLICATION OF PHYSICO-CHEMICAL METHODS IN MONITORING OF THE ANTHROPOGENIC POLLUTION BY AN EXAMPLE OF *Ceratophyllum demersum* L.

N. V. ILYASHENKO, L. A. KURBATOVA, M. B. PETROVA,
N. V. PAVLOVA, AND P. M. PAKHOMOV

CONTENTS

12.1   INTRODUCTION

The effect of anthropogenic pollution on the chemical composition and anatomic structure of bioindicator plants (hydrophytes) was studied with the aid of Fourier transform IR spectroscopy (FTIR), scanning electron microscopy (SEM), and X-ray microanalysis. A correlation between the changes existing in the IR spectrum of the plant samples and anthropogenic pollution of the plant inhabitation is established. Deformation and epidermis cell disruption were revealed in the samples from polluted sites.

This chapter is aimed at the experimental study of the chemical composition and anatomic structure of polluted higher aquatic plants making use of combined physical methods of characterization by FTIR, SEM, and X-ray microanalysis.

Recent studies demonstrated that aquatic plants are highly sensitive tools for assessing the status of water bodies [1-4]. Because they quickly establish themselves with their environment. Further usage of plants as bioindicators requires accumulation of knowledge and development of appropriate methods of their characterization.

Higher hydrophytic plants possess high accumulative properties and find their use in the determination of anthropogenic chemical loads on the water body in the system of environment state biomonitoring [1-4].

The IR spectroscopic data on the chemical composition changes in bioindicator plants may be informative for the estimation of hydrosphere pollution in industrial regions. The exact identification of the types of compounds formed in the plant as a result of accumulation of various pollutants enables the use of FTIR for biomonitoring of acid pollutions (sulfur and nitrogen dioxides) and petroleum products and also organic compounds [5-7].

As far as bioindicator plants are subject to change at chemical and anatomic levels due to the action of anthropogenic environment pollution, these transformations may be effectively monitored with the aid of FTIR spectroscopy, SEM, and energy-dispersive X-ray spectroscopy (EDX) [8,9].

Combination of SEM and EDX makes it possible to determine the element composition in a volume of ~1 cubic micrometer by means of registration the X-ray characteristic irradiation occurring during the primary electron interaction with the sample surface. The EDX is an analytical technique used for the chemical characterization of a sample. Its characterization capabilities are due largely to the fundamental principle that each element has a unique atomic structure allowing X-rays that are characteristic of an element's atomic structure to be identified uniquely from one another. The number and energy of the X-rays emitted from a specimen are measured by an energy-dispersive spectrometer. As the energy of the X-rays are characteristic of the difference in energy between atomic shells, and of the atomic structure of the element from which they were emitted, this allows the elemental composition of the specimen to be measured.

The aim of the present chapter is to study the effect of pollutants on the chemical composition and anatomic structure of hydrophytic plant *Ceratophyllum demersum* L. (Hornwort deep green) deep green by means of FTIR spectroscopy, SEM, and EDX.

12.2 EXPERIMENTAL

Ceratophyllum demersum L. occurs in Tver region in stagnant reservoirs with slowly flowing water, ponds, peaceful river backwater, cutoff meanders, and is able to grow both in clear and contaminated inhabitants [1,10]. Collection of the plants and water intake for chemical analysis were performed in water bodies of the city of Tver and Tver region classified by the factor of proximity to the sources of contamination in two groups—control and polluted (Table 1).

In parallel the chemical analysis of the water from the sites of *C. demersum* growth was made to ensure the proper interpretation of the element composition and IR spectra of the aquatic plants from industrial regions. Chemical analysis was performed with the aid of a spectrofluorimeter "Fluorat-02-Panorama" and capillar electrophoresis system «Kapel-105» (Lumex). Determination of the contents of inorganic anions, surfactants, petroleum products, and phenols in water was made in accordance with standard methods described in [11-14].

The IR spectra of the samples under study were recorded in the range of 400–4000 cm^{-1} by a standard method with potassium bromide [15] making use of the Fourier transform spectrometer «Equinox 55» (Bruker).

TABLE 1   Sites of *C. demersum* collection

No	Water body	Ecological status / Source of pollution [16]		Main pollutants [17]
I	Mezha river, Big Fyodorov village, Nelidov district of Tver region	SPNT* (control area), Central-Forest National Natural Biosphere Park		
II	Coolant lake Udomlya, Udomlya, Tver region	Ecologically stressed nodes	Kalinin nuclear energy station	Petroleum products, small radiation doses
III	Cunette, Redkino township, Konakov district, Tver region		Public corporation «Redkino pilot plant»	Petroleum products, inorganic anions, ammonia, aromatic compounds, anionic surfactants

* SPNT – Special Protected Natural Territory

Scanning electron microscope studies were made in high vacuum regime with JEOL 6610LV SEM (Japan), EDX was performed with the INCA Energy system (OXFORD INSTRUMENTS, UK). The morphology of the epidermal surfaces was examined at the magnifications of × 500 and × 1000. Samples of the plants for SEM and EDX studies were dried at 30–40°C and fixed by a graphite adhesive tape [8, 9].

12.3   DISCUSSION AND RESULTS

12.3.1   CHEMICAL ANALYSIS OF THE WATER IN RESERVOIRS UNDER STUDY

Results of the water analysis of the sites specified in Table 1 confirmed the existence of pollutants listed in Table 2.

Main water indices (Cl^-, NO_2^-, NO_3^-, SO_4^{2-}, phenols, and anionic surfactants) in samples II, III are much higher as compared with the control (I).

In the water sample III, the content of phenols and inorganic anions is by a factor of several tens larger than in sample I, thus indicating discharges of chemical pollutants from the Redkino pilot plant (Table 2).

The existence of various pollutants in the water reservoirs should be taken into account because aquatic plants are able to accumulate the contaminants (Table 2) [18-20].

TABLE 2   Chemical analysis of the water samples from the reservoirs under study

Index	Content, mg/l		
	I (control)	II	III
Cl–	1.2	6.2	64.0
NO2–	< 0.2	0.88	5.0
SO42–	2.8	7.4	42.3
NO3–	< 0.2	1.2	< 0.2
PO43–	< 0.2	< 0.2	< 0.2
Petroleum products	0.17	-	0.16
Phenols (total)	0.001	-	0.38
Anionic surfactants	0.11	-	0.09

12.3.2   THE IR SPECTRAL ANALYSIS OF C. DEMERSUM SAMPLES

Shown in Figure 1 are the IR spectra of C. *demersum* samples from different reservoirs.

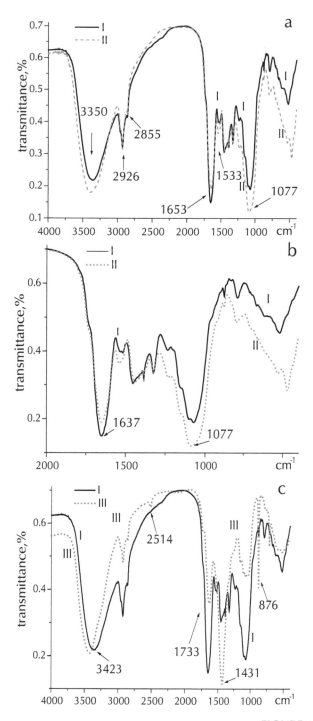

FIGURE 1 *(Continued)*

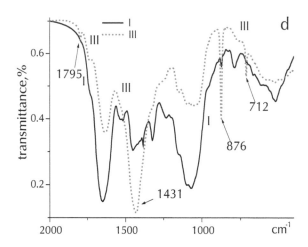

FIGURE 1   The IR spectra of *C. demersum* samples:

I (control), II (lake Udomlya), and III (Cunette, Redkino township).
The IR spectral analysis shows that all samples under study have absorption bands corresponding to the main chemical components of the plant: carbohydrates ~56% (of the absolute dry weight), proteins ~18%, and fats~1% [1]. The existence of carbohydrates in the plant is testified by absorption bands due to stretching vibration of CH$_2$-groups at a frequency of ~2925 cm^{-1} and OH-groups at ~3400 cm^{-1} [5,21-22]. The existence of proteins is evidenced by the absorption bands at ~1640 (Amid I), ~1535 (Amid II), and ~1235 cm^{-1} (Amid III) [7, 22]. The presence of fat may be judged by the existence of absorption bands at ~1735 ($v_{C=O}$), ~1446 (δ_{CH2-}) (Table 3) [6].

TABLE 3   Assignment of IR absorption bands of C. demersum samples

Band position, cm-1			Assignment	Reference
I	II	III		
3350	3393	3423	v(OH)	[4,7,21]
2926	2925	2925	vas(CH2)	
2855	2855	2854	vs(CH3)	[6,21]
–	–	2514	v(S-H)	[22]
–	–	1795	v(C=O)	[22]
1733	1738	1735	v(C=O)	[4,6,21]
1653	1653	1637	Amid I v(C=O)	[7,22]
1533	1542	–	Amid II v(O-C-N)	[21]

TABLE 3   *(Continued)*

–	1438	1431	vas(SO2), δ(CH2), δ(N-H)	[22]
1385	1385	–	δ(OH)	[5,22]
1325	1325	1325	δ(CH2)	[22]
1233	1230	1265	Amid III δ(N-H)	[7,22]
1201	1202	1204	v (C–O), δ(OH)	[15,21]
1098	1153	1101	v (C–O), vas(COC)	[4,21]
1071	1078	1077	vas(COC) chain	
–	–	1052	vas(COC) chain	[5,21]
–	–	1025	v(OH)	
874	874	876	S-O-C	[6,7,22]
–	–	712	v (C-S-C)	[22]

Comparison of the IR spectra of *C. demersum* samples taken from control and contaminated reservoirs demonstrates considerable changes in the chemical composition of the plants. It is important that most significant changes of the absorption bands intensity correspond to the samples collected in the regions of industrial contamination. In samples collected from the reservoirs not subjected to direct contamination the band intensity conforms to the control values.

Spectra of plant samples from contaminated sites (Figure 1) demonstrate essential changes at the following frequencies: ~2514 cm⁻¹ (III), due to stretching vibration of S-H groups, ~1794 cm⁻¹ because of stretching vibration of C=O groups (III), ~876 cm⁻¹ owing to stretching symmetrical vibration of the S-O-C groups (III), and ~712 cm⁻¹ by virtue of stretching vibration of C-S-C groups (III) (Table 3) [7,22].

The absorption band at ~1431 cm⁻¹, attributed to $v_{as}(SO_2)$, δ(N-H), observed in all spectra of samples from contaminated sites but it is mostly intensive in the spectrum of sample III on account of high concentration of sulfur-bearing anions absorbed from the water by the plant (Figure 1 and Table 3).

Comparative analysis of *C. demersum* from different sites enables to propose that the intensity increase of the mentioned absorption bands of the samples collected from contaminated reservoirs may be attributed to the accumulation and adoption of chemical compounds not inherent to the plant containing thiol-, carbon-, and nitrogen-bearing groups [5,23].

12.3.3   EXAMINATION OF HYDROPHYTIC PLANT SAMPLES WITH THE AID OF EDX AND SEM

Electron images and X-ray spectra of *C. demersum* samples studied with the aid of SEM and EDX are presented in Figure 2.

It is seen that in samples II and III collected in polluted sites, there are deformation and destruction of epidermis resulting in violation of the epidermal cell layer integrity.

In samples II, III the attention is attracted by the high density of diatoms (Diatomea sp.) at the epidermal layer of the hydrophytic plant under study. Such kind of hornwort leaves encrustation by diatoms (fouling) is to all appearance stipulated by the favorable conditions for their vital activity due to heightened thermal regime of the Udomlya lake of the Kalinin nuclear plant [24,25].

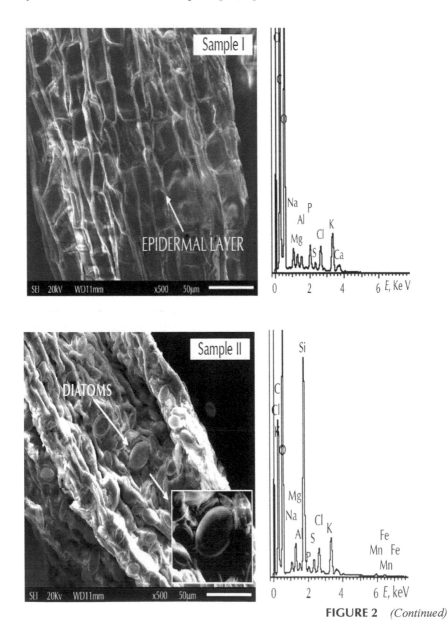

FIGURE 2   (Continued)

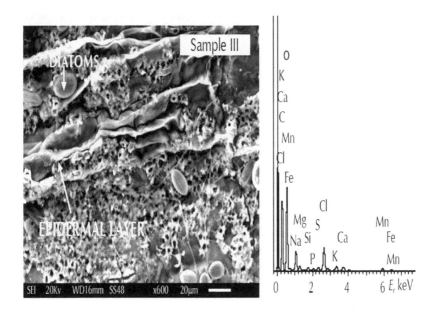

FIGURE 2   The SEM images and energy X-ray spectra of *C. demersum* leaf samples I, II, and III at magnification × 500.

The elemental composition and quantitative chemical analysis of *C. demersum* plants was performed with the aid of EDX. Given (Table 4) are the data for elemental composition and occurrence of various chemical elements in the tissues of *C. demersum* from some water bodies of Tver region.

TABLE 4   Elemental chemical composition of C. demersum

Sample	Chemical element	C	O	Na	Mg	Al	P	S	Cl	K	Ca	Si	Mn	Fe	Total
I		56.35	40.90	0.60	0.27	0.18	0.41	0.10	0.40	0.67	0.12	–	–	–	100
II	Wt%	48.34	45.31	0.34	0.68	0.11	0.12	0.23	0.50	0.75	–	3.48	0.07	0.06	100
III		53.44	41.92	1.93	0.27	–	0.05	0.14	1.25	0.25	0.26	0.09	0.26	0.11	100

The EDX data reflect the overall chemical composition of *C. demersum* [1]. However in samples II and III the manganese, iron, sulfur and chlorine content is higher than in the control sample, which is in accordance with the corresponding chemical analysis data of the water given in Table 2. In addition, high content of silicon in sample II should be mentioned. Apparently the silica-containing frustules of diatoms observed on the hornwort foliage serve as a source of silicon.

12.4 CONCLUSION

By the use of FTIR, it was proved that higher aquatic plants have a capability to respond actively on the water chemical composition changes by the increase of absorption bands intensity related to contaminants. The results of the study show that the Fourier IR spectroscopy may be recommended for the effective application in biomonitoring of contaminated water bodies.

With the aid of SEM by an example of biomonitoring *C. demersum* plants anatomic changes in the leaf platelets were noticed. Deformation and destruction of epidermis cells were observed thus giving an evidence of high accumulation and adoption properties of aquatic plants. The elemental composition of plants under study was determined by means of EDX. Heavy metals were detected in samples collected from the sites of anthropogenic pollution.

The described physical methods of analysis may be effectively employed in biomonitoring of the ambient environment.

KEYWORDS

- **Anthropogenic pollution**
- **Fourier IR spectroscopy**
- **Hydrophytic plants**
- **Scanning electron microscopy**
- **X-ray microanalysis**

ACKNOWLEDGMENT

This work was performed within the Federal Target Program "Scientific and Scientific–Pedagogical Personnel of Innovation Russia for 2009–2013", contract № 14.740.11.1281 and supported by the grant of foundation for development of small enterprises in scientific–technical sphere under the program "Participant of the Youth Scientific–Innovation Competition", contract № 8754p /14008. Permission to use the equipment and support of the sharing service center of Tver State University is greatly acknowledged.

REFERENCES

1. Sadchikov, A. P. and Kudryashov, M. A. *Hydrobotany: Coastal Vegetation* (in Russian), Moscow, (2005).
2. Chukina, N.V. *Effect of water contamination on the chemical composition of hydrophytic leafs* (in Russian), Proc. XIII Intern. Youth School, Rybinsk, pp. 290–296 (2007).

3. Goldovskaya, L. F. *Chemistry of Ambient Environment* (in Russian), Mir, Moscow, (2005).
4. Ilyashenko, H. V., Ilyashenko, V. D., Dementieva, S. M., Khizhnyak, S. D., and Pakhomov, P. M. *Vestnik TvGU. Ser. Biologiya i Ekologiya* (in Russian), (13), 211–221 (2009).
5. Duraees, N., Bobos, I., and Ferreira Da Silva, E. *Chemistry and FT-IR spectroscopic studies of plants from contaminated mining sites in the Iberian Pyrite Belt*, Portugal Mineralogical Magazine. **72**(1), 405–409 (2008).
6. Ilyashenko, N. V., Khizhnyak, S. D., and Pakhomov, P. M. Effect of anthropogenic factor on the content of biologically active compounds in Bidens tripartita **L**. And Potentilla erecta L, Intern. Conf. *"Renewable Wood and Plant Resources: Chemistry, Technology, Pharmacology, Medicine"*, Book of abstracts, Saint-Petersburg, pp. 265–266 (2011).
7. Meisurova, A. F., Khizhnyak, S. D., and Pakhomov, P. M. IR spectral analysis of the chemical composition of the lichen Hypogymnia physodes to assess atmospheric pollution. *Journ. Applied Spectroscopy*, 76, 420–426 (2009).
8. Pathan, A. K., Bond, J., and Gaskin, R. E. *Sample preparation for SEM of plants surfaces*. *Materials Today*, **12**, 32–43 (2009).
9. Goldstein, J. I., Newbury, D. E., Echlin, P., Joy, D. C., Fiori, C., and Lifshin, E. *Scanning Electron Microscopy and X-ray Microanalysis*, Plenum Press, New York and London, (1981).
10. National Atlas of Russia, Vol. 2: *Nature and Ecology* (in Russian), Moscow, (2008).
11. *Determination of inorganic anions in water*. Russian federal standard M 01-30-2009 PND F 14.1:2:4, pp. 157–99 (2009).
12. *Determination of anionic surfactants in water,* Russian federal standard M 01-06-2009 PND F 14.1:2:4.158-2000 GOST P 51211-98, (2009).
13. *Determination of petroleum in water*, Russian federal standard M 01-05-2007 PND F 14.1:2:4.128-98 MYK 4.1.1262-03 (2007).
14. *Determination of phenols in water*, Russian federal standard MVI M 01-07-2006 PND F 14.1:2:4.182-02 MYK 4.1.1263-03 (2006).
15. Smith, A. *Applied IR Spectroscopy – Bases, Techniques, and Analytical Applications* (1982).
16. A. S. Kurbatov (Ed.), *City Ecology*, Moscow, 2004 (in Russian).
17. Federal Report "On the Status and Ambient Environment Protection in Russian Federation in 2009" (in Russian). Moscow (2009).
18. Ilkun, G. M. *Air Filtering of Pollutants with the Aid of Woody Plants* (in Russian). Tallin, (1982).
19. Ilkun, G. M. *Atmosphere Contamination and Plants* (in Russian). Naukova dumka, Kiev (1978).
20. Smolov, A. P. and Opanasenko, V. K. Ammonium factor in the vital activity of plant cell. *Cytology*, **51**, 358–366 (2009).
21. Bazarnova, N. G., Karpova, E. V., and Katrakov, I. B. *Methods of Study of Wood and its Derivatives* (in Russian). Altai State University, Barnaul (2002).
22. G. Socrates (Ed.), *Infrared Characteristic Group Frequencies. Tables and Charts*. John Wiley & Sons, London (1994).
23. Garty, J., Kunin, P., Delarea, J., and Weiner, S. *Calcium oxalate and sulphate-containing structures on the thallial surface of the lichen Ramalina lacera: response to polluted air and simulated acid rain, Plant, Cell and Environment*, **25** 1591–1604 (2002).
24. *Ecological monitoring of the terrestrial and water ecosystems of the Kalinin nuclear energy plant region* (in Russian). Atom Energy Institute Report, Moscow (2005)

25. Barinova, S. S., Anisimova, O. V., and Medvedeva, L. A. Algae Indicators in the Estimation of the Environment Quality (in Russian). All-Union Research Institute of Nature, Moscow, p. 150 (2000).

CHAPTER 13

COMPLEXITIES IN NANOMATERIALS— CASE STUDIES

M. NIDHIN, K. J. SREERAM, B. U. NAIR, J. JAKUBOWICZ,
G. ADAMEK, S. E. KAREKAR, B. A. BHANVASE, S. H. SONAWANE,
SUMAN SINGH, V. K. MEENA, D. V. S. JAIN, and M. L. SINGLA

CONTENTS

13.1 INTRODUCTION

Recently, nanotechnology has attracted much attention because of its application in chemistry, physics, materials science, and biotechnology to create novel materials that have unique properties. The combination of a high surface area, flexibility, and superior directionality makes nano structures suitable for many applications. Difference in properties at bulk and nano-level has provided new insight and direction to the researchers to meet the ever increasing demand of smaller and more efficient devices. Nanotechnology offers potential solutions to many problems using emerging nano techniques. Depending on the strong inter disciplinary character of nanotechnology there are many research fields and several potential applications that involves nanotechnology.

This chapter is the collection of four case studies on nanomaterials: Synthesis, characterization and application The main objectives of these case studies are to promote interdisciplinary research on synthesis, characterization and application of nanoparticles, their composites.

13.2 CASE STUDY I: SYNTHESIS AND APPLICATION OF INORGANIC NANOPARTICLES

13.2.1 INTRODUCTION

Iron oxide nanoparticles are undoubtedly one of the most investigated nanoparticles owing to their importance in industrial and medical applications. Hematite nanoparticles are comparatively more stable and therefore have wider applications as well. While several methods are available for the synthesis of nanoparticles, the green synthesis routes are preferred, more so in biological applications. It has been demonstrated that natural products such as polysaccharides can be employed as effective templates, both through sacrificial and otherwise routes to generate mono-dispersed nanoparticles of below 20 nm. Such methods are easy to adopt for other ferrite nanoparticles as well. Incidentally, some of the polysaccharide methods are facile and possibly simple for replication at industrial scales.

13.2.2 INCREASING RELEVANCE OF NANOTECHNOLOGY IN SCIENCE AND INDUSTRY

Innovations in modern technology are directed towards smaller, cheaper, faster and smarter products than that is currently available in the market. These properties are interconnected with size reduction. For instance, size reduction to 10^{-9} meters, considerably changes the physical and chemical properties of the material. "Metals" or "Metal oxides" are old materials. However, their fabrication in new forms, more specifically in the nano-scale has aroused tremendous interest [39]. Nanotechnology is a broad interdisciplinary area of R&D and industrial activity. It involves manufacturing, processing and application of materials with size below 100 nm. Research areas in nanotechnology can be broadly grouped under nanomedicine, nanofabrication, nanometrology, nanomaterials and nanoparticles [2,3]. Through the years, we have seen increasing relevance for nanoparticles and nanomaterials in fundamental and applied science. These nanoparticles and nanostructures exhibit different physical and chemical properties compared to bulk. They have been of keen interest to science and technology, especially in miniaturized devices. Nanotechnology based industries are expected to have a market of USD 1 trillion by 2015. Aitken *et al.*, has provided a detailed review on the potential of nanotechnology [2]. He opines that an impediment to the quantum jump in growth of this industry is the issue of scale-up from lab scale to industrial level production. There is a need to develop or reorient the synthesis and functional strategies of industrially relevant metal or metal oxide nanoparticles.

The synthesis of nanoparticles of iron oxides have been reported in recent times (especially α-Fe_2O_3) by using different chemical methods [103]. Iron oxide nanoparticles are usually synthesized by the coprecipitation of ferrous (Fe^{2+}) and ferric (Fe^{3+}) ions by a base. Other synthetic methods include the thermal decomposition of an alkaline solution of an Fe^{3+}chelate in the presence of hydrazine and the sonochemical decomposition of an Fe^{2+}salt followed by thermal treatment [83]. Uniformly sized iron oxide nanoparticles were synthesized by the high-temperature reaction of $Fe(acac)_3$ in octyl ether and oleic acid or lauric acid or a mixture of four solvents and ligands, namely, 1,2-hexadecanediol, phenyl ether, oleic acid, and oleylamine. But the major drawbacks of these synthetic procedures are the low dispersion in solvents and wide particle size distribution [45]. However, the uniformity of the size and shape of the particles of these nanomaterials was rather poor. These methods usually involve synthesizing a precursor gel of iron, followed by decomposing the gel or precursor into the designed crystalline iron oxide phase at an elevated temperature [47]. They require expensive and often toxic reagents, complicated synthetic steps, and are not biocompatible. To understand the environmental implications of these nanoparticles and facilitate their potential applications, it is important to develop a simple, green, and generic method for the preparation of Fe_2O_3 nanoparticles [48]. Green synthesis of nanosized Fe_2O_3 with uniform size is one of the important issues of the present research for understanding the fundamental properties of nanomaterials and utilizing their nano scaled properties for various applications.

One of the industrially most relevant and fascinating properties of iron oxide is its magnetism. To a large extent, magnetism is a nano phenomenon [48]. When particle dimensions are reduced to nanosize and shape, new properties such as exchange magnetic moments,[7] exchanged coupled dynamics, [21] quantization of spin waves [35] and giant magneto-resistance [5] have been reported. These new properties lead to applications in permanent magnets, data storage devices, magnetic refrigeration, catalysis, targeted drug delivery and MRI. Superparamagnetic iron oxide nanoparticles (SPIONs) possess high transverse (spin-spin) relaxivity and are therefore good T_2 contrast agents, enabling them to produce darker images when accumulated in certain areas of the body [88].

13.2.3 HEMATITE: ABUNDANT TUNABLE INDUSTRIAL APPLICATIONS

Almost all processes or environment have iron oxide in it. This includes the depths of Earth, the surfaces of Mars, factories, high-tech magnetic recorders, brains of birds and magnetotactic bacteria. Nanotechnology has made the applications of iron oxide more interesting and increasing. Iron oxides are technologically important transition metal oxides. Sixteen pure phases of iron oxides, viz., oxides, hydroxides or oxy-hydroxides are known till date [14,94]. These include $Fe(OH)_3$, $Fe(OH)_2$, $Fe_3HO_8.4H_2O$, Fe_3O_4, FeO, five polymorphs of FeOOH and four polymorphs of Fe_2O_3 (α, β, γ, ϵ) etc. This complexity has brought about a limitation in the knowledge of structural details, thermodynamics and reactivity of iron oxides [70]. The physical and chemical properties associated with these particles are dependent on size and degree of hydration and therefore there is a larger emphasis on the synthesis strategies employed, more so in the case of α-Fe_2O_3. α-Fe_2O_3 (Hematite), is an n-type semiconductor, and the most prevalent metal oxide on Earth. It is stable, corrosion resistant and non-toxic under ambient conditions. Historically, one of the first uses of α-Fe_2O_3 was in the production of red pigments [14,94]. Relatively small bandgap (2.2 eV; 564 nm) and related visible light absorption that (d $\rightarrow$ d type) [84] are not strongly absorptive, low cost and stability under deleterious chemical conditions make α-Fe_2O_3 an ideal candidate for several industrial applications such as in catalysis, gas sensors, solar cells, anode material for lithium ion batteries, field emission devices, magnetic materials, hydrogen storage, pigments and water splitting [10,50,79,108,119]. However, challenges such as poor conductivity and short hole-diffusion length (2-4 nm) are some of the drawbacks which need to be overcome prior to practical usage [13,96].

α-Fe_2O_3 has a hexagonal unit cell and entirely octahedrally coordinated Fe^{3+} atoms (corundum structure).[14,94]. Crystal structure consists of alternating iron and oxygen layers stacked along the [001] axis of the hexagonal unit. At temperatures below the Neel temperature (T_N – 955 K), the magnetization directions of neighboring Fe layers become antiparallel due to antiferromagnetic ordering of the layers. A first-order spin-reorientation transition (Morin transition) occurs at 263 K. Below the Morin transition temperature (T_M), moments of any two magnetic sublattices are antiparallel and aligned along the rhombohedra [111] axis and above morin transition temperature (T_M), moments lie in the basal plane, resulting in a weak net magnetic moment [58]. The polymorphs of Fe_2O_3 and their nanoparticles can in general, be synthesized through natural and synthetic thermal transformations of iron-bearing materials in an oxidizing atmosphere [113]. For instance, iron bacteria coated with ferrihydrite is known to precipitate iron as α-Fe_2O_3 [86]. A large host of literature exists on synthetic α-Fe_2O_3 with crystallographic surfaces, tunable through synthesis parameters such as temperature, pressure, additives, pH etc [14]. Many of these variables are also known to affect the sorption reactivity, surface area, particle size, point of zero charge and color [14]. These variables also affect size and shape of the particles [47]. For instance, a surfactant mediated hydrothermal synthesis results in uniform nanocrystalline α-Fe_2O_3[55].

The iron oxide polymorphs, more particularly, their nanoparticles, have unique magnetic, catalytic, optical, sorption and other properties [61]. Decreasing size to less than 20 nm, makes α-Fe_2O_3 nanoparticles smaller than a single magnetic domain and shows superparamagnetic behavior [16,82]. Decrease in size also brings down the T_M and T_N, with T_M completely vanishing between 8 – 20 nm [110]. Above 950 K (T_N of α-Fe_2O_3), α-Fe_2O_3 loses its magnetic ordering and becomes a paramagnet [61].

Nanoscale particles of α-Fe$_2$O$_3$ have been reported to trigger different toxicological reaction pathways rather than microscale particles [6].

13.2.4 SIZE TUNING IN HEMATITE

It is a big challenge to develop simple and reliable synthesis strategies for nanostructures with controlled morphologies, which strongly influences the application and specific properties of the metal oxide [29]. α-Fe$_2$O$_3$ nanostructures compared to the bulk, have been used in heterogeneous catalysis, alkylation of phenols, selective catalysis of cyclohexane oxidation, dye solar cells, electrolysis of water, gas sensors, photo catalysts, transistors, controlled drug delivery and so forth [1].α-Fe$_2$O$_3$ nanoparticles coated with gold nanoparticles are used as catalysts in CO oxidation [118]. Until now, only a few reports are concerned with the synthesis of α-Fe$_2$O$_3$ nanoparticles, especially through methodologies with a potential for scale up. Colloidal chemistry has offered scope to create nanostructures with high monodispersity, well-controlled sizes, unique shapes and complex structures. Compared to top-down approaches like lithography, bottom up approaches such as those based on colloidal chemistry have advantages to achieve small particle size and unmatched structural complexity[49]. Based on bottom up approach, α-Fe$_2$O$_3$ nanostructures have been synthesized in various shapes such as nanowires, nanorods, nanotubes, hollow fibere, nanorings, and cubes [1,33,34,107,109].

Methods of synthesis includes, but not limited to a) chemical precipitation such as mixing of ferric or ferrous salts with alkali to form goethite nanoparticles and subsequent conversion into α-Fe$_2$O$_3$ nanoparticles through thermal transformation (20 Fan et al. 2005), b) forced hydrolysis of FeCl$_3$ using HCl, resulting in direct transformation of amorphous iron oxides to crystalline α-Fe$_2$O$_3$ [106] microemulsion method based on ferrihydride and Fe(II) catalyst [24] through sol-gel process, such as with ethylene oxide and FeCl$_3$ as starting material [19,121]. However, the uniformity of size of the nanoparticles obtained by most of these methods was relatively poor.

Modification to these methods includes a) microwave irradiation based hydrolysis of iron salt using urea [15] or ethylenediamine [72], b) hydrothermal treatment followed by calcinations [11], c) alcoholysis of ferric ion under solvothermal condition [46], d) reverse microemulsion technique employing water, chloroform, 1-butanol and a surfactant [25], e) employing hydrothermal methods with amino acids as morphology control agent [105], and so on. These modifications in the synthesis routes can provide size reduction as well as improved physiochemical properties of the nanoparticles.

13.2.5 SHAPE TUNING IN HEMATITE

Materials with directional properties are opening new avenues in nanomaterial research. Synthesis of nanomaterial's with anisotropic morphologies so as to incorporate them directly into devices has enthused researchers worldwide [23]. Particle anisotropy produces great changes in properties that are difficult to obtain simply by size tuning of spherical nanoparticles. While 0D or spherical particles cannot be easily tuned owing to the confinement of electrons to the same extent in all dimensions, 1, 2, or 3D nanostructures where the electron motions are possible in more than one dimension are easily tunable [89]. Understanding the interactions between solids, interfacial reactions and kinetics and solution or vapor chemistry is essential in developing methodologies for particle shape control. Shape controllable synthesis of nanomaterials is of great interest and is actively pursued. Efforts on precise control of shape is poised towards potential scale-dependent applications in catalysis, drug delivery, active material encapsulation, ionic intercalation, light weight fillers, surface functionalization, energy storage, and so on [45]. Shape controlled synthesis involves processes such as seed-mediated [69] (synthesis of seed → growth of seed in the presence of shaping agent), polyol synthesis [42] (reduction of inorganic salt by polyol at elevated temperature), biological [27] (growth under constrained environments such as peptides), hydro/solvothermal synthesis [41] (synthesis in hot water/solvent in autoclave under high pressure), galvanic replacement reactions [17] (spontaneous reduction in the absence of an electric field), photochemical reactions [36] (use of radiolytic or photochemical methods for reduction), electrochemical methods [28] (bulk metal as sacrificial anode is oxidized and the metal cations migrate to cathode and get reduced in the presence of a stabilizing agent) and template mediated synthesis [23] (easy fabrication, low cost, high through-put and adaptability). Significantly, large numbers of these methods are best suited for noble metal nanoparticles, with limited scope for metal oxide nanoparticles, resulting in lower options for their morphology control.

Based on the above methods, researchers have developed many approaches for the preparation of α-Fe$_2$O$_3$ with different morphologies.

These include:

a) Hierarchical mesoporous microspheres,
b) Acicular nanoparticles,
c) Nanocubes,
d) Tube-in-tube structures,
e) Hollow spheres,
f) Cigar shapes,
g) Dendric forms,
h) Urchin like,
i) Flower like,
j) Nanorods,
k) Nanowires,
l) Nanobelts,
m) Nanochains,
n) Spindles, and so on. [32,44,59,60,66,71,100,114]

Most of these approaches advantageously utilize the phase transfer process that occurs when akaganeite (β-FeOOH) synthesized by hydrolysis of FeCl$_3$ is transformed to α-Fe$_2$O$_3$ by heat treatment.Introduction of an organic or inorganic additive during the phase transfer process enables the confinement of growth of the products in specific directions through selective adsorption on specific crystal surfaces [31]. For instance, sodium dodecyl benzene sulfonate can cooperate with Fe^{3+} through electrostatic interactions to form a yellow Fe^{3+}-DBS complex in water solution, which then decomposes during hydrothermal process. The free DBS groups adsorb preferentially on certain surface planes of the freshly formed nanocrystals to confine the growth progress, thus modifying the geometry and size [112]. However, as in the case of size, it is still a challenge to produce large homogeneous quantities of various shapes of nanoparticles with similar morphologies.Ionic liquid assisted hydrothermal synthesis of α-Fe$_2$O$_3$ in various morphologies has been reported by, [45] recently. In this method, an ionic liquid, 1-*n*-butyl-3-methylimidazolium chloride was used as soft template, in a two part reaction, whose mechanism was illustrated as:

$$CH_3COO^- + H_2O \Leftrightarrow CH_3COOH + OH^- \tag{1}$$

$$Fe^{3+} + 3OH^- \rightarrow Fe\frac{(OH)_3}{FeOOH} \xrightarrow[\text{Ionic liquid template}]{\text{hydrothermal condition}} \alpha - Fe_2O_3 \tag{2}$$

$$\alpha - Fe_2O_3 + 6CH_3COOH \Leftrightarrow 2Fe^{3+}6CH_3COO^- + 6H_2O \tag{3}$$

While in the second step, α-Fe$_2$O$_3$ nanoparticles were formed, in the third step, to reduce the total surface energy, the core dissolves in the presence of excess CH$_3$COOH, while a shell of α-Fe$_2$O$_3$ nanoparticles keeps forming on the dissolving core, resulting in hollow spheres.

13.2.6   TEMPLATE LESS SYNTHESIS

During the past few years, template-free methods such as hydrothermal/solvothermal synthesis, based on mechanisms such as oriented attachments [58,114], Kirkendall effects [57] and Ostwald ripening have been developed to synthesize nanoparticles. Hydrothermal synthesis based on FeCl$_3$ and CH$_3$COONH$_4$ has been reported. In this method by appropriate choice of reagent concentration, reaction temperature and time, nanocubes of Fe$_2$O$_3$ could be generated as per the following mechanism.

$$3CH_3COONH_4 + FeCl_3 \rightarrow Fe(CH_3COO)_3 + 3NH_4Cl$$
$$Fe(CH_3COO)_3 + 2H_2O \rightarrow Fe(OH)_2(CH_3COO) \downarrow +2(CH_3COOH)$$
$$Fe(OH)_2(CH_3COO) \rightarrow \beta - FeOOH + CH_2COOH$$
$$2\beta - FeOOH \rightarrow \alpha - Fe_2O_3 + H_2O$$

Overall reaction can be summarized as:

$$Fe^{3+}(l) \rightarrow Fe(OH)_2 CH_3COO(S) \rightarrow \beta - FeOOH(s) \rightarrow \alpha - Fe_2O_3(s)$$

Solvothermal reaction of ethanolic solution of iron acetate in the presence of L-Lycine to produce magnetite, which can subsequently be converted by heat treatment to α-Fe_2O_3 hollow spheres, has been reported. Such hollow spheres have applications in catalysis, gas sensors and lithium ion batteries [37]. However in many cases the synthesis of nanoparticles in the absence of template not provided good control over the morphology of nanoparticles.

13.2.7 TEMPLATE SYNTHESIS

Substrates that have their surfaces modified to have active sites which can induce nanoparticle deposition are generally called as templates. Examples include porous alumina, polymer nanotubes and patterned catalysts to control the oriented growth [9,26,87]. They can be smaller or larger in size than the nanoparticle deposited. It can also be considered as a scaffold wherein the particles can be arranged into a structure with a morphology that is complementary to that of the template [22]. Examples of templates include single molecules, microstructures or block copolymers. They are classified as soft and hard [76]. In a soft-template method, growth occurs by chemical or electrochemical reduction, usually in the presence of a surfactant or structure directing molecule.

The template governs growth of certain faces or structures. Polymer-surfactant complexes are promising templates. At first, the surfactant associated with the polymer chain form micellar aggregates at a critical association concentration (much lower than the critical micelle concentration of the surfactant in water). Secondly, both polymer and surfactant are good crystal stabilizers, preventing their aggregation and at the same time connecting them together into loose flocs through a bridging mechanism [43].

Finally, through electrostatic, hydrophobic and van der Waals interactions between polymer and the surfactant, stable structures are generated, which serve as ideal soft templates, providing monodisperse nanoparticles. Soft template methods include electrochemical reduction, seeding followed by chemical reduction, redox reactions, selective etching etc.

Soft-template technique offers advantage of scalability [39]. In hard-template method, a porous membrane of inorganic or polymeric material serves as a rigid mold for chemical or electrochemical replication of structure. This method provides an easy manner for production of 1-D nanostructures, but with difficulties of scale up. Hard templates such as silica or carbon spheres are also ideal for synthesis of hollow structures (11 Chen et al. 2003). Classical examples where the template enables the control of morphology of α-Fe_2O_3 nanoparticles can be found in literature (Table 1).

TABLE 1   Examples of template enabled morphology control in α-Fe_2O_3 nanoparticles

Template	Features
Carbon nanotubes [58]	Nanochains – weakly ferromagnetic in room temperature, antiferromagnetic below TM
Amino acids [8]	Nanocubes – double hydrophilic functional groups (-NH2 and –COOH to control growth
Carbon spheres [30]	Hollow spheres to nanocups – variable magnetic properties – calcination before gel stage results in spheres and calcination after thixotropic gel formation in cups.
Sulfonatedpolystyrene microspheres [116]	large scale monodisperse urchin - like hollow microspheres - high remanent magnetization
Anionic surfactant as a rod-like template [63]	Porous structure - weak ferromagnetic behavior
TiO2 nanotubes	One-dimensional Fe2O3 – TiO2nanorod -nanotube arrays
Polycarbonate/Alumina [119]	Nanotubes to amorphous hydroxides
Ferritin protein cage [38]	Boiling aqueous solution by refluxing the ferritin protein cage/ferrihydrite composite
PVA [62]	Oriented growth
PEG [99]	Necklace shaped – sacrificial template
PEG + CTAB [85]	Shuttle like structures
ZnO nanowire arrays [53]	Iron oxide-based nanotube arrays – sacrificial template
MCM-48 type silica support and wide-pore silica gel [102]	Ordered porous structure
FDU-1 type cubic ordered mesoporous silica [65]	Sacrificial template
DNA [90]	Chain like structures - antiferro to ferromagnetic transition at a temperature similar to 240 K
Polymethyl methacrylate as imprint template [115]	Hierarchically mesoporous and macroporous hematite Fe2O3
Protonated triethylenetetramine [120]	Sacrificial template – nanoribbons
Greek wood and mill scale wastes [4]	Iron oxide ceramics
Cellulose films [54]	Sacrificial, looped structures, replicates template

However, there are also reports that the template synthesis results in nanoparticles with low crystalinity. From Table 1 it can be seen that the template is likely to be present as an impurity or as a part of a composite with α-Fe_2O_3, other than in cases where template is selectively sacrificed. Even when sacrificed, it would leave a residue on top of the nanoparticle.

13.2.8 GREEN SYNTHESIS

Green nanotechnology transforms existing processes and products to enhance environmental quality, reduce pollution and conserve natural and non-renewable resources. Nano-synthesis involving a) cost effective and non-toxic precursors, b) negligible quantities of carcinogenic reagents or solvents, c) few number of reagents, d) lesser number of reaction steps and hence lesser waste generation, reagent use and power consumption, e) little or no byproducts, and f) room temperature synthesis under ambient conditions are considered ideal under green nanotechnology principles. Such processes also need to be efficient in terms of scale-up [64]. Molten-salt synthesis method is one of the most versatile and cost effective green synthesis approaches. With NaCl as the reaction medium, large-scale production of single-crystalline α-Fe_2O_3 rhombohedra have been obtained from relatively polydisperse, polycrystalline and/or amorphous, commercially available starting precursor materials [81]. Reactor conditions where the autogenous pressure far exceeds ambient pressure, allows solvents to be brought to temperatures much above their boiling points. When chemical reactions are performed under such conditions, they are referred to as solvothermal synthesis. Reaction temperature, time, pH, solvent choice and concentration, additives, autoclave geometries etc. influence the size and shape of the synthesized nanoparticles [67]. Template assisted green synthesis is one of the environmentally progressive methodologies for nanoparticle synthesis. Most of these reactions can be run under ambient conditions, with reliable control over shape and dimensionality. A potential to develop generalized protocol for synthesis, capping and functionalization exists with template synthesis, which is more biologically relevant when biopolymers are employed as templates.

13.2.9 POLYSACCHARIDE TEMPLATES

Polysaccharides are a class of biopolymers with repeating units of mono- or disaccharides linked by glycosidic bonds. They can be linear or branched, have high variability of building block composition and physico-chemical properties [18]. They have predominant applications in the area of biomaterials [85]. For a nanotechnologist, they are attractive candidates as stabilizing agents, functionalization moieties, drug delivery vehicles and reducing cum capping agents in the synthesis of noble metal nanoparticles [56,97]. Simple monosaccharide such as sucrose is a ready source of carbon in the synthesis of carbide nanoparticles [111]. Fibre like morphology of cellulose ($1\rightarrow4$, glycoside linked β-glucose) is advantageously employed for replicating the fiber morphology on to the nanoparticles [117]. Similarly, the branched structure of dextran enables formation of nanowires [40]. Starch, by itself, is a complex glucose polymer with inherent molecular anisotropy and the same is transferrable to other materials with ease [95].

Alginate is composed of blocks of poly-guluronate and poly-mannuronate and their ratio and distribution varies with source of seaweed [93]. By introducing metal cations to a solution of alginate, metallic species are preferentially taken up by poly-G species leading to controlled nucleation and growth. The cocooning effect of alginate prevents nanoparticle coalescence leading to uniform and homogeneous dispersion of nanoparticles [78]. Finally, chitosan being cationic, forms stable complexes with anionic inorganics, which on calcination forms networked structures [77].

In a series of works the ability of polysaccharide to aid nanoparticle synthesis when employed as templates were reported. Polysaccharides with varying charge such as cationic chitosan, anionic pectin/alginate and neutral starch were chosen to form iron–polysaccharide complexes which were calcined to generate iron oxide nanoparticles. The crystallographic phases matched well with that of hematite (α-Fe$_2$O$_3$). While Dynamic light scattering measurements indicated an average intensity average diameter of around 270 nm, the SEM images suggested spherical morphology for the nanoparticles generated on chitosan, alginate and starch [75]. In the case of pectin, a linear aggregation of spherical nanoparticles resulting in a rod like morphology was observed. Narrow particle size distribution was observed with starch. At the concentration range investigated in that study, the spatial separation of iron(II) centres during the complexation process is expected to have provided the monodispersity to the nanoparticles.

Thermogravimetric analysis suggested that the iron-polysaccharide complexes were stable to degradation at high temperature, when compared to polysaccharides alone and that they were not completely removed even at temperatures of 800°C, leading to the presence of a residual carbon shell around the nanoparticle core. The authors by employing the Peniche method estimated the percentage of nanoparticle content in the core-shell as 5.4%. The low size and monodispersity of starch templated nanoparticles, conferred a higher level of water dispersible character when compared to other nanoparticles. The nanoparticles were not cytotoxic in the concentration range of 50 to 200 µM, possibly due to the carbonaceous shell. The nanoparticles were easily dispersed into medium employed in surface coating, leather finishing and plastic coloration, where a uniform distribution of the color at low concentrations of the nanoparticles was observed. The ability of the nanoparticles to disperse well into polymeric matrices, thus conferring thermal stability to the same was found advantageous in automobile coatings. Dispersibility was attributed to the carbon coating of polysaccharide template over the nanoparticle.

The study was also extended to CoO nanoparticles [73] employing a sacrificial starch template. This method provided uniform shape and size, and was easy to perform at bulk levels. The size distribution of the nanoparticles was in the range of 15-30 nm, with a crystallite size of 18 nm. The low size and aggregation free character of the nanoparticles resulted in a NIR reflectance of above 75%, in spite of a black color (where NIR reflectance values are expected to be less than 30%). This provided feasibility for employing CoO nanoparticles as components of cool coatings on leather, automobile surfaces, and so on.

With an aim to further reduce the size of the nanoparticles and to prove that the spatial distribution between the iron(II) centers was responsible for the aggregation free synthesis of nanoparticles, the nanoparticles are synthesized on green templates such as chitosan polyion and blended films. Through electrostatic interaction between cationic chitosan and anionic alginate/pectin, self-assembled films were prepared. Similarly chitosan supported starch films were also obtained. Iron(II) in solution interacted with the functional groups in the film, following their adsorption on to polysaccharide films, which were subsequently calcined. Based on pores and the pore distribution, aligned α-Fe$_2$O$_3$ nanostructures were observed in the case of chitosan-alginate film template, rhombohedra shaped nanoparticle in the case of chitosan-pectin and spherical nanoparticles in the case of chitosan-starch film template. The weight percentage of the nanoparticles in core-shell dropped to 2.0%, indicating a reduction in size to as low as 2 nm (in the case alginate film). Here again the biocompatibility of the nanoparticles was established using MTT assay.

The above two methods of synthesis of hematite nanoparticles on polysaccharide template alone and chitosan polyion and blended films were high energy consuming calcination reactions, where after the template removal, potential for aggregation cannot be ruled out. The synthesized nanoparticles by the above methods had also demonstrated super paramagnetic behavior ideal for biological applications (chitosan-alginate film template). The synthesis methodology was modified to a reflux process as against calcination so as not to remove the polysaccharide template by calcination. Nanoparticles were found to grow into rhombohedra with reflux duration. Possibly based on the ions adsorbed from the medium, the nanoparticles demonstrated fluorescence properties and were found ideal for stabilizing collagen in solution. Starch mediated interaction with amino acids in collagen, resulting in H-bonded networks provided a 3.1°C increase in denaturation temperature, over the native collagen (37.2°C).

The study was extended to cobalt ferrite, where incorporation of a surfactant such as CTAB, provided an opportunity for the assembly of the nanoparticles into flower shaped nanostructures with a particle size of around 25 ± 3 nm. This method also provided an opportunity to systematically study its morphology, crystallinity, particle size, magnetic properties, biocompatibility, cytotoxicity and their biomedical applications such a contrast agents for MRI. The salient observation from this study was that the assembling of the nanoparticles into nanoflowers was essential to utilize them as T_2 contrast for MR imaging, while the nanoparticles in the absence of assembling process were neither a T_1 nor T_2 contrast agents. The reflux method of synthesis of anisotropic ferrite nanoparticles with uniform size and shape was found to be simple, low in reaction temperature, high yielding and low in cost of inorganic precursors.

13.2.10 SURFACE MODIFICATION OF HEMATITE NANOPARTICLES

Unique and advanced properties observed in nanoparticles are usually associated with bare (uncoated) nanoparticles. However, bare nanoparticles possess excessive surface energy and should be protected by way of capping or surface functionalization. In liquid media, unprotected metal and metal oxide nanoparticles are thermodynamically unstable and tend to spontaneously coalesce. In nanostructured hematite materials, long range magnetic dipole interactions can have strong influence on magnetic dynamics. When in close proximity, exchange bias which manifests as shift in hysteresis curves after field cooling, is generally observed. In α-Fe$_2$O$_3$, suppression of superpara-magnetic relaxation is observed when particles are in close proximity, which is more in the case of uncoated particles than coated particles [68].

A large number of synthesis methods have been developed to prevent the aggregation of nanoparticles. These include sonication by ultrasound, capping with polymers and surfactants etc. In several instances, equilibrium is established between nanoparticles, aggregates, dispersed particles and reaggregated particles, resulting in a polydispersed system of particles. One of the drawbacks of these methods is their limited stability in aqueous solutions. Particles with a polymer coating, more so polymeric shell have more stability against aggregation because of large decrease in surface energy. Functionalization of hematite core with polymeric materials can result in three types of structures, *viz.*, core-shell, matrix and shell-Core-shell. The choice of solvent for functionalization plays a crucial role in achieving sufficient repulsive interactions to prevent agglomeration. For instance, when functionalized with surfactants, the functionalized nanoparticles can be divided as oil-soluble, water-soluble and amphiphilic. This division is based on the nature of interaction of the surfactant with the solvent employed.

Accordingly with water as solvent, fatty acid or alkyl phenol functionalized nanoparticles are oil-soluble, polyol or lysine coated particles are water-soluble and sulfuric lysine coated particles are amphiphilic. Stabilization and functionalization of nanoparticles in an environmentally benign manner are among one of the goals of green nanotechnology. Such agents, which are from a natural/renewable source, can function as reducing/capping/stabilizing agents, can be effectively used for functionalization. Interaction of nanoparticles with variety of compounds found in nature which provides good control over the size and morphology are highly desirable for various applications of nanotechnology [104].

One of the best ways by which aggregation can be avoided is surface passivation [3]. The largest class of compounds explored in nanosynthesis for passivation is biopolymers. They are produced by living organisms and consist of simple biological compounds. They are generally renewable, non-toxic and biodegradable [92,101]. Metal and metal oxide nanoparticles when dispersed into biopolymers can overcome many of the short comings of bare nanoparticles such as aggregation and reduction in surface area to volume ratio, without affecting the properties of parent nanoparticles. Appropriate choice of biopolymers would also provide functional groups and enhancement of properties [91]. Donor or acceptor species that are bound to the hematite nanoparticle surface such as polysaccharides are finding increasing relevance in surface passivation. A detailed biotechnological perspective on the applications of iron oxide nanoparticles modified with polysaccharides can be found in the work of 18 Dias et al. 2011. Polysaccharides, can be neutral (agarose, dextran, pullalan, and starch), negative (alginate, carrageenans, gum Arabic and heparin) or positively charged (chitosan). They carry functional groups such as OH, COO$^-$, OSO$_3^-$, COO$^-$, and NH$_3^+$. They act as recognition markers in several biological processes (80 Park et al. 2008). Chemical co-precipitation is the preferred method of *in-situ* coating, while encapsulation, microemulsion, hydrothermal treatment, covalent bonding, adsorption or sonication is the preferred post-synthesis coating methods. The biomedical applications of such coated iron oxide nanoparticles include cell tracking, drug delivery, detection of emboli, liver tumor, liver lesion etc. by MRI, cardiovascular applications and hyperthermia.

REFERENCES-CASE STUDY I

1. Agarwala, S., Lim, Z. H., Nicholson, E., and Ho, G. W. Probing the morphology-device relation of Fe_2O_3 nanostructures towards photovoltaic and sensing applications. *Nanoscale*, **4**(1), 194–205 (2012).
2. Aitken, R. J., Chaudhry, M. Q., Boxall, A. B. A., and Hull, M. Manufacture and use of nanomaterials: current status in the UK and global trends. *Occupational Medicine-Oxford*, **56**(5), 300–306 (2006).
3. Bakunin, V. N., Suslov, A. Y., Kuzmina, G. N., and Parenago, O. P. Synthesis and application of inorganic nanoparticles as lubricant components - a review. *Journal of Nanoparticle Research*, **6**(2–3), 273–284 (2004).
4. Bantsis, G., Betsiou, M., Bourliva, A., Yioultsis, T., and Sikalidis, C. Synthesis of porous iron oxide ceramics using Greek wooden templates and mill scale waste for EMI applications. *Ceramics International*, **38**(1), 721–729 (2012).
5. Berkowitz, A. E., Mitchell, J. R., Carey, M. J., Young, A. P., Zhang, S., Spada, F. E., Parker, F. T., Hutten, A., and Thomas, G. Giant magnetoresistance in heterogeneous Cu-Co alloys. *Physical Review Letters*, **68**(25), 3745–3748 (1992).
6. Bhattacharya, K., Hoffmann, E., Schins, R. F. P., Boertz, J., Prantl, E. M., Alink, G. M., Byrne, H. J., Kuhlbusch, T. A. J., Rahman, Q., Wiggers, H., Schulz, C., and Dopp, E. Comparison of Micro- and Nanoscale Fe^{3+}-Containing (Hematite) Particles for Their Toxicological Properties in Human Lung Cells In Vitro. *Toxicological Sciences*, **126**(1), 173–182 (2012).
7. Bucher, J. P., Douglass, D. C., and Bloomfield, L. A. Magnetic properties of free cobalt clusters. *Physical Review Letters*, **66**(23), 3052–3055 (1991).
8. Cao, H. Q., Wang, G. Z., Warner, J. H., and Watt, A. A. R. Amino-acid-assisted synthesis and size-dependent magnetic behaviors of hematite nanocubes. *Applied Physics Letters*, **92**(1), 3 (2008).
9. Cao, H. Q., Xu, Z., Sang, H., Sheng, D., and Tie, C. Y. Template Synthesis and Magnetic Behavior of an Array of Cobalt Nanowires Encapsulated in Polyaniline Nanotubules. *Advanced Materials*, **13**(2), 121–123 (2001).
10. Cao, M., Liu, T., Gao, S., Sun, G., Wu, X., Hu, C., and Wang, Z. L. Single-Crystal Dendritic Micro-Pines of Magnetic α-Fe_2O_3: Large-Scale Synthesis, Formation Mechanism, and Properties. *Angewandte Chemie International Edition*, **44**(27), 4197–4201 (2005).
11. Chen, D., Chen, D., Jiao, X., and Zhao, Y. Hollow-structured hematite particles derived from layered iron (hydro)oxyhydroxide-surfactant composites. *Journal of Materials Chemistry*, **13**(9), 2266–2270 (2003).
12. Chen, J. S., Zhu, T., Yang, X. H., Yang, H. G., and Lou, X. W. Top-Down Fabrication of a-Fe_2O_3 Single-Crystal Nanodiscs and Microparticles with Tunable Porosity for Largely Improved Lithium Storage Properties. *Journal of the American Chemical Society*, **132**(38), 13162–13164 (2010).
13. Cherepy, N. J., Liston, D. B., Lovejoy, J. A., Deng H., and Zhang, J. Z. Ultrafast Studies of Photoexcited Electron Dynamics in α- and β-Fe_2O_3 Semiconductor Nanoparticles. *The Journal of Physical Chemistry B*, **102**(5), 770–776 (1998).
14. Cornell, R. M. and Schwertmann, U. *The Iron Oxides: Structure, Properties, Reactions, Occurrences and Uses*. John Wiley & Sons (2007).

15. Daichuan, D., Pinjie, H., and Shushan, D. Preparation of uniform a-FeO(OH) colloidal particles by hydrolysis of ferric salts under microwave irradiation. *Materials Research Bulletin*, **30**(5), 537–541 (1995).

16. del Monte, F., Morales, M. P., Levy, D., Fernandez, A., Oca, M., Roig, A., Molins, E., O'Grady, K., and Serna, C. J. Formation of a-Fe$_2$O$_3$ Isolated Nanoparticles in a Silica Matrix. *Langmuir*, **13**(14), 3627–3634 (1997).

17. Dement'eva, O. V. and Rudoy, V. M. Colloidal synthesis of new silver-based nanostructures with tailored localized surface plasmon resonance. *Colloid Journal*, **73**(6), 724–742 (2011).

18. Dias, A. M. G. C., Hussain, A., Marcos, A. S., and Roque, A. C. A. A biotechnological perspective on the application of iron oxide magnetic colloids modified with polysaccharides. *Biotechnology Advances*, **29**(1), 142–155 (2011).

19. Dong, W. and Zhu, C. Use of ethylene oxide in the sol-gel synthesis of α-Fe$_2$O$_3$ nanoparticles from Fe(iii) salts. *Journal of Materials Chemistry*, **12**(6), 1676–1683 (2002).

20. Fan, H., Song, B., Liu, J., Yang, Z., and Li, Q. Thermal formation mechanism and size control of spherical hematite nanoparticles. *Materials Chemistry and Physics*, **89**(2–3), 321–325 (2005).

21. Fullerton, E. E., Jiang, J. S., Sowers, C. H., Pearson, J. E., and Bader, S. D. Structure and magnetic properties of exchange-spring Sm-Co/Co superlattices. *Applied Physics Letters*, **72**(3), 380–382 (1998).

22. Grzelczak, M., Vermant, J., Furst, E. M., and Liz-Marzaìn, L. M. Directed Self-Assembly of Nanoparticles. *ACS Nano*, **4**(7), 3591–3605 (2010).

23. Hall, S. R. Biotemplated syntheses of anisotropic nanoparticles. *Proceedings of the Royal Society a-Mathematical Physical and Engineering Sciences*, **465**(2102), 335–366 (2009).

24. Han, L. H., Liu, H., and Wei, Y. In situ synthesis of hematite nanoparticles using a low-temperature microemulsion method. *Powder Technology*, **207**(1â€"3), 42–46 (2011).

25. Housaindokht, M. R. and Pour, A. N. Precipitation of hematite nanoparticles via reverse microemulsion process. *Journal of Natural Gas Chemistry*, **20**(6), 687–692 (2011).

26. Huang, M. H., Mao, S., Feick, H., Yan, H. Q., Wu, Y. Y., Kind, H., Weber, E., Russo, R., and Yang, P. D. Room-temperature ultraviolet nanowire nanolasers. *Science*, **292**(5523), 1897–1899 (2001).

27. Huang, X., Neretina, S., and El-Sayed, M. A. Gold Nanorods: From Synthesis and Properties to Biological and Biomedical Applications. *Advanced Materials*, **21**(48), 4880–4910 (2009).

28. Huang, X., Qi, X., Huang, Y., Li, S., Xue, C., Gan, C. L., Boey, F., and Zhang, H. Photochemically Controlled Synthesis of Anisotropic Au Nanostructures: Platelet-like Au Nanorods and Six-Star Au Nanoparticles. *ACS Nano*, **4**(10), 6196–6202 (2010).

29. Huber, D. L. Synthesis, Properties, and Applications of Iron Nanoparticles. *Small*, **1**(5), 482–501 (2005).

30. Jagadeesan, D., Mansoori, U., Mandal, P., Sundaresan, A., and Eswaramoorthy, M. Hollow Spheres to Nanocups: Tuning the Morphology and Magnetic Properties of Single-Crystalline α-Fe$_2$O$_3$ Nanostructures. *Angewandte Chemie International Edition*, **47**(40), 7685–7688 (2008).

31. Jia, B. and Gao, L. Growth of Well-Defined Cubic Hematite Single Crystals: Oriented Aggregation and Ostwald Ripening. *Crystal Growth & Design*, **8**(4), 1372–1376 (2008).

32. Jia, C. J., Sun, L. D., Yan, Z. G., Pang, Y. C., You, L. P., and Yan, C. H. Iron Oxide Tube-in-Tube Nanostructures. *The Journal of Physical Chemistry C*, **111**(35), 13022–13027 (2007).

33. Jia, C. J., Sun, L. D., Yan, Z. G., You, L. P., Luo, F., Han, X. D., Pang, Y. C., Zhang, Z., and Yan, C. H. Single-Crystalline Iron Oxide Nanotubes. *Angewandte Chemie International Edition*, **44**(28), 4328–4333 (2005).

34. Jones, N. O., Reddy, B. V., Rasouli, F., and Khanna, S. N. Structural growth in iron oxide clusters: Rings, towers, and hollow drums. *Physical Review B*, **72**(16), 165411 (2005).

35. Jung, S., Watkins, B., DeLong, L., Ketterson, J. B., and Chandrasekhar, V. Ferromagnetic resonance in periodic particle arrays. *Physical Review B*, **66**(13), 132401 (2002).

36. Khomutov, G. B. Interfacially formed organized planar inorganic, polymeric and composite nanostructures. *Advances in Colloid and Interface Science*, **111**(1–2), 79–116 (2004).

37. Kim, H. J., Choi, K. I., Pan, A., Kim, I. D., Kim, H. R., Kim, K. M., Na, C. W., Cao, G., and Lee, J. H. Template-free solvothermal synthesis of hollow hematite spheres and their applications in gas sensors and Li-ion batteries. *Journal of Materials Chemistry*, **21**(18), 6549–6555 (2011).

38. Klem, M. T., Young, M., and Douglas, T. Biomimetic synthesis of photoactive alpha-Fe_2O_3 templated by the hyperthermophilic ferritin from Pyrococus furiosus. *Journal of Materials Chemistry*, **20**(1), 65–67 (2010).

39. Kline, T. R., Tian, M., Wang, J., Sen, A., Chan, M. W. H., and Mallouk, T. E. Template-Grown Metal Nanowires. *Inorganic Chemistry*, **45**(19), 7555–7565 (2006).

40. Kong, R., Yang, Q., and Tang, K. A Facile Route to Silver Nanowires. *Chemistry Letters*, **35**(4), 402–403 (2006).

41. Kumar, S. and Nann, T. Shape control of II-VI semiconductor nanomateriats. *Small*, **2**(3), 316–329 (2006).

42. Lee, G., Cho, Y. S., Park, S., and Yi, G. R. Synthesis and assembly of anisotropic nanoparticles. *Korean Journal of Chemical Engineering*, **28**(8), 1641–1650 (2011).

43. Leontidis, E., Kyprianidou-Leodidou, T., Caseri, W., Robyr, P., Krumeich, F., and Kyriacou, K. C. From Colloidal Aggregates to Layered Nanosized Structures in Polymer-Surfactant Systems. 1. Basic Phenomena. *The Journal of Physical Chemistry B*, **105**(19), 4133–4144 (2001).

44. Li, L., Chu, Y., Liu, Y., and Dong, L. Template-Free Synthesis and Photocatalytic Properties of Novel Fe2O3 Hollow Spheres. *The Journal of Physical Chemistry C*, **111**(5), 2123–2127 (2007).

45. Lian, J., Duan, X., Ma, J., Peng, P., Kim, T., and Zheng, W. Hematite (α-Fe_2O_3) with Various Morphologies: Ionic Liquid-Assisted Synthesis, Formation Mechanism, and Properties. *ACS Nano*, **3**(11), 3749–3761 (2009).

46. Lian, S. Y., Li, H. T., He, X. D., Kang, Z. H., Liu, Y., and Lee, S. T. Hematite homogeneous core/shell hierarchical spheres: Surfactant-free solvothermal preparation and their improved catalytic property of selective oxidation. *Journal of Solid State Chemistry*, **185**, 117–123 (2012).

47. Liang, X., Wang, X., Zhuang, J., Chen, Y. T., Wang, D. S., and Li, Y. D. Synthesis of nearly monodisperse iron oxide and oxyhydroxide nanocrystals. *Advanced Functional Materials*, **16**(14), 1805–1813 (2006).

48. Lin, X. M. and Samia, A. C. S. Synthesis, assembly and physical properties of magnetic nanoparticles. *Journal of Magnetism and Magnetic Materials*, **305**(1), 100–109 (2006a).

49. Lin, X. M. and Samia, A. C. S. Synthesis, assembly and physical properties of magnetic nanoparticles. *Journal of Magnetism and Magnetic Materials*, **305**(1), 100–109 (2006b).

50. Lin, Y. J., Xu, Y., Mayer, M. T., Simpson, Z. I., McMahon, G., Zhou, S., and Wang, D. W. Growth of p-Type Hematite by Atomic Layer Deposition and Its Utilization for Improved Solar Water Splitting. *Journal of the American Chemical Society*, **134**(12), 5508–5511 (2012).

51. Liu, B. and Zeng, H. C. Fabrication of ZnO Dandelions via a Modified Kirkendall Process. *Journal of the American Chemical Society*, **126**(51), 16744–16746 (2004a).

52. Liu, B. and Zeng, H. C. Mesoscale Organization of CuO Nanoribbons. *Journal of the American Chemical Society*, **126**(26), 8124–8125 (2004b).

23. Liu, J. P., Li, Y. Y., Fan, H. J., Zhu, Z. H., Jiang, J., Ding, R. M., Hu, Y. Y., and Huang, X. T. Iron Oxide-Based Nanotube Arrays Derived from Sacrificial Template-Accelerated Hydrolysis: Large-Area Design and Reversible Lithium Storage. *Chemistry of Materials*, **22**(1), 212–217 (2010).

54. Liu, S., Tao, D., and Zhang, L. Cellulose scaffold: A green template for the controlling synthesis of magnetic inorganic nanoparticles. *Powder Technology*, **217**(0), 502–509 (2012).

55. Liu, X. H., Qiu, G. Z., Yan, A. G., Wang, Z., and Li, X. G. Hydrothermal synthesis and characterization of α-FeOOH and α-Fe$_2$O$_3$ uniform nanocrystallines. *Journal of Alloys and Compounds*, **433**(1–2), 216–220 (2007).

56. Liu, Z. H., Jiao, Y. P., Wang, Y. F., Zhou, C. R., and Zhang, Z. Y. Polysaccharides-based nanoparticles as drug delivery systems. *Advanced Drug Delivery Reviews*, **60**(15), 1650–1662 (2008).

57. Lu, H. B., Liao, L., Li, J. C., Shuai, M., and Liu, Y. L. Hematite nanochain networks: Simple synthesis, magnetic properties, and surface wettability. *Applied Physics Letters*, **92**(9), 3 (2008).

58. Lu, H. M. and Meng, X. K. Morin Temperature and Neĩel Temperature of Hematite Nanocrystals. *The Journal of Physical Chemistry C*, **114**(49), 21291–21295 (2010).

59. Lv, B., Liu, Z., Tian, H., Xu, Y., Wu, D., and Sun, Y. Single-Crystalline Dodecahedral and Octodecahedralα-Fe$_2$O$_3$ Particles Synthesized by a Fluoride Anion–Assisted Hydrothermal Method. *Advanced Functional Materials*, **20**(22), 3987–3996 (2010).

60. Lv, B., Xu, Y., Wu, D., and Sun, Y. Single-crystal [small alpha]-Fe2O3 hexagonal nanorings: stepwise influence of different anionic ligands (F- and SCN- anions). *Chemical Communications*, **47**(3), 967–969 (2011).

61. Machala, L., Zboril, R., and Gedanken, A. Amorphous Iron(III) OxideA Review. *The Journal of Physical Chemistry B*, **111**(16), 4003–4018 (2007).

62. Mahmoudi, M., Simchi, A., Imani, M., Stroeve, P., and Sohrabi, A. Templated growth of superparamagnetic iron oxide nanoparticles by temperature programming in the presence of poly(vinyl alcohol). *Thin Solid Films*, **518**(15), 4281–4289 (2010).

63. Mandal, S. and Muller, A. H. E. Facile route to the synthesis of porous α-Fe$_2$O$_3$ nanorods. *Materials Chemistry and Physics*, **111**(2–3), 438–443 (2008).

64. Mao, Y., Park, T. J., Zhang, F., Zhou, H., and Wong, S. S. Environmentally Friendly Methodologies of Nanostructure Synthesis. *Small*, **3**(7), 1122–1139 (2007).

65. Martins, T. S., Mahmoud, A., da Silva, L. C. C., Cosentino, I. C., Tabacniks, M. H., Matos, J. R., Freire, R. S., and Fantini, M. C. A. Synthesis, characterization and catalytic evaluation of cubic ordered mesoporous iron-silicon oxides. *Materials Chemistry and Physics*, **124**(1), 713–719 (2010).

66. Matijevic, E. Preparation and properties of uniform size colloids. *Chemistry of Materials*, **5**(4), 412–426 (1993).

67. Michailovski, A. and Patzke, G. R. Hydrothermal Synthesis of Molybdenum Oxide Based Materials: Strategy and Structural Chemistry. Chemistry. *A European Journal*, **12**(36), 9122–9134 (2006).

68. Morup, S., Hansen, M. F., and Frandsen, C. Magnetic interactions between nanoparticles. *Beilstein Journal of Nanotechnology*, **1**, 182–190 (2010).

69. Murphy, C. J., San, T. K., Gole, A. M., Orendorff, C. J., Gao, J. X., Gou, L., Hunyadi, S. E., and Li, T. Anisotropic metal nanoparticles: Synthesis, assembly, and optical applications. *Journal of Physical Chemistry B*, **109**(29), 13857–13870 (2005).

70. Navrotsky, A., Mazeina, L., and Majzlan, J. Size-driven structural and thermodynamic complexity in iron oxides. *Science*, **319**(5870), 1635–1638 (2008).

71. Ngo, A. T. and Pileni, M. P. Assemblies of cigar-shaped ferrite nanocrystals: orientation of the easy magnetization axes. *Colloids and Surfaces a-Physicochemical and Engineering Aspects*, **228**(1–3), 107–117 (2003).

72. Ni, H., Ni, Y. H., Zhou, Y. Y., and Hong, J. M. Microwave-hydrothermal synthesis, characterization and properties of rice-like alpha-Fe2O3 nanorods. *Materials Letters*, **73**, 206–208 (2012).

73. Nidhin, M., Sreeram, K. J., and Nair, B. U. Green synthesis of rock salt CoO nanoparticles for coating applications by complexation and surface passivation with starch. *Chemical Engineering Journal*, **185–186**(0), 352–357 (2012a).

74. Nidhin, M., Sreeram, K. J., and Nair, B. U. Polysaccharide films as templates in the synthesis of hematite nanostructures with special properties. *Applied Surface Science*, **258**(12), 5179–5184 (2012b).

75. Nidhin, M., Indumathy, R., Sreeram, K. J., and Nair, B. U. Synthesis of iron oxide nanoparticles of narrow size distribution on polysaccharide templates. *Bulletin of Material Science*, **31**, 93–96 (2008).

76. Nie, Z., Petukhova, A., and Kumacheva, E. Properties and emerging applications of self-assembled structures made from inorganic nanoparticles. *Nat Nano*, **5**(1), 15–25 (2010).

77. Ogawa, K., Oka, K., and Yui, T. X-ray study of chitosan-transition metal complexes. *Chemistry of Materials*, **5**(5), 726–728 (1993).

78. Ozin, G. A., Arsenault, A. C., Cademartiri, L., and Chemistry, R. S. O. Nanochemistry: a chemical approach to nanomaterials. *Royal Society of Chemistry* (2009).

79. Park, S. J., Kim, S., Lee, S., Khim, Z. G., Char, K., and Hyeon, T. Synthesis and Magnetic Studies of Uniform Iron Nanorods and Nanospheres. *Journal of the American Chemical Society*, **122**(35), 8581–8582 (2000).

80. Park, S., Lee, M. R., and Shin, I. Chemical tools for functional studies of glycans. *Chem Soc Rev*, **37**(8), 1579–91 (2008).

81. Park, T. J. and Wong, S. S. As-Prepared Single-Crystalline Hematite Rhombohedra and Subsequent Conversion into Monodisperse Aggregates of Magnetic Nanocomposites of Iron and Magnetite. *Chemistry of Materials*, **18**(22), 5289–5295 (2006).

82. Parker, F. T., Foster, M. W., Margulies, D. T., and Berkowitz, A. E. Spin canting, surface magnetization, and finite-size effects in a-Fe$_2$O$_3$ particles. *Physical Review B*, **47**(13), 7885–7891 (1993).

83. Perez, J. M., Simeone, F. J., Saeki, Y., Josephson, L., and Weissleder, R. Viral-Induced Self-Assembly of Magnetic Nanoparticles Allows the Detection of Viral Particles in Biological Media. *Journal of the American Chemical Society*, **125**(34), 10192–10193 (2003).

84. Quinn, R. K., Nasby, R. D., and Baughman, R. J. Photoassisted electrolysis of water using single crystal a-Fe$_2$O$_3$ anodes. *Materials Research Bulletin*, **11**(8), 1011–1017 (1976).

85. Rinaudo, M. Main properties and current applications of some polysaccharides as biomaterials. *Polymer International*, **57**(3), 397–430 (2008).

86. Robbins, E. I. and Iberall, A. S. Mineral remains of early life on earth -on mars. *Geomicrobiology Journal*, **9**(1), 51–66 (1991).

87. Routkevitch, D., Bigioni, T., Moskovits, M., and Xu, J. M. Electrochemical Fabrication of CdS Nanrrowire Aays in Porous Anodic Aluminum Oxide Templates. *The Journal of Physical Chemistry*, **100**(33), 14037–14047 (1996).

88. Rudzka, K., Delgado, Ã. n. V., and Viota, J. n. L. Maghemite Functionalization for Antitumor Drug Vehiculization. *Molecular Pharmaceutics* (2012).

89. Sajanlal P. R., Sreeprasad T. S., Samal A. K. and Pradeep T. (2011). Anisotropic nanomaterials: structure, growth, assembly, and functions

90. Sarkar D., Mandal K. and Mandal M. (2011). Synthesis of Chainlike alpha-Fe2O3 Nanoparticles in DNA Template and Their Characterization. Nanoscience and Nanotechnology Letters, 3(2), 170–174.

91. Sarkar S., Guibal E., Quignard F. and SenGupta A. K. (2012). Polymer-supported metals and metal oxide nanoparticles: synthesis, characterization, and applications. Journal of Nanoparticle Research, 14(2), 24.

92. Satyanarayana K. G., Arizaga G. G. C. and Wypych F. (2009). Biodegradable composites based on lignocellulosic fibers-An overview. Progress in Polymer Science, 34(9), 982–1021.

93. Schnepp Z. A. C., Wimbush S. C., Mann S. and Hall S. R. (2008). Structural Evolution of Superconductor Nanowires in Biopolymer Gels. Advanced Materials, 20(9), 1782–1786.

94. Schwertmann U. and Cornell R. M. (2000). Iron Oxides in the Laboratory: Preparation and Characterization. Wiley-VCH

95. Shi W., Liang P., Ge D., Wang J. and Zhang Q. (2007). Starch-assisted synthesis of polypyrrole nanowires by a simple electrochemical approach. Chemical Communications, (23), 2414–2416.

96. Sivula K., Zboril R., Le Formal F., Robert R., Weidenkaff A., Tucek J., Frydrych J. and Graȋtzel M. (2010). Photoelectrochemical Water Splitting with Mesoporous Hematite Prepared by a Solution-Based Colloidal Approach. Journal of the American Chemical Society, 132(21), 7436–7444.

97. Sreeram K. J., Nidhin M. and Nair B. U. (2008). Microwave assisted template synthesis of silver nanoparticles. Bulletin of Materials Science, 31(7), 937–942.

98. Sreeram K. J., Nidhin M. and Nair B. U. (2009a). Synthesis of aligned hematite nanoparticles on chitosan-alginate films. Colloids and Surfaces B: Biointerfaces, 71(2), 260–267.

99. Sreeram K. J., Nidhin M. and Unni Nair B. (2011). Formation of necklace-shaped haematite nanoconstructs through polyethylene glycol sacrificial template technique. Journal of Experimental Nanoscience, 1–13.

100. Suber L., Fiorani D., Imperatori P., Foglia S., Montone A. and Zysler R. (1999). Effects of thermal treatments on structural and magnetic properties of acicular α-Fe$_2$O$_3$ nanoparticles. Nanostructured Materials, 11(6), 797–803.

101. Sundar S., Kundu J. and Kundu S. C. (2010). Biopolymeric nanoparticles. Science and Technology of Advanced Materials, 11(1).

102. Surowiec Z., Gac W. and Wiertel M. (2011). The Synthesis and Properties of High Surface Area Fe$_2$O$_3$ Materials. Acta Physica Polonica A, 119(1), 18–20.

103. Tartaj P., Morales M. P., Gonzalez-Carreno T., Veintemillas-Verdaguer S. and Serna C. J. (2005). Advances in magnetic nanoparticles for biotechnology applications. Journal of Magnetism and Magnetic Materials, 290–291, Part 1(0), 28–34.

104. Virkutyte J. and Varma R. S. (2011). Green synthesis of metal nanoparticles: Biodegradable polymers and enzymes in stabilization and surface functionalization. Chemical Science, 2(5), 837–846.

105. Wang G. H., Li W. C., Jia K. M. and Lu A. H. (2011). A Facile Synthesis of Shape and SizeControlled α-Fe$_2$O$_3$ Nanoparticles Through Hydrothermal Method. Nano, 6(5), 469–479.

106. Wang S.–B., Min Y.-L. and Yu S.-H. (2007). Synthesis and Magnetic Properties of Uniform Hematite Nanocubes. The Journal of Physical Chemistry C, 111(9), 3551–3554.

107. Wen X., Wang S., Ding Y., Wang Z. L. and Yang S. (2004). Controlled Growth of Large-Area, Uniform, Vertically Aligned Arrays of α-Fe2O3 Nanobelts and Nanowires. The Journal of Physical Chemistry B, 109(1), 215–220.

108. Wheeler D. A., Wang G., Ling Y., Li Y. and Zhang J. Z. (2012). Nanostructured hematite: synthesis, characterization, charge carrier dynamics, and photoelectrochemical properties. Energy & Environmental Science, 5(5), 6682–6702.

109. Woo K., Lee H. J., Ahn J. P. and Park Y. S. (2003). Sol–Gel Mediated Synthesis of Fe_2O_3 Nanorods. Advanced Materials, 15(20), 1761–1764.

110. Xue D. S., Gao C. X., Liu Q. F. and Zhang L. Y. (2003). Preparation and characterization of haematite nanowire arrays. Journal of Physics: Condensed Matter, 15(9), 1455.

111. Yang Z., Xia Y. and Mokaya R. (2004). High Surface Area Silicon Carbide Whiskers and Nanotubes Nanocast Using Mesoporous Silica. Chemistry of Materials, 16(20), 3877–3884.

112. Yanyan X., Shuang Y., Guoying Z., Yaqiu S., Dongzhao G. and Yuxiu S. (2011). Uniform hematite a-Fe_2O_3 nanoparticles: Morphology, size-controlled hydrothermal synthesis and formation mechanism. Mater Lett, 65(12), 1911-1914.

113. Zboril R., Mashlan M. and Petridis D. (2002). Iron(III) Oxides from Thermal Proces-sesSynthesis, Structural and Magnetic Properties, Mossbauer Spectroscopy Characterization, and Applicationsâ€ Chemistry of Materials, 14(3), 969–982.

114. Zeng S. Y., Tang K. B., Li T. W., Liang Z. H., Wang D., Wang Y. K., Qi Y. X. and Zhou W. W. (2008). Facile route for the fabrication of porous hematite nanoflowers: Its synthesis, growth mechanism, application in the lithium ion battery, and magnetic and photocatalytic properties. Journal of Physical Chemistry C, 112(13), 4836–4843.

115. Zhang X. J., Hirota R., Kubota T., Yoneyama Y. and Tsubaki N. (2011). Preparation of hierarchically meso-macroporous hematite Fe_2O_3 using PMMA as imprint template and its reaction performance for Fischer-Tropsch synthesis. Catalysis Communications, 13(1), 44–48.

116. Zhang Y. P., Chu Y. and Dong L. H. (2007). One-step synthesis and properties of urchin-like PS/alpha-Fe_2O_3 composite hollow microspheres. Nanotechnology, 18(43), 5.

117. Zheng Z., Huang, Ma, Zhang, Liu, LiuLiu Z., Wong K. W. and Lau W. M. (2007). Biomimetic Growth of Biomorphic $CaCO_3$ with Hierarchically Ordered Cellulosic Structures. Crystal Growth & Design, 7(9), 1912–1917.

118. Zhong Z., Ho J., Teo J., Shen S. and Gedanken A. (2007). Synthesis of Porous α-Fe_2O_3 Nanorods and Deposition of Very Small Gold Particles in the Pores for Catalytic Oxidation of CO. Chemistry of Materials, 19(19), 4776–4782.

119. Zhou H. and Wong S. S. (2008). A Facile and Mild Synthesis of 1-D ZnO, CuO, and α-Fe_2O_3 Nanostructures and Nanostructured Arrays. ACS Nano, 2(5), 944–958.

120. Zhou, Y. X., Yao, H. B., Yao, W. T., Zhu, Z., and Yu, S. H. Sacrificial Templating Synthesis of Hematite Nanochains from [Fe18S25](TETAH)14 Nanoribbons: Their Magnetic, Electrochemical, and Photocatalytic Properties. Chemistry-a European Journal, 18(16), 5073–5079 (2012).

121. Zhu, L. P., Xiao, H. M., Liu, X. M., and Fu, S. Y. Template-free synthesis and characterization of novel 3D urchin-like [small alpha]-Fe2O3 superstructures. Journal of Materials Chemistry, 16(19), 1794–1797 (2006).

13.3   CASE STUDY II: FORMATION OF POROUS NANOCRYSTALLINE TI-6AL-4V AND (AL, V)-FREE ALLOYS WITH COMPOSITION OF THE TI-15ZR-4NB AND TI-6ZR-4NB

13.3.1   INTRODUCTION

Formation of Ti-6Al-4V, Ti-15Zr-4Nb and Ti-6Zr-4Nb porous nanocrystalline bioalloys was described. The alloys were prepared by mechanical alloying followed by pressing, sintering and subsequent anodic electrochemical etching in 1M H_3PO_4 + 2% HF electrolyte at 10 V for the time of 30-300 min. We show that ultrafine structure improves etching process. The electrolyte penetrates sinters through the empty spaces and grain boundaries, which results in an effective grain removal and pore formation. The porosity of the Ti-15Zr-4Nb is larger than of the Ti-6Al-4V. The macropore diameter reaches up to 60 μm, while the average micropore size is in the range from 3.5 to 5.46 nm for Ti-15Zr-4Nb and Ti-6Al-4V electrochemically etched alloys, respectively.

The mechanism of the bioactive ceramic Ca-P layer formation on the porous etched surface was investigated independently. The Ca-P compounds were cathodically deposited at a range of different potentials (from −0.5 to −10 V), using a solution of 0.042M $Ca(NO_3)_2$ + 0.025M $(NH_4)_2HPO_4$ + 0.1M HCl. The Ca and P ions penetrate preferentially the pores inside, which results in improved bonding of the bioceramic layer to the metallic background. Changes in the depositing potential result in different morphology, porosity, composition and thickness of the growing Ca-P layer. We propose the electric field enhancement mechanism of the electrolytic ions flow resulting in Ca-P growth on the surface irregularities, such as pores and surrounding hillocks.

The corrosion resistance of the alloys was investigated in Ringer's solution. Electrochemical etching improves the corrosion resistance of the nanocrystalline alloys.

The prepared porous nanocrystalline Ti-6Al-4V, Ti-15Zr-4Nb and Ti-6Zr-4Nb alloys with electrochemically biofunctionalized surface could be a possible candidate for hard tissue implant applications.

Ti alloys are widely used in medical applications, because of theirs excellent corrosion resistance, biocompatibility and mechanical properties [12]. In comparison to the pure Ti, the Ti-alloys have better mechanical properties. The elements such as Al and V are main alloying additives in the Ti-alloys. Unfortunately, these elements exhibit high cytotoxicity and may induce senile dementia [3,9]. Replacing of Al and V by Nb and Zr leads to excellent biocompatibility, because these elements belong to vital group in the tissue reaction [14]. The most popular Ti-6Al-4V alloy is still in use in biomedical as well as other technical applications [18,20], but recently new Ti alloys containing Zr have begun to be more frequently used [21]. The latter have properties promoting them in the hard tissue implant applications, for example: low density, high strength, relatively low elastic modulus, high wear resistance, no toxic behavior, good corrosion resistance and excellent biocompatibility. Additionally these alloys show long lifetime in human body, which extends the time between implant replacements and thus between surgery operations.

The problem regarding the metallic materials for hard tissue implant applications is the mismatch of Young's modulus between the Ti-type implant (100-120 GPa) and the bone (10-30 GPa), which is unfavorable for bone healing and remodeling [10]. An introduction of pores into the alloy microstructure results in smaller modulus [15]. Decreasing the grain size to a nanoscale could also lead to a decrease of Young's modulus of the Ti alloys. The nanostructure could be generated in a large scale by mechanical methods, for example by mechanical alloying (MA) [5,11]. Thus, in the alloys for hard tissue implant applications, the best solution seems to be a combination of nanostructure with porosity.

Although, pores in the whole volume of the implant material can lead to lower mass and Young's modulus, unfortunately the lower mechanical strength is expected. So, the electrochemical etching seems to be a very promising method for the surface pores formation. In this case, the bulk background provides good mechanical properties, while the porous surface morphology improves the bone fixing [7]. In the nanocrystalline alloys the large volume of the grain boundaries improves the electrochemical etching [1,6]. The pore formation (material removing) is easy and effective, due to the large grains surface area. Additionally, the surface etched at positive (vs. OCP) potentials is oxidized, resulting in thick TiO_x layer, which improves corrosion resistance [25].

On the porous surface bioceramic material could be deposited, which improves the bioactivity of the metallic implant [13]. Usually, a bioactive hydroxyapatite (HA) is required on the surface. Hydroxyapatite, which is a Ca-P compound with chemical composition of $Ca_{10}(PO_4)_6(OH)_2$ and Ca/P ratio of 1.67, can easily bond to living tissues, improving biocompatibility of the implant, while the metallic background provides required mechanical properties. Coating of the implant by Ca-P compounds results in good osseointegration and short convalescence of the patient after the surgery implantation [8,22].

Adhesion and proliferation of the human osteoblast cells is accelerated by the topography of the implant. For example, number of the adhered cells to the nano-rough TiO_2 nanotubes surface increases by 400% in comparison to microcrystalline Ti [17]. The calcium and phosphorus are components of the bone, and these elements are crucial for good osseointegration. Depositing of the HA layer on the rough and porous implant background improves fixation of the layer, as well as fixation of the implant with the human body, in comparison to the flat implant surface [16]. The HA can be deposited using electrochemical cathodic deposition method, where the electrolyte consist of Ca and P ions flowing into the negatively charged Ti-type surface [19,24].

Below, we show recent achievements in the formation of nanocrystalline/ultrafine Ti-typed alloys prepared by mechanical alloying with electrochemically biofunctionalized surface intended for hard tissue implant applications.

13.3.2   MECHANICAL ALLOYING OF THE TI-AL-V AND TI-ZR-NB ALLOYS

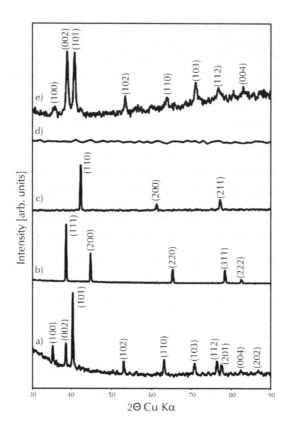

FIGURE 1   XRD data for pure Ti (a), Al (b),V (c), Ti-6Al-4V after 48 h MA (d) and after sintering (e).

The mechanical alloying is a process, where the pure elemental powders are mixed and cold welded together, resulting in amorphous or nanocrystalline phases [1,5,6,11]. The powders are hit by balls of the mill, resulting in effective grain size reduction. The weight ratio of the balls to the powders is 10: 1. In this work a SPEX 8000 mixer mill, equipped with hardened steel container with steel balls inside, was applied. The protective atmosphere is very important for high quality material preparation, so all powders handling was done in Labmaster glove box from MBraun, with argon atmosphere inside and controlled levels of oxygen and humidity. The argon atmosphere was ensured also inside the mill container. A mixture of powders of Ti, Al, V, Zr, and Nb (Figure 1a-c and Figure 2a-c), with stoichiometric ratio required for obtaining the Ti-6Al-4V, Ti-15Zr-4Nb and Ti-6Zr-4Nb alloys, was loaded into the container. All powders have initial size below 300 mesh.

After 48 h of mechanical alloying the powders are composed from amorphous and/ or nanocrystalline phases (Figure 1d, Figure 2d,f). In the mechanically alloyed powders high plastic deformations are generated resulting in high density of dislocation lines and subsequently subgrains formation, which finally leads to amorphisation [1]. When the powders are trapped between balls, the high energy of the balls hits leads to powders deformation, cold welding and finally lamellar structure formation (Figure 3) [1].

The lamellar or onion-type morphology is well visible in large plastically deformed particles (Figure 3a). In these particles the layers do not strictly adhere to each other and large volume of small pores occurs (Figure 3b). The continuous isotropic deformations of the particles between balls and walls of the milling container and relatively small stress in the large particles (in comparison to smaller ones) leads to gaps and pores existing between the layers.

The long time milling leads finally to amorphous (Figure 4) or nanocrystalline (Figure 5) particles. After mechanical alloying usually agglomerates are formed, which are composed mainly of the smaller amorphous (Figure 4) or nanocrystalline (Figure 5) areas [1]. Figure 4 shows amorphous agglomerate particle of the mechanically alloyed Ti-6Al-4V alloy.

The amorphous structure is unstable and under temperature rising (for example during sintering) a crystallization takes place. Due to a high stress and strong plastic deformations observed in the mechanical alloying process, a nanocrystalline areas in the larger powder particles could be formed directly in the process (Figure 5). So, the mixture of amorphous and nanocrystalline areas could coexists together. In these crystalline large particles, which are formed more often in the Ti-15Zr-4Nb than in the Ti-6Al-4V, the nanostructure is built from small subgrains with size below 20 nm (visible in the upper part of the Figure 5).

The subgrains formation is related to high plastic deformations ratio and high density of dislocation lines (visible in the lower part of the Figure 5) formed in the process.

The subsequent uniaxial pressing at 500 MPa leads to formation of green compacts, which were sintered at 1000°C for 1h. The sinters have crystallographic structure (Figure 1e, Figure 2e,g) corresponding to commercial alloy and density of 3.6 g/ cm^3, which is about 80% of the theoretical value for the bulk commercial microcrystalline alloy.

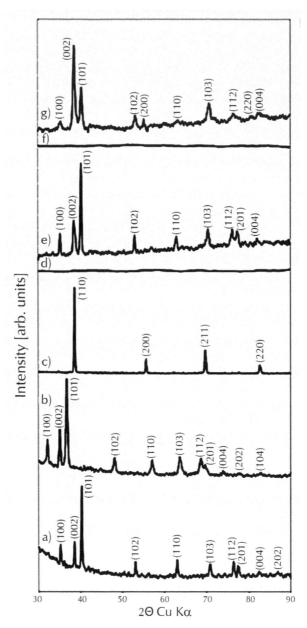

FIGURE 2   XRD data for pure Ti (a), Zr (b), Nb (c), Ti-15Zr-4Nb alloy after 48 h MA (d) and after sintering (e), as well as Ti-6Zr-4Nb alloy after 48 h MA (f) and after sintering (g)

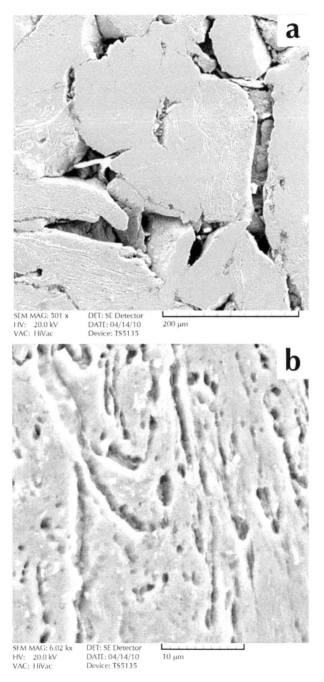

FIGURE 3 Cross section SEM images of the lamellar (onion-type) structure formed in the mechanical alloying process of the Ti-15Zr-4Nb alloy particles (a, b – different magnifications)

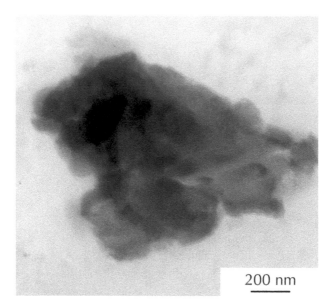

FIGURE 4   TEM image of the mechanically alloyed amorphous Ti-6Al-4V alloy agglomerate
particle

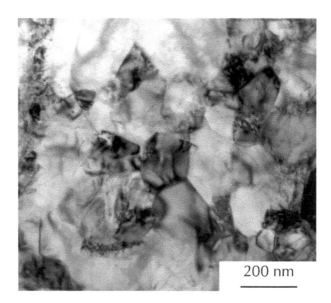

FIGURE 5   TEM image of the nanocrystals in the mechanically alloyed large particle of the
Ti-15Zr-4Nb alloy

The high temperature sintering leads to increase of the grain size, more in case of the Ti-6Al-4V than the Ti-15Zr-4Nb alloy (Figure 6). For the Ti-6Al-4V and the Ti-15Zr-4Nb alloy the grain size after sintering is in the range of 0.2-2 □m and 0.1-1 □m, respectively. In case of the Ti-6Al-4V, the grains have more uniform isotropic shape (Figure 6a). For the Ti-15Zr-4Nb the grains have more anisotropic (lamellar) shape (Figure 6b). Because of relatively low pressing pressure, the sintering a large volume of free spaces remains in the whole volume of the sinters (Figure 6). Hence the above mentioned sinters density stays at the level of about 80% of the theoretical value. The presented pores are useful in the next electrochemical etching stage of the surface bio-functionalization.

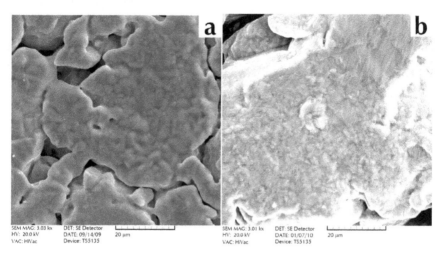

FIGURE 6   SEM images of the Ti-6Al-4V (a) and the Ti-15Zr-4Nb (b) sinters rough surface.

13.3.3   ELECTROCHEMICAL BIOFUNCTIONALIZATION OF THE TI-AL-V AND TI-ZR-NB ALLOYS

ELECTROCHEMICAL ANODIC ETCHING

The Ti-type nanocrystalline sinters made by mechanical alloying and powder metallurgy have initial porosity useful in decreasing of Young's modulus as well as in improving the etching process.

The electrochemical etching was done by using potentiostat connected to respective electrodes (graphite rod – counter electrode, SCE – reference electrode and Ti-alloy sinter – working electrode) in the electrochemical etching cell. The magnetic stirring was applied for the uniform etching, etching products removing as well as removing of the hydrogen bubbles released from the etched surface. As an etching electrolyte the solution of 1M H_3PO_4 + 2% HF in distilled water was applied. The applied etching potential was fixed on 10 V vs. OCP and the etching time was changed from 30 to 300 min.

Large volume of the pores and grain boundaries leads to penetration of these features by electrolyte (Figure 7, Figure 8). The large pores, remained after sintering (Figure 7a, Figure 8a), increase during the surface etching and the small grains are released from the bulk by electrolyte penetration through the grain boundaries (Figure 7b, Figure 8b). The largest pores have diameter up to 60 μm (Figure 7a, Figure 8a), while the smallest ones have diameter corresponding to the nanograins diameter (Figure 7b, Figure 8b). The etching time in the range of 30-300 min does not significantly affect the pore size. Until recently it was commonly accepted that for the effective tissue growth the pores should be close in size to 100-400 μm (macropores) [4]. However, the latest research shows that smaller micro pores also support living cells growth [23]. So the useful range of pores in the implants is very broad, from few nanometers up to 400 μm, supporting growth of different living cells, leading to shortening of the osseo integration time and stronger bonding of the implant to the bone. The stronger bonding of the tissue to the implant extends a lifetime of the implant in the human body.

The initial porosity of the sinters is well visible on the Figure 7c and Figure 8c, where the sinters cross section is shown. Depending of the etching time the thickness of the etched layer could be controlled. After 30 min (Figure 7c,d) and 300 min (Figure 8c,d) etching time, the porous etched layer thickness increases to 50 μm and 150 μm, respectively. During the etching connected and open channels are formed, resulting in sponge surface morphology (Figure 7d, Figure 8d). The comparable results of surface etching were achieved for the Ti-Zr-Nb alloy.

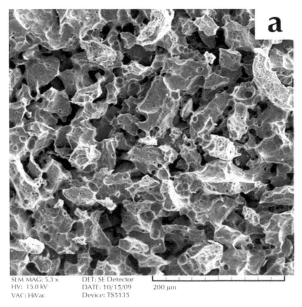

SEM MAG: 5.3 x   DET: SE Detector
HV: 15.0 kV   DATE: 10/15/09   200 μm
VAC: HiVac   Device: TS5135

FIGURE 7   *(Continued)*

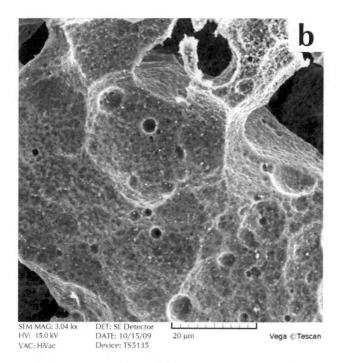

SEM MAG: 3.04 kx   DET: SE Detector
HV: 15.0 kV   DATE: 10/15/09   20 µm
VAC: HiVac   Device: TS5135   Vega ©Tescan

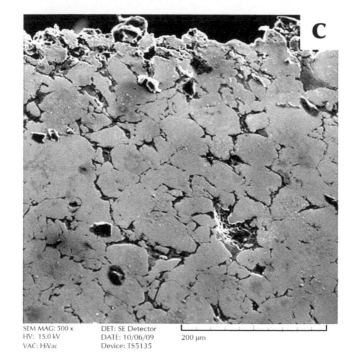

SEM MAG: 500 x   DET: SE Detector
HV: 15.0 kV   DATE: 10/06/09   200 µm
VAC: HiVac   Device: TS5135

FIGURE 7 *(Continued)*

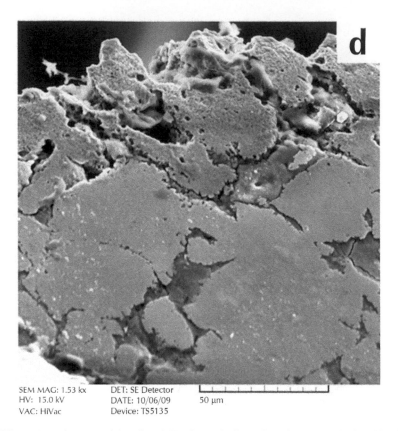

SEM MAG: 1.53 kx   DET: SE Detector
HV: 15.0 kV   DATE: 10/06/09   50 µm
VAC: HiVac   Device: TS5135

FIGURE 7   SEM images of the Ti-6Al-4V sintered alloy after electrochemical etching in 1M H_3PO_4 + 2% HF electrolyte at 10 V for 30 min: (a) and (b) – surface morphology (different magnifications), (c) and (d) – cross section of the surface (different magnifications).

The chemical composition after etching is shown on Figure 9. Because during the mechanical alloying some amount of powders strongly stick into the milling container walls and materials looses during the etching, the Al content is slightly different from the stoichiometric 6% in the Ti-6Al-4V alloy. The more important is that during the etching, phosphorus is introduced into the surface (from the phosphoric acid). The phosphorus is one of the bone components, so its presence on the surface is highly recommended for osseo integration improvement or HA growth. The rest of the composition (Cl, Fe, and Mg) is the contamination introduced during the process.

The porosity of the nanocrystalline Ti-6Al-4V and Ti-15Zr-4Nb alloys was investigated with an image analysis software (Figure 10) and Accelerated Surface Area and Porosimetry Analyzer (ASAP) (Figure 11, Table 1) [1].

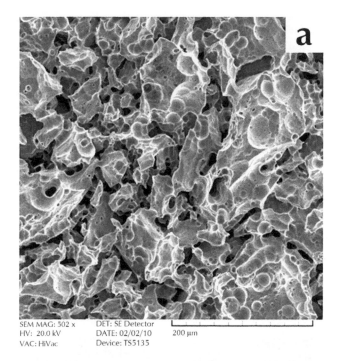

SEM MAG: 502 x   DET: SE Detector
HV: 20.0 kV   DATE: 02/02/10   200 µm
VAC: HiVac   Device: TS5135

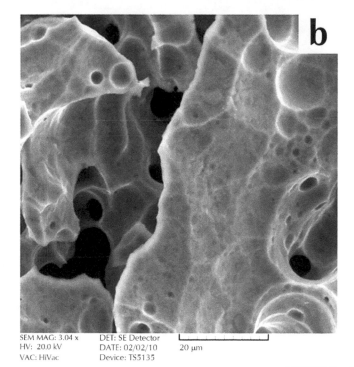

SEM MAG: 3.04 x   DET: SE Detector
HV: 20.0 kV   DATE: 02/02/10   20 µm
VAC: HiVac   Device: TS5135

FIGURE 8   *(Continued)*

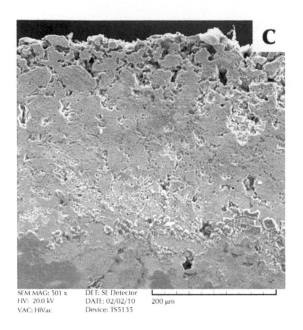

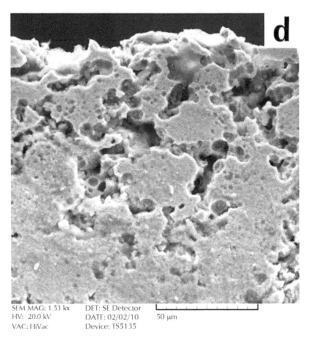

FIGURE 8   SEM images of the Ti-6Al-4V sintered alloy after electrochemical etching in 1M H3PO4 + 2% HF electrolyte at 10 V for 300 min: (a) and (b) – surface morphology (different magnifications), (c) and (d) – cross section of the surface (different magnifications

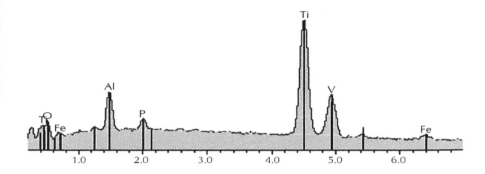

Element	Wt%	At%
Ti	74.46	57.7
Al	6.3	8.68
V	3.4	2.47
P	1.39	1.67
Cl	1.72	1.81
Mg	0.45	0.68
O	11.38	26.39
Fe	0.91	0.6
Total	100	100

FIGURE 9 EDS spectrum of the Ti-6Al-4V sintered alloy after electrochemical etching in 1M H3PO4 + 2% HF electrolyte at 10 V for 30 min.

Because large isolated pores are located inside the sinters, the good method for porosity analysis is a software method (Figure 10). For the Ti-6Al-4V (Figure 10a,b,c) and Ti-15Zr-4Nb (Figure 10d,e,f) after sintering the pore volume reaches 22.4% and 27.3%, respectively. The results are consistent with those from density measurements, mentioned earlier. It is very promising that the porosity of the Ti-15Zr-4Nb is higher than the Ti-6Al-4V. For example Oh et al. found that 30% porosity results in Young modulus close to those of the human cortical bone [15]. The implants with such porosity should induce smaller stress and suppress damage at the implant/bone interface.

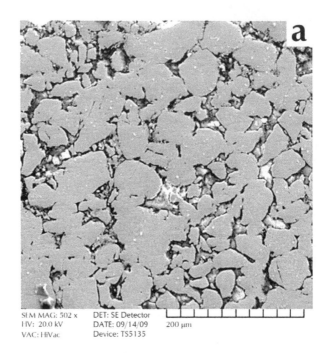

SEM MAG: 502 x   DET: SE Detector
HV: 20.0 kV   DATE: 09/14/09   200 μm
VAC: HiVac   Device: TS5135

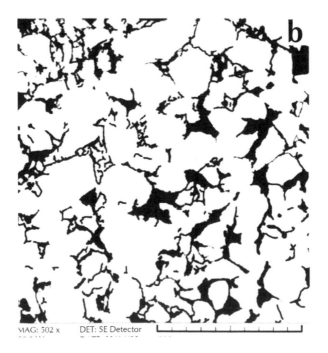

MAG: 502 x   DET: SE Detector

FIGURE 10   *(Continued)*

C

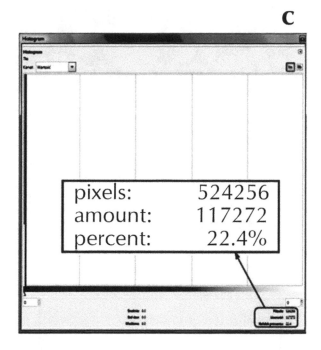

pixels: 524256
amount: 117272
percent: 22.4%

d

100 μm

FIGURE 10 *(Continued)*

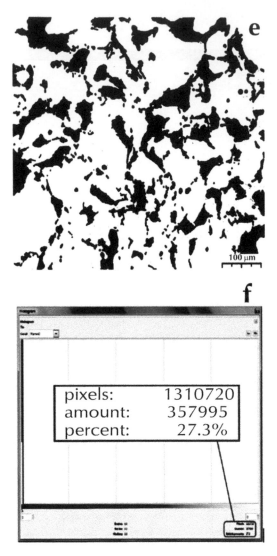

FIGURE 10   Overall porosity of the Ti-6Al-4V (a, b, c) and Ti-15Zr-4Nb (d, e, f) nanocrystalline sinters; microscope images (a, d), dual tone images (b, e), porosity content spectrum (c, f).

The porosity results after additional electrochemical etching are shown on Figure 11 and in Table 1. Because of the interconnected pore formation during the etching, the ASAP is a more accurate method of the porosity measurements. For the Ti-6Al-4V alloy increase of the etching time from 30 to 300 min results in increase of the average pore size (micro- and mesopores with size <100 nm) and total volume of the pores, while the BET surface, micropore surface and volume decreases. In the case of the Ti-15Zr-4Nb alloy two orders higher porosity parameters were achieved in comparison to the Ti-6Al-4V alloy, with additionally smaller average micropore size, too.

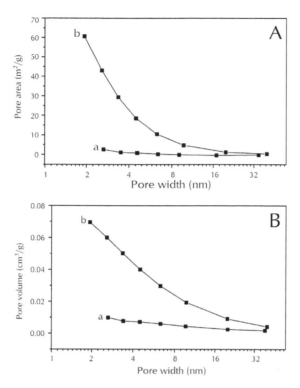

FIGURE 11 Pore surface area (A) and pore volume (B) of the electrochemically etched (1M H3PO4 + 2% HF; 10 V, 30 min) Ti-6Al-4V (a) and Ti-15Zr-4Nb (b) alloys.

The SEM image sample surface analysis and porosity measurements lead to the conclusion that smaller micro- and mesopores are present on the macropore walls (see Figure 7 and Figure 8). The ASAP analysis made on the unetched samples did not reveal micro- and mesopores with diameter <100 nm – so the surface is relatively continuous and flat (in the ASAP analysis the large pores observed on Figure 6 were not taken into account).

TABLE 1 BET/BJH (BET: Brunauer Emmett Teller; BJH: Barrett Joyner Halenda) porosity parameters for sintered nanocrystalline alloys, etched at 10 V for 30 min and 300 min in 1M H3PO4 + 2% HF electrolyte

Alloy	Etching time [min]	BET surface area [m²/g]	Micropore surface area [m²/g]	Micropore volume [cm³/g]	Total volume of pores (micro- and mesopores <100 nm) [cm³/g]	Average pore size [nm]
Ti-6Al-4V	30	0.8503	0.8347	0.000424	0.00116	5.46
Ti-6Al-4V	300	0.3087	0.2826	0.000142	0.00130	16.87
Ti-15Zr-4Nb	30	94.8676	87.3278	0.081295	0.08297	3.50

ELECTROCHEMICAL CATHODIC CA-P DEPOSITION

The next stage of the Ti-alloys surface bio functionalization, with respect to the implant applications, is a deposition of the calcium-phosphate compounds. The calcium-phosphate was cathodically deposited on the porous etched surface for the time of 60 min at the potential –0.5 V (Figure 12), –1.5 V (Figure 13), –3.0 V (Figure 14), –5.0 V (Figure 15) and –10 V (Figure 16) vs. OCP using a mixture of 0.042M Ca(NO3)2 + 0.025M (NH4)2HPO4 + 0.1M HCl electrolyte. Depositing of the Ca-P layer at –0.5V (Figure 12) slightly changes the surface morphology. Relatively low cathodic potential and low current density lead to low stream of calcium and phosphate ions flowing into the surface. The formed calcium-phosphate layer has a scaffold morphology, with pore diameter in the range from 0.1 up to 15 µm. So, the pores in the etched alloy surface are only partly filled with Ca-P. The increase of the depositing potential to –1.5 V (Figure 13) leads to more continuous, but slightly cracked layer, which fully covers the metallic background. Increase of the depositing potential to –3.0 V (Figure 14) leads to significant cracks in the deposited layer. The cracking is related to higher stress at the Ti/Ca-P interface. Depositing of the Ca-P at –5.0 V (Figure 15) results in transformation of the layer to highly cracked lamellar grains with high roughness. At –10 V the Ca-P layer is more continuous and homogenous (Figure 16). Unfortunately the layer has less adhesion to the surface and part of the layer falls down during removing of the sample from the electrolyte and drying it in the stream of nitrogen. Weak adhesion of the Ca-P layer could be related to a significant hydrogen emission at the electrode/electrolyte interface and large thickness of the deposited material.

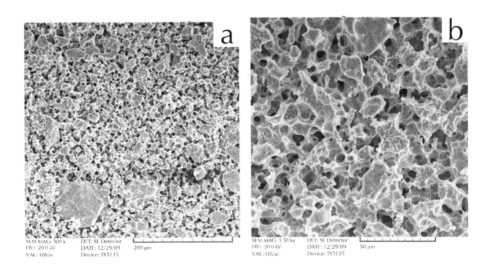

FIGURE 12   *(Continued)*

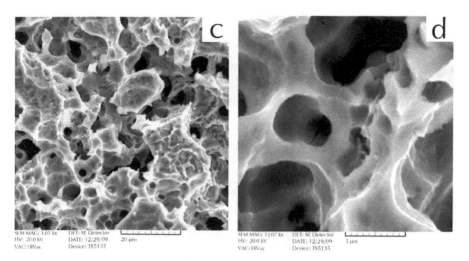

FIGURE 12 The SEM images of the Ca-P deposited (−0.5 V, 60 min) on the nanocrystalline porous Ti-6Al-4V (a→d: sequence of images with increasing magnification).

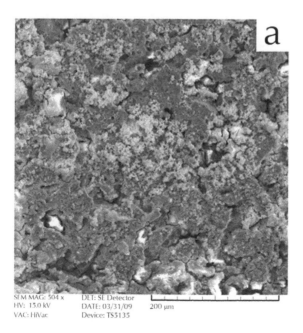

FIGURE 13 *(Continued)*

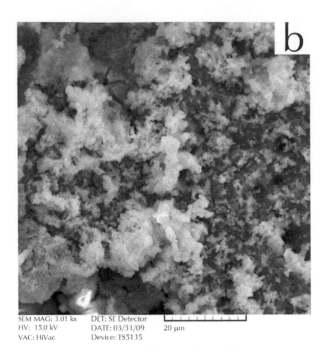

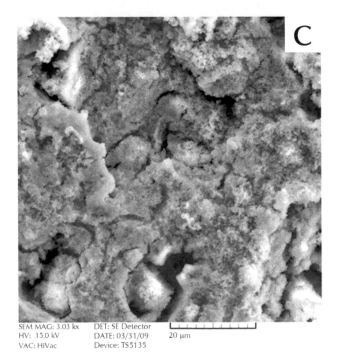

FIGURE 13   *(Continued)*

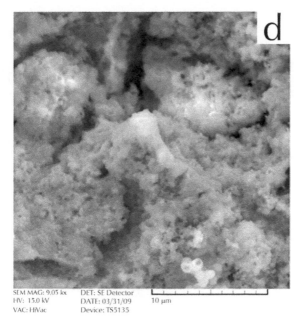

FIGURE 13   SEM images of the Ca-P deposited (–1.5 V, 60 min) on the nanocrystalline porous Ti-6Al-4V (a→d: sequence of images with increasing magnification).

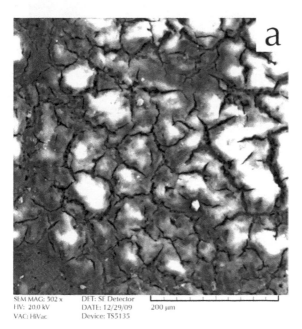

FIGURE 14   (Continued)

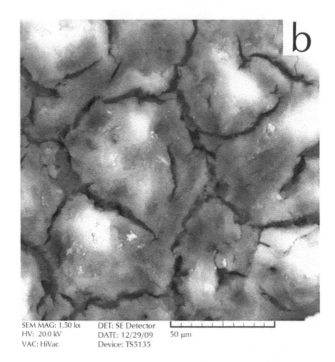

SEM MAG: 1.50 kx DET: SE Detector
HV: 20.0 kV DATE: 12/29/09 50 μm
VAC: HiVac Device: TS5135

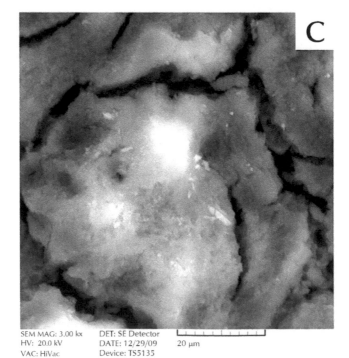

SEM MAG: 3.00 kx DET: SE Detector
HV: 20.0 kV DATE: 12/29/09 20 μm
VAC: HiVac Device: TS5135

FIGURE 14 *(Continued)*

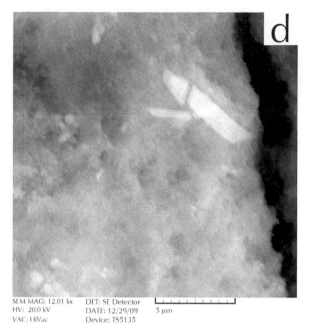

SEM MAG: 12.01 kx   DET: SE Detector
HV: 20.0 kV   DATE: 12/29/09   5 μm
VAC: HiVac   Device: TS5135

FIGURE 14   SEM images of the Ca-P deposited (–3.0 V, 60 min) on the nanocrystalline porous Ti-6Al-4V (a→d: sequence of images with increasing magnification).

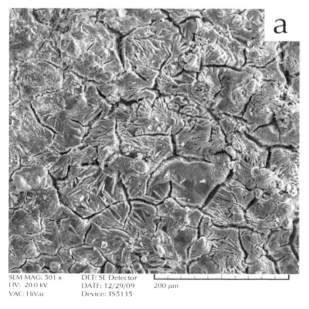

SEM MAG: 501 x   DET: SE Detector
HV: 20.0 kV   DATE: 12/29/09   200 μm
VAC: HiVac   Device: TS5135

FIGURE 15   *(Continued)*

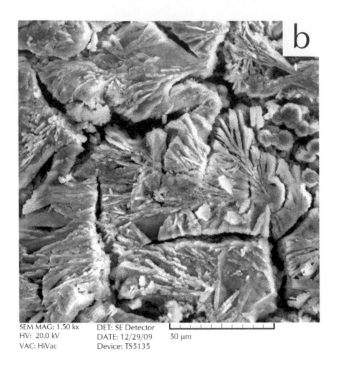

SEM MAG: 1.50 kx   DET: SE Detector
HV: 20.0 kV   DATE: 12/29/09   50 µm
VAC: HiVac   Device: TS5135

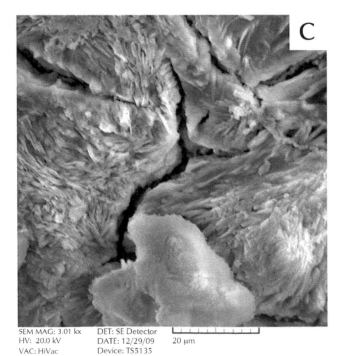

SEM MAG: 3.01 kx   DET: SE Detector
HV: 20.0 kV   DATE: 12/29/09   20 µm
VAC: HiVac   Device: TS5135

FIGURE 15   *(Continued)*

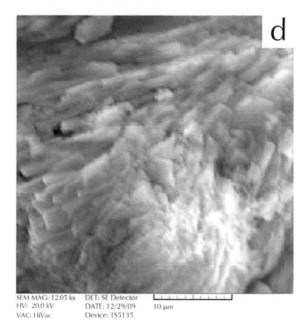

FIGURE 15 SEM images of the Ca-P deposited (−5.0 V, 60 min) on the nanocrystalline porous Ti-6Al-4V (a□d: sequence of images with increasing magnification).

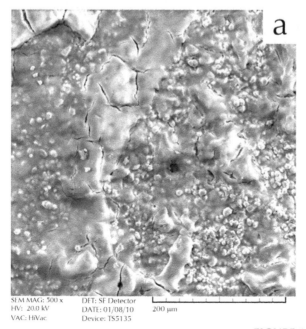

FIGURE 16 *(Continued)*

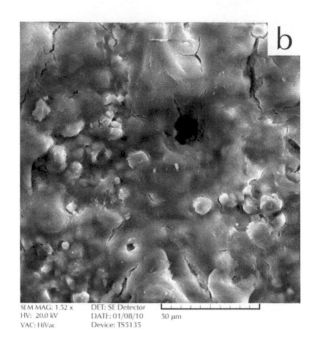

SEM MAG: 1.52 x   DET: SE Detector
HV: 20.0 kV   DATE: 01/08/10   50 μm
VAC: HiVac   Device: TS5135

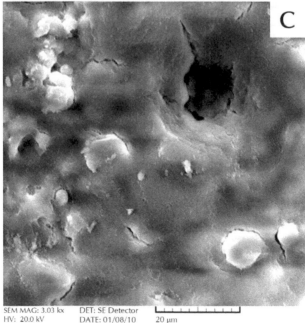

SEM MAG: 3.03 kx   DET: SE Detector
HV: 20.0 kV   DATE: 01/08/10   20 μm
VAC: HiVac   Device: TS5135

FIGURE 16   *(Continued)*

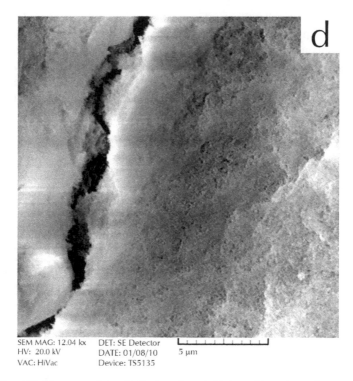

SEM MAG: 12.04 kx   DET: SE Detector
HV: 20.0 kV   DATE: 01/08/10   5 µm
VAC: HiVac   Device: TS5135

FIGURE 16 SEM images of the Ca-P deposited (–10.0 V, 60 min) on the nanocrystalline porous Ti-6Al-4V (a→d: sequence of images with increasing magnification).

The chemical composition analysis of the Ca-P deposited at different potentials (according to Figures. 12-16) on the Ti-6Al-4V porous surface is shown on Figure 17.

The Ca/P ratio in the ceramic layer changes with the depositing potential. After deposition at –0.5 V, –1.5 V, –3 V, –5 V, and –10 V, the Ca/P atomic ratio is 0.24, 1.28, 1.12, 1.30 and 1.71, respectively, leading to hydroxyapatite layer composition (Ca/P = 1.66) deposited at the potential range between –5 V and –10 V. Only in the (a) spectrum (sample with Ca-P deposited at –0.5V) the significant amount of Ti, Al and V from the metallic background was recorded, which is in agreement with the data observed on Figure 12. At low depositing potential the Ca-P do not fully cover the Ti-6Al-4V alloy surface. The other element traces are the contaminations introduced from the reagents of the depositing solutions or from the sample holder's silicone seal.

Using the image analysis software the surface porosity was determined (Figure 18 – columns). The initial surface porosity after electrochemical etching is estimated to be 49.3%. After additional Ca-P deposition, the surface porosity decreases with increasing depositing cathodic potential (increasing flowing charge density and amount of the Ca-P), reaching of about 4.3% porosity for –10 V depositing potential. The significant differences in porosity appear between –0.5 V and –3 V (as observed on Figures. 12-14). The Ca-P layer thickness, measured on the samples cross sections, increases with increasing depositing potential (Figure 18 – solid line). For –10 V the

Ca-P layer thickness reaches 90 μm. The Ca-P growing rate is estimated to be 9 μm/V. The shift in the depositing potential to more negative value results in thicker Ca-P layer, but simultaneously more hydrogen emission occurs (hydrogen bubbles formed at the electrode/electrolyte interface). It is obvious that the increase of potential to more negative value results in higher current and more flowing charge densities, leading to larger amount of the deposited Ca-P compounds. Because of the low adhesion and flat morphology of the Ca-P deposited at −10 V, the most promising layer is deposited below −10 V, for example at −5 V.

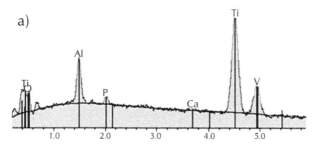

Element	Wt%	At%
Ti	78.63	61.06
Ca	0.29	0.27
P	0.92	1.1
O	11.33	26.34
AL	7.36	10.15
V	1.47	1.07
Total	100	100

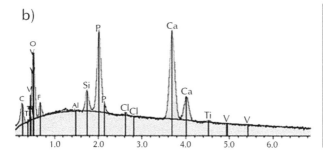

Element	Wt%	At%
Ti	0.23	0.10
Ca	26.30	12.98
P	15.83	10.11
O	37.01	45.76
AL	0.07	0.05
V	0.00	0.00
Si	1.54	1.09
F	1.98	2.06
Cl	0.19	0.11
C	16.85	27.76
Total	100	100

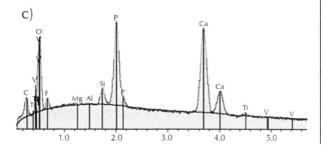

Element	Wt%	At%
Ti	0.24	0.10
Ca	24.89	12.24
P	17.17	10.92
O	38.24	47.11
AL	0.05	0.04
V	0.00	0.00
Si	1.19	0.83
F	1.68	1.74
Cl	16.37	26.86
Mg	0.19	0.15
Total	100	100

FIGURE 17   *(Continued)*

d)

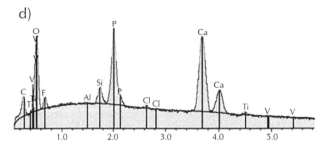

Element	Wt%	At%
O	39.46	44.43
Ti	0.22	0.1
V	0.00	0.00
Al	0.05	0.03
P	0.05	10.00
Ca	25.55	13.03
C	15.85	22.77
Si	1.44	9.58
Cl	0.13	0.06
Total	100	100

e)

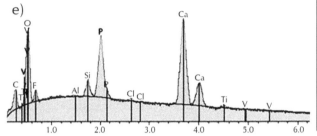

Element	Wt%	At%
O	39.51	43.98
Ti	0.00	0.00
V	0.03	0.00
Al	0.05	0.03
P	10.3	5.92
Ca	22.85	10.15
C	26.71	39.57
Si	0.44	0.28
Cl	0.11	0.05
Total	100	100

FIGURE 17 EDS analysis of the Ca-P deposited on porous Ti-6Al-4V surface at: –0.5 V (a), –1.5 V (b), –3 V (c), –5 V (d) and –10 V (e); the spectra (a-e) corresponds to Figures 12-16, respectively

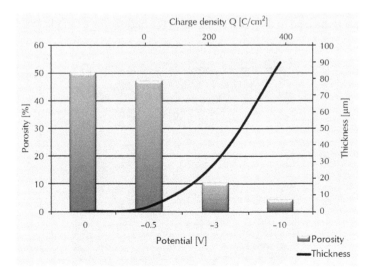

FIGURE 18 Ca-P surface porosity and thickness as a function of the Ca-P depositing potential and charge density for the nanocrystalline Ti-6Al-4V alloy; porous sample, after electrochemical etching without Ca-P is indicated by column at 0 V.

Comparable results were obtained for the porous etched nanocrystalline Ti-15Zr-4Nb alloys. The Ca-P layer deposited at −1.5 V is built from lamellas, resulting in rough surface morphology (Figure 19). The deposition of the Ca-P layer on the flat polished Ti-15Zr-4Nb nanocrystalline alloy surface results in significantly different morphology in comparison to the previously shown results (Figures 20). On the polished surface mainly loose-lying particles (mainly spherical) are deposited. The Ca-P particles do not have a possibility to anchor to the flat surface, so the adhesion of the Ca-P layer is rather poor. The Ca-P effectively grows inside the pores and pits, which are present on the surface (Figure 21). For example, after Ca-P deposition at −0.5 V, the Ca-P layer thickness (layer on the flat part of the alloy) is estimated to be about 2-5 μm, while the pores are almost filled (but still some free space is observed) and the local thickness of the Ca-P in the pores is up to 60 μm (depth of the pores). It means that the pores are preferentially filled by Ca-P in the electrochemical deposition process. The Ca-P mass transport is enhanced at the surface pits and pores.

We suggest, that a possible mechanism of the Ca-P (HA) growth is related to the electric field enhancement, occurring on the surface irregularities, such as pores and surrounding hillocks (Figure 22). The electric field lines are concentrated on these specific surface features, enhancing ions flow to these places (Figure 22a). Pore walls, which are perpendicular to the surface, are rather unaffected at the initial stage of the deposition process, especially if they are smooth, without irregularities. If the pore walls consist of protrusions, then the Ca-P grows on them. Increasing of the Ca-P (HA) cathodic deposition potential (as well as time) leads then to the pore filling and semi-continuous cracked layer formation (Figure 22b). The more pronounced increase of the depositing potential leads to the fully filled pores and formation of homogenous Ca-P (HA) layer (Figure 22c). The significant shift in the cathodic potential to the more negative value leads to electric field enhancement, resulting in enhancing of the $Ca2+$ and the $HPO42-$ (or $PO43-$) ions flow to the specific surface features, where the chemical reactions lead to the Ca-P compound formation, for example hydroxyapatite.

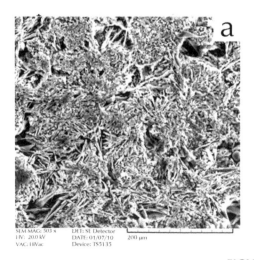

FIGURE 19   *(Continued)*

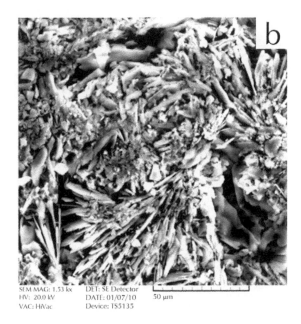

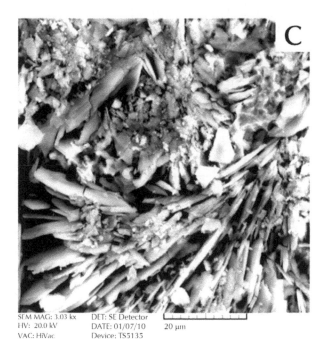

FIGURE 19 *(Continued)*

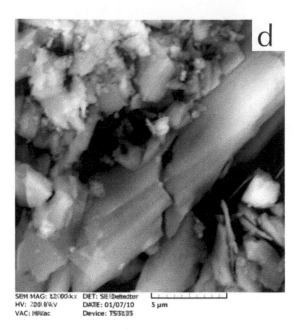

SEM MAG: 12000kx   DET: SE Detector
HV: 20.0 kV   DATE: 01/07/10   5 µm
VAC: HiVac   Device: TS5135

FIGURE 19   SEM images of the Ca-P deposited (−1.5 V, 60 min) on the nanocrystalline porous Ti-15Zr-4Nb (a→d: sequence of images with increasing magnification).

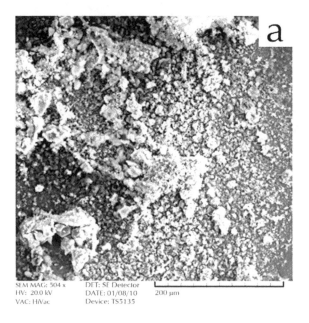

SEM MAG: 504 x   DET: SE Detector
HV: 20.0 kV   DATE: 01/08/10   200 µm
VAC: HiVac   Device: TS5135

FIGURE 20   *(Continued)*

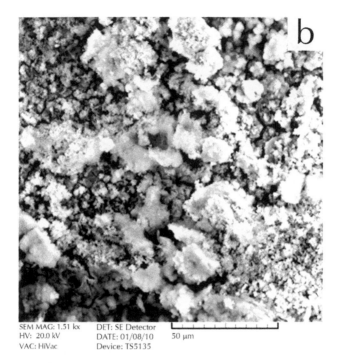

SEM MAG: 1.51 kx   DET: SE Detector
HV: 20.0 kV   DATE: 01/08/10   50 μm
VAC: HiVac   Device: TS5135

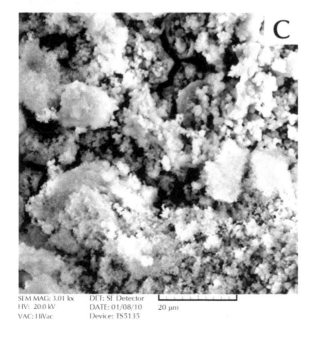

SEM MAG: 3.01 kx   DET: SE Detector
HV: 20.0 kV   DATE: 01/08/10   20 μm
VAC: HiVac   Device: TS5135

FIGURE 20 *(Continued)*

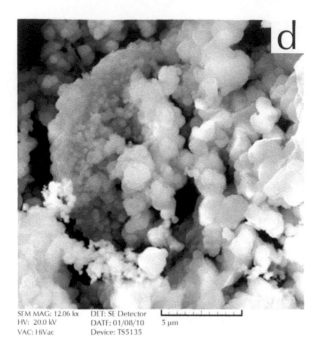

FIGURE 20 SEM images of the Ca-P deposited (−1.5 V, 60 min) on the nanocrystalline polished Ti-15Zr-4Nb (a→d: sequence of images with increasing magnification).

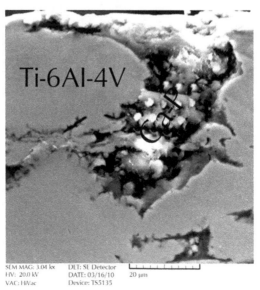

FIGURE 21 Cross section SEM image of the porous nanocrystalline Ti-6Al-4V alloy after Ca-P (HA) deposition for 60 min at −0.5 V.

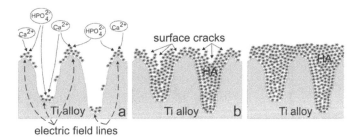

FIGURE 22 Scheme of the Ca-P (HA) layer growth mechanism on the porous surface during electrochemical cathodic deposition; a→c – stages of the HA growth

13.3.4 CORROSION RESISTANCE OF THE TI-AL-V AND TI-ZR-NB ALLOYS

The corrosion resistance of the biomaterials is a very important factor, which can decide about application of the material in the aggressive human body environment. In our tests we have applied potentiodynamic method, where the samples were immersed in Ringer's electrolyte containing significant amount of chloride (simulated body fluid with composition: NaCl: 9g/l, KCl: 0.42 g/l, CaCl2: 0.48 g/l, NaHCO3: 0.2 g/l). The sample was charged from –1 to 2.5 V vs. OCP with scan rate 0.5 mV/s. From the corrosion curves (Figure 23), using the Tafel extrapolations, the corrosion current density (Icorr) and corrosion potential (Ecorr) were estimated (Table 2). The corrosion resistance of the untreated Ti-6Al-4V nanocrystalline alloy is lower than the Zr-containing alloys (Icorr is higher for the Ti-6Al-4V). After electrochemical etching the corrosion current decreases (except Ti-15Zr-4V alloy). The subsequent Ca-P deposition does not significantly change the corrosion current, even at different depositing potential (probably due to layer's properties: the Ca-P layer is cracked and belongs to isolating materials). Taking into account all our corrosion experiments, we can generally say that the corrosion resistance increases after anodic etching and does not significantly change after the Ca-P deposition.

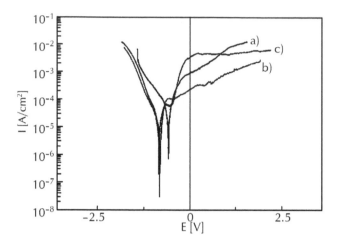

FIGURE 23 Corrosion curves for the nanocrystalline Ti-6Al-4V (a), nanocrystalline Ti-15Zr-4Nb (b) and nanocrystalline Ti-6Zr-4Nb (c); all samples without etching and Ca-P deposition.

TABLE 2 Corrosion current density Icorr and corrosion potential Ecorr of the nanocrystalline Ti-alloys before and after electrochemical etching (1M H3PO4 + 2% HF; 10 V, 30 min) as well as after additional Ca-P deposition (60 min) at different potentials.

Material	I_{corr} [A/cm²]	E_{corr} [V]
Ti-6Al-4V before etching	9.19×10^{-6}	−0.58
Ti-6Al-4V after etching	4.28×10^{-6}	−0.36
Ti-6Al-4V after etching with Ca-P (−0.5 V)	9.35×10^{-6}	−0.94
Ti-6Al-4V after etching with Ca-P (−1.5 V)	1.88×10^{-5}	−0.85
Ti-6Al-4V after etching with Ca-P (−3.0 V)	7.57×10^{-6}	−0.94
Ti-6Al-4V after etching with Ca-P (-5.0 V)	7.14×10^{-6}	−0.85
Ti-6Al-4V after etching with Ca-P (−10.0 V)	1.88×10^{-5}	−0.85
Ti-15Zr-4V before etching	1.85×10^{-6}	−0.84
Ti-15Zr-4V after etching	3.12×10^{-5}	−0.85
Ti-15Zr-4V after etching with Ca-P (−1.5 V)	1.10×10^{-5}	−0.99
Ti-6Zr-4V before etching	2.27×10^{-6}	−0.82
Ti-6Zr-4Nb after etching	4.43×10^{-6}	−0.75
Ti-6Zr-4Nb after etching with Ca-P (−1.5 V)	5.74×10^{-6}	−0.89

13.3.5 CONCLUSION

In this article formation of porous nanocrystalline Ti-6Al-4V and (Al, V)-free alloys with composition of the Ti-15Zr-4Nb and Ti-6Zr-4Nb prepared by mechanical alloying, pressing, sintering and subsequent electrochemical biofunctionalization was described. Mechanically alloyed powders are composed from nanograins and subgrains, formed by high plastic deformations, or even amorphous areas. The sintering process done at high temperature results in grain size growth, but due to the composition and refinement effect of the Zr, the Ti-15Zr-4Nb and Ti-6Zr-4Nb alloys are more resistant to grain growth than the Ti-6Al-4V. The ultrafine structure affects the etching process, improving porosity, which is two order of magnitude larger for the Ti-15Zr-4Nb in comparison to the Ti-6Al-4V. The electrolyte penetrates sinters through the grain boundaries, resulting in effective material removing and pores formation.

On the porous background bioactive Ca-P layer was deposited. Increase of the deposition potential leads to more calcium and phosphate ions flow to the surface, which results in significant different Ca-P layer morphologies. At low depositing potential –0.5V the Ca-P is porous, thin and does not fully cover the metallic background, while the increase of the depositing potential by a few volts (up to –10V) results in thick continuous Ca-P layer. We suggest, that the electric field enhancement on the surface pores and surrounding hillocks plays a key role in the Ca-P growth, leading to preferential Ca-P deposition.

Due to obtained properties, the prepared porous biofunctionalized nanocrystalline Ti-type alloys could be a possible candidate for the hard tissue implant applications.

REFERENCES –CASE STUDY II

1. Adamek, G. and Jakubowicz, J. Microstructure of the mechanically alloyed and electro-chemically etched Ti-6Al-4V and Ti-15Zr-4Nb nanocrystalline alloys. Materials Chemistry and Physics, 124, 1198–1204 (2010).
2. Adamek, G. and Jakubowicz, J. Mechanoelectrochemical synthesis and properties of porous nano-Ti-6Al-4V alloy with hydroxyapatite layer for biomedical applications. Electrochemistry Communication, 12, 653–656 (2010).
3. Davidson, J. A., Mishra, A. K., Kovacs P., and Poggie, R.A. New surface-hardened, low-modulus, corrosion resistant Ti-13Nb-13Zr alloy for total hip arthroplasty. Biomed. Mater. Eng., 4, 231–243 (1994).
4. Itälä, A.I., Ylänen, H.O., Ekholm, C., Karlsson, K.H., and Aro, H.T. Pore diameter of more than 100 μm is not requisite for bone ingrowth in rabbits. J. Appl. Biomater., 58, 679–683 (2001).
5. Jakubowicz, J. and Adamek, G. Preparation and properties of mechanically alloyed and electrochemically etched porous Ti-6Al-4V. Electrochemistry Communication, 11, 1772–1775 (2009).
6. Jakubowicz, J., Jurczyk, K., Niespodziana, K., and Jurczyk, M. Mechanoelectrochemical synthesis of porous Ti-based nanocomposite biomaterials. Electrochemistry Communication, 11, 461–465 (2009).
7. Kim, H.M., Miyaji, F., Kokubo, T., and Nakamura, T. Preparation of bioactive Ti and its alloys via simple chemical surface treatment. J. Biomed. Mater. Res., 32, 409–417 (1996).

8. Kuo, M.C.and Yen, S.K. The process of electrochemical deposited hydroxyapatite coatings on biomedical titanium at room temperature. Mater. Sci. Eng. C, 20, 153–160 (2002).
9. Lugowski, S.J., Smith, D.C., McHugh, A.D., and Loon, V. Release of metal ions from dental implant materials in vivo: determination of Al, Co, Cr, Mo, Ni, V, and Ti in organ tissue. J. Biomed. Mater. Res., 25, 1443–1458 (1991).
10. Niinomi, M. Cyto-toxicity and fatigue performance of low rigidity titanium alloy, Ti–29Nb–13Ta–4.6Zr, for biomedical applications. Biomaterials, 24, 2673–2683(2003).
11. Niespodziana, K., Jurczyk, K., Jakubowicz, J., and Jurczyk, M. Fabrication and properties of titanium-hydroxyapatite nanocomposites. Mater. Chem. Phys., 123, 160–165 (2010).
12. Noort, R.V. Titanium: The implant material of today. J. Mater. Sci., 22, 3801–3811 (1987).
13. Narayanan, R., Seshadri, S.K., Kwon, T.Y., and Kim, K.H. Electrochemical nano-grained calcium phosphate coatings on Ti-6Al-4V for biomaterial applications. Scripta Mater., 56, 229–232 (2007).
14. Okazaki, Y., Rao, S., Ito, Y., and Tateishi, T. Corrosion resistance, mechanical properties, corrosion fatigue strength and cytocompatibility of new Ti alloys without Al and V. Biomaterials, 19, 1197–1215 (1998).
15. Oh, I. H., Nomura, N., Masahashi, N.and Hanada, S. Mechanical properties of porous titanium compacts prepared by powder sintering. Scripta Mater., 49, 1197–1202 (2003).
16. Oh, I. H., Nomura, N., Chiba, A., Murayama, Y., Masahashi, N., Lee, B. T., and Hanada, S. Microstructures and bond strengths of plasma-sprayed hydroxyapatite coatings on porous titanium. J. Mat. Sci.: Mater. in Med., 16, 635–640 (2005).
17. Oh, S. and Jin, S. Titanium oxide nanotubes with controlled morphology for enhanced bone growth. Mater. Sci. Eng. C, 26, 1301–1306 (2006).
18. Rack, H.J.and Quazi, J.I. Titanium alloys for biomedical applications.Mater. Sci. Eng. C, 26, 1269–1277(2006).
19. Raja, K.S., Misra, M., and Paramguru, K. Deposition of calcium phosphate coating on nanotubular anodized titanium. Mater. Lett., 59, 2137–2141 (2005).
20. Santos, L.V., Trava-Airoldi, V.J., Corat, E.J., Nogueira, J., and Leite, N.F. DLC cold welding prevention films on a Ti6Al4Valloy for space applications. Surf. Coat. Techn., 200, 2587–2593 (2006).
21. Saji, V.S. and Choe, H.Ch. Electrochemical corrosion behaviour of nanotubular Ti–13Nb–13Zr alloy in Ringer's solution. Corr. Sci., 51, 1658–1663 (2009).
22. Sgambato, A., Cittadini, A., Ardito, R., Dardeli, A., Facchini, A., Pria, P.D., and Colombo, A. Osteoblast behavior on nanostructured titanium alloys. Mater. Sci. Eng. C, 23, 419–423 (2003).
23. Webster, T.J. and Ejiofor, J.U. Increased osteoblast adhesion on nanophase metals: Ti, Ti6Al4V and CoCrMo. Biomaterials, 25, 4731–4739 (2004).
24. Xiao, X. F., Liu, R.F., and Zheng, Y.Z. Hydoxyapatite/titanium composite coating prepared by hydrothermal–electrochemical technique. Mater. Lett., 59, 1660–1664 (2005).
25. Yang, B., Uchida, M., Kim, H. M., Zhang, X., and Kokubo, T. Preparation of bioactive titanium metal via anodic oxidation treatment. Biomaterials, 25, 1003–1010 (2004).

13.4   CASE STUDY III:MULTILAYERED ASSEMBLY OF ZINC PHOSPHOMOLYBDATECONTAINERS

13.4.1   INTRODUCTION

The current chapter consists of synthesis of zinc phosphomolybdate (ZMP) by ultrasound assisted sonochemical synthesis and ZMP nanocontainer by the formation of multilayered structure of oppositely charged species on the surface of ZMP nano pigment by ultrasonic emulsion synthesis. To ease the application of deposition of oppositely charged layers ZMP nanoparticles were functionalized with myristic acid (MA). Benzotriazole is a good corrosion inhibitor for copper containing alloys so it is adsorbed in between two polyelectrolyte layers (polyaniline and polyacrylic acid). The average particle size of ZMP nanoparticles which are synthesized by sonochemically assisted emulsion method was found to be 68.4275 nm. Lesser particle size obtained which is due to efficient mixing and maximum amplitude of ultrasound horn. ZMP nanocontainer and ZMP nanoparticles characterized by XRD, PSD, FTIR, zeta potential, and TEM which are evident to show successful formation of multilayered structure of ZMP nanoparticles with ZMP nanoparticles at the core. UV-vis spectroscopic analysis was used to estimate the release rate, release flux and diffusivity of benzotriazole in water with respect to time and at varying pH environment. The obtained results showed that the benzotriazole release is found to be decreased with an increase in pH value. Also the estimated diffusivity is found to be higher at lower pH i.e. 2.78 $\times$ 10-14cm2/min at pH 2.From above results it can be predicted that the multilayered structure of ZMP nanoparticle can be use an good anticorrosive coating formulation in paint. The ability of ZMP nanocontainer to sustain against the corrosive environment was tested by dispersing ZMP nanocontainers in alkyd resin. From these results, it is observed that the corrosion resistance ability of ZMP nanocontainer is dependent upon the amount incorporating it in alkyd resin. It is found that with an increase in the loading of ZMP nanocontainer (0 to 5 wt % total paint volume composition) in alkyd resin, the corrosion rate was found decreased significantly. Also, it is observed that the corrosion rate was higher in 5% NaOH solution compared to 5% HCl and NaCl solution which indicates that ZMP nanocontainer has more corrosion resistant ability in HCl solution. Tafel plot results shows nanocontainer has improvement in the anticorrosive behavior.

The corrosion occurs due various phenomena's e.g. reaction of metal surface with oxygen and water, contact of metal surface with acid, base, salt or due to any electrochemical reaction. There is no such industry which doesn't have corrosion; mainly corrosion takes place in chemical and petrochemical industries. Any metal surface can be protected from corrosion by using three various mechanisms such as cathodic protection, anodic protection (passivation technique) and barrier mechanism [29].Almost in every industry barrier technique is more popular due to ease of application and main thing is that it forms a barrier of anticorrosive coatings in between metal surface and corrosive environment which also results in low penetration of corrosive chemicals on metal surface [31].A corrosion inhibiting compound mainly used as a barrier coating which forms a barrier film on surface of metal by attaching on the surface of metal, which results in preventing the corrosive reaction from occurring and can be easily added into paint coating formulation. Borisova et al. put forth that ACP i.e. active corrosion protection is a system in which anticorrosive liquids can be introduced into the coatings. Murphy et al.predicted that the direct application of anticorrosive liquid in anticorrosive coatings is not preferable which have an effect on the performance formed coatings [22].Bare corrosion inhibitor pigment can be encapsulatedbyformation of multilayered structure of different anticorrosive species for the responsive release with respect to outside environment and which offers good corrosion resistance performance than alone bare corrosion inhibitor pigment [8,10,23,37,39,40,43,44]. Polyelectrolyte encapsulated nanocontainer can be used in variety of applications such as biomedical [3], controlled release drugs [41],reaction rate acceleratorcatalyst [9], textile industries, etc. The polyelectrolyte layer is a pH dependent formulation which forms a matrix layer on encapsulated material thus can be applicable in number of fields and to release species in responsive manner. Tedim et al. invented the Synthesis of nanocontainer and encapsulation of core corrosion inhibiting pigment [1,8,14,17, 23,24,34,35,37,39,43,44].The multilayered structure of ZMP nanocontainer is useful because of its responsive and sustained release of active anticorrosion species.Liquid corrosion inhibitors generally react with the coating when it is directly used, therefore it can be used in the form of layered structure of micro/nanocontainer. Encapsulated corrosion inhibiting species can release in responsive manner but which dependent upon pH, temperature, mechanical stimulation and their layers combinations.

Nanocontainer can be used for corrosion inhibition when the following conditions are precisely maintained:

(1) Nanocontainer should be dispersed in paint formulation uniformly and it should remain inactive when not in application,

(2) Nanocontainer should be compatible with paint formulation for uniform formation of coating.

(3) Nanocontainer which incorporated in paint coating formulation should not disturb other properties of coatings such as adherence, and so on,

(4) Self healing is an important function of nanocontainer so for that nanocontainer show fast release kinetics and responsive release and

(5) nanocontainer should be added in required quantity to maintain pigment volume concentration (PVC ratio).

Shchukin et al. put-forth that Layer by layer (LbL) synthesis of nanocontainer by encapsulated with polyelectrolyte [35]. PAA is a matrix like structure which encapsulates the corrosion inhibiting pigment in multilayered synthesis of nanocontainer which shows responsive and controlled release based on outside environment. A multifunctional shell assembly of nanocontainer can be prepared using deposition of oppositely charged species (polyelectorlytes and nanoparticles) on the surface of the core material. Shchukin et al. have used inorganic nanotubes called as halloysite nanoclay of inside diameter 25 nm for encapsulation of corrosion inhibitor that is benzotriazole the inhibitor [35]. Above nanoclays are pH based basically made up of kaolin on upper surface of which negative and inside which positive charge exist at pH 8. This halloysite nanoclay has an ability to absorb species inside the lumen and shows release according to external environmental pH conditions. Shchukin et al. used halloysite nanoclay with inside lumen loaded with corrosion inhibitor such as benzotriazole [34]. Thesehalloysite tubes were encapsulated with polyelectrolyte by using Sol–gel method, which shows responsive and sustained release. Shchukin et al. reported core and shell structure of halloysite nanoclay using emulsion method [33,34]. Responsive release of halloysite nanocontainer depends upon external pH conditions [34]. Corrosion is a destructive attack of metal by external environmental conditions and which occurs due to formation of carbonic acid (H3O). When any metal surface comes in contact with oxygen and water corrosion occurs due to formation of carbonic acid, therefore zinc phosphomolybdate forms a passive layer in contact with water and oxygen, this results in inhibition of the flow of electrons from anode to cathode [7,13] as well as metal surface get protected from rusting. Some chromate and lead bearing pigments are environmentally restricted due to their toxicity to nature but Zn2PMoO7 is efficient pigment because of optimum cost and environmental friendly behavior. From all such a characteristics Zn2PMoO7 is used as an inorganic core in the synthesis of layered structure of ZMP nanocontainer and to avoid rusting at edge of metal and polyaniline and PAA are important because these are liquid corrosion inhibitors which attaches to metal surface and inhibit the corrosion even if other film is removed from the surface.A Zn2PMoO7 act as an anodic-type inhibitor for metal surface which was encapsulated in between multilayered structure of ZMP nanocontainer also it is excellent nano-filler.

In the present book chapter Zn2PMoO7 nanocontainers were formed by deposition of polyaniline/benzotriazole/polyacrylic acid layers incorporated in alkyd resin and which act as excellent corrosion inhibitor for metal surface. Benzotriazole is encapsulated in between the two polyelectrolytes (polyaniline and PAA) layers. Liquid corrosion inhibitor such as polyaniline PANI used as a polyelectrolyte which has inhibiting effect due to the creation of a compact iron/dopant complex assembly at the metal-coating interface, which acts as a passive protective layer having a redox potential to go through a continuous charge transfer reaction at the metal-coating boundary, in which PANI is reduced from emeraldine salt form (ES) to an emeraldine base (EB) [12]. The above synthesized nanocontainer has an excellent corrosion inhibiting efficiency. Estimation of release of benzotriazole at various pH was investigated using UV spectroscopic analysis. Corrosion inhibiting ability of Zn2PMoO7 on MS has been confirmed by various analysis as well as responsive release of benzotriazole at different pH was seen. ZMP nanoparticles can be incorporated in alkyd resin and can be used as excellent anticorrosive coating for MS.

13.4.2 EXPERIMENTAL

MATERIALS

For the synthesis of ZMP nanocontainer, analytical grade sodium molybdate (Na2MoO4), zinc sulphate (ZnSO4), potassium dihydrogen phosphate (KH2PO4), 16 N nitric acid, 50% caustic lye ,ammonium persulphate (APS, (NH4)2S2O8) as an initiator and sodium dodecyl sulfate (SDS, NaC12H25SO4) as a surfactant were procured from S.D. Fine Chem. and used as received without further purification. HCl, NaOH, NaCl, benzotriazole, and ethanol these are all of analytical grade chemicals procured from Sigma Aldrich and were used as received. Polyacrylic acid (PAA, Mw= 50000 g mol–1) procured from Sigma Aldrich and were used as received. The monomer aniline (analytical grade, M/s Fluka) was distilled two times before actual use. Millipore water was used as a medium throughout the all experimentation.

PREPARATION OF ZINC PHOSPHOMOLYBDATENANOCONTAINERS

Zn2PMoO7 as an anticorrosive pigment in nanoform was synthesised using so-nochemical irradiation method. At first aqueous solutions of sodium molybdate dihy-drate, zinc sulphate heptahydrate and potassium dihydrogen phosphate were prepared separately by adding 153.5 g sodium molybdate dihydrate in 50 ml distilled water, 100 g zinc sulphate heptahydrate in 50 ml distilled water and 8.0 g potassium dihydrogen phosphate in 20 ml distilled water. The solutions of zinc sulfate heptahydrate and potassium dihydrogen phosphate were added drop wise simultaneously to the sodium molybdate dihydrate solution under ultrasonic irradiation within time span of 10 min. After complete addition of zinc sulfate heptahydrate and potassium dihydrogen phos-phate solution, 16 N nitric acid is added as an oxidizing agent under sonication. After addition of nitric acid the aqueous clear reaction mixture was turned to yellowish initially. During the continuation of the reaction the pH of solution was adjusted to 7 by the addition of 50 % caustic lye solution under constant sonication which results into dense white precipitate. The total reaction time was about 1 h under sonication at room temperature (24 ± 2 0C). To separate out the synthesized product after comple-tion of reaction, the reaction mixture was kept in water bath at 150 0C for 10 min which results into formation of ZMP (whitish in nature) nanoparticles. Synthesis of Zn2PMoO7 nanocontainers have been carried out in a subsequent steps, which are summarized below:

(1) Doping of PANI layer on nano Zn2PMoO7 by sonochemical emulsion polym-erization method: ZMP nanocontainer was synthesized using Zn2PMoO7 as a core material due to its multifunctional and environmental legislation accept-able nature and excellent anticorrosive characteristics. Hydrophobic property incorporated in ZMP nanoparticles (5 g in 100 ml water) for that these were initially functionalised with 0.5 g myristic acid solution in 10 ml methanol at 60 0C under sonication at 60 min. After modification of ZMP with myristic acid it will create negative charges onto the surface of ZMP nanoparticles as C13H27COO– functional group adsorbed on ZMP nanoparticle surface. ZMP nanoparticles having negative charge on the surface and which are func-tionalized with the help of myristic acid were encapsulated by polyaniline (PANI, positively charged) layer by sonochemical microemulsion polymer-ization method reported by 4Bhanvase and Sonawane (2010). According to method proposed by Bhanvase and Sonawane (2010), at first surfactant solu-tion of sodium dodecyl sulfate was prepared by adding 3 g of SDS and 0.2 g of Zn2PMoO7 (on basis of monomer aniline) in 50 ml water, which was directly transferred to sonochemical reactor (Hielscher Ultrasonics GmbH, 22 KHz frequency 240W power). APS i.e. ammonium persulfate used as an initiator for the polymerization reaction which solution was prepared separately with the addition of 3.5 g APS in 20 ml of deionized water and which was mixed to the reactor. Aniline in amount of 5 g was added in semibatch mode within time span of 30 min, after completion of addition the reaction was kept continued further for 1 h (total reaction tine 1.5 h) in the presence of ultrasound at 4 0C

temperature. The synthesized product was separated by using centrifugal sedimentation method (Remi Instruments Supply 220/230 V, 50Hz, 1 AC) at 8000 rpm for 10 min and further washed with ultrapure water to take out un-reacted material and impurities. The separated product was dried with the help of oven for and at 60 0C.

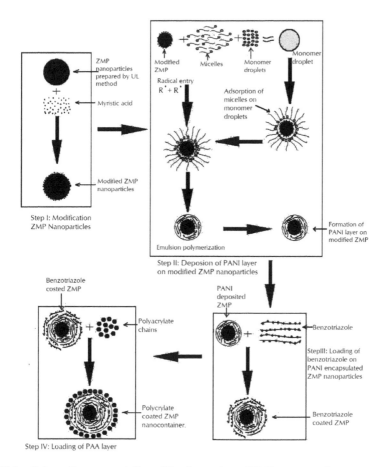

FIGURE 1   Schematic representation of the formation of ZMP nanocontainer.

(2) Loading of Benzotriazole (Corrosion Inhibitor) Layer on PANI Encapsulated Zn2PMoO7:Benzotriazolelayer loaded in between two polyelectrolyte layers i.e. PANI and PAA layer. This is carried out by adding 2 g PANI loaded Zn2PMoO7 (synthesized in step 1) in 0.1 N NaCl solution prepared within 100 ml water. Positively charged benzotriazole (third layer) layer deposited on the surface of PANI loaded Zn2PMoO7 nanoparticles by adding 2 mg ml-1 of benzotriazole in acidic media at pH 3 in sonochemical assisted culture for 20 min. Synthesized product was separated by using centrifugal sedimentation

method at 8000 rpm continued for 10 min, dried at 60 0C for a time period of 3 h and which was used again for loading of PAA layer.

(3) Deposition of PAA Polyelectrolyte Layer on Benzotriazole Loaded PANI Encapsulated Zn2PMoO7 Nanoparticles: PAA polyelectrolyte layer was adsorbed on benzotriazole loaded PANI encapsulated ZMPnanoparticles to attain sustained and responsive release of corrosion inhibitor at varying pH environment and to make ZMP nanocontainer suitable to prepare alkyd resin based paint formulation. For the adsorption of PAA layer, 2 mg ml-1 of concentration solution of PAA was prepared in 0.1 N NaCl solution by using of sonochemical irradiation for time duration of 20 min. The prepared product was separated by centrifugal sedimentation at 8000 rpm and for time span of 10 min, washed with deionized water and dried in oven at 60 oC for 48 h. Formed nanocontainer has multilayered structure which contain myristic acid functionalized ZMP nanoparticles as a core and benzotriazole corrosion inhibitor was entrapped in between two polyelectrolyte layers i.e. PANI and PAA layer. Sonochemical microemulsion method polymerization was used to make sure the complete coverage of ZMP nanoparticles surface with PANI. The formation mechanism of ZMP nanocontainer is reported in figure 1.

PREPARATION OF ZINC PHOSPHOMOLYBDATENANOCONTAINER/ALKYD MULTIFUNCTIONAL COATINGS

Pigment muller was used to incorporate ZMP nanocontainer particles in alkyd resin. Zn2PMoO7 nanocontainers/alkyd multifunctionalcoatings has been prepared by dispersing ZMP nanocontainer with an concentration of 0 to 5.0 wt % of total paint volume concentration (10 g) in alkyd resin with help of pigment Muller (Sheen Instruments at 400 RPM). To make uniform coating and to relieve the procedure of application above prepared coating was carefully mixed with acetone solution by using bar coater on MS panels having dimensions 50 × 40 × 1 mm.

CHARACTERIZATION OF ZINC PHOSPHOMOLYBDATE NANOCONTAINER AND ZINC PHOSPHOMOLYBDATENANOCONTAINER/ALKYD COATINGS

To identify the kinds of phases exist, crystal size in Zn2PMoO7 nanoparticles and Zn2PMoO7 nanocontainer X-ray diffractometer (Rigaku Mini-Flox, USA).The morphology of Zn2PMoO7 nanocontainer was performed by using transmission electron microscopy (TEM) (Technai G20 working at 200 kV). Responsive release of (corrosion inhibitor) benzotriazole was measured at various pH values of 2, 4, 7, and 10 using UV-vis spectrophotometer (SHIMADZU 160A model). To observe the kinds of bonding present in molecules infrared spectroscopic analysis of samples was carried out using SHIMADZU 8400S analyzer in the spectral range of 4000–500 cm−1. To find out potential present at the surface of the molecule that is zeta potential and to observe particle size distribution present in molecules Malvern Zetasizer Instrument (Malvern Instruments, Malvern, UK) was used.

To test corrosion inhibiting ability of ZMP nanocontainer MS plate (density 7.86 gm/cm3) was used. Coating film of ZMP nanocontainer was maintained in the thickness of 50 μm on the surface of MS plate and further these plates were used to carry out dip test in acid. Above plates were dipped in acid, base and salt (5 wt % for each) solution for nearly 750 h. Ability to resist corrosion of ZMP nanocontainer with respect to loss in weight of MS was estimated by calculating the corrosion rate (Vc) in cm/year for each one of the samples [4] by using an fundamental expression:

$$V_C = \frac{\Delta g}{Atd} \tag{1}$$

Where Δg is the loss in weight of each species in gram calculated by difference in weight of each species before dipping and after dipping, A is the exposed area of the sample in cm2 in corrosive media, t is the time of exposure in years under dip test, and d is the density of the metallic species in g/cm3 (mild steel i.e. 7.86 gm/cm3).

Loss in weight of each species was measured by washing the samples with ultra-pure water after 750 h and removing loosen material from species and then drying was carried in oven at 60 0 C (±1) out to obtain equilibrium weight by removing moisture present at surface of species because of washing. Electrochemical corrosion potential i.e. Tafel plot analysis (log |I| vs. E) of 0%,2%,4% (wt% of total paint composition) ZMP nanocontainer coated plate of M.S and was carried out in 5 % NaCl solution as an electrolyte at room temperature (25 oC) and this characterization was performed on computerized electrochemical analyzer (supplied by Autolab Instruments, Netherlands).

Three different MS plates of Zn2PMoO7nanocontainer having composition 0, 2.0, and 4.0 wt % of coated with alkyd resin containing were used as working electrode, as Pt and Ag/AgCl were used as counter and reference electrodes respectively. The plate area used for characterization was about 1 cm2 for sample testing. The electrochemical window was −1.0 V to +1 V with 2mV/s scanning rate.

13.4.3 RESULTS AND DISCUSSIONS

FORMATION MECHANISM, ZETA POTENTIAL AND PARTICLE SIZE DISTRIBUTION OF ZINC PHOSPHOMOLYBDATENANOCONTAINER

Figure 1 explains the formation mechanism of ZMP nanocontainer. ZMP nanoparticles were prepared by sonochemical irradiation method explained in figure 1. ZMP nanoparticles were functionalized by using myristic acid (C13H27COO–) with the help sonochemical micro emulsion method of to make ZMP nanoparticles hydrophobic in nature for efficient encapsulation of ZMP nanoparticles in PANI layer. MA has ability to improve the hydrophobicity and to create negative potential at surface molecule by adsorption of negatively charged C13H27COO– functional group. After modification step functionalised ZMP nanoparticles were doped by PANI layer and which was attained by sonochemical micro-emulsion polymerization process. PANI layer on ZMP nanoparticles was successfully adsorbed because of hydrophobic nature of ZMP nanoparticles due to myristic acid modification step and presence of negative charge on ZMP nanoparticles, The adsorption of next layer i.e. benzotriazole (corrosion inhibitor) was successfully carried out on PANI/MA/ZMP particles. Finally, above Benzotriazole/ PANI/MA/ZMP structure was encapsulated by using deposition of the negatively charged PAA layer (forth layer) and which was carried out after the formation of the layer of benzotriazole by sonochemical microemulsion polymerization method. As sonochemical irradiation method is used for synthesis of ZMP nanocontainer which results into distinct reduction in particle size and formation of ZMP nanoparticles without agglomeration.

Nanocontainer is a multilayered structure, due to deposition of number of layers ionic strength of liquid changes which affects stability of nanocontainer. Zeta potential gives the value of the charge present at surface after each layer deposition. The value of Zeta ( ) potential after each layer deposition of synthesized ZMP nanocontainer reported in Figure 2a. The zeta potential value of bare ZMP nanoparticles is −21.2 mV. After modification of ZMP nanoparticles with myristic acid the negative value of zeta potential increases and becomes -25.7 mV, because myristic acid which is having C13H27COO– functional group get adsorbed on bare ZMP nanoparticles. After deposition of PANI layer negative value of zeta potential decreases slightly and becomes 24.3 mV. It is seems that the negative value of Zeta potential decreases after loading PANI layer. Zeta potential shows negative value due to presence of sodium dodecyl sulfate (SDS), which was used during emulsion polymerization though PANI layer introduces positive charges on the surface. During emulsion polymerization, aniline comes in contact with the micelle and occupies the space near to the head group of the adsorbed SDS rather than the hydrophobic tail region [2].Due to this reduction in the value of zeta potential takes place after loading into PANI layer. The phenyl moiety is located within the hydrophobic region and the polar group remains between the SDS head groups put forth by Kim et al [16]. After deposition of PANI layer loading of benzotriazole (corrosion inhibitor) takes place which results in decrease in the negative value of zeta potential that is. 19.9 mV.

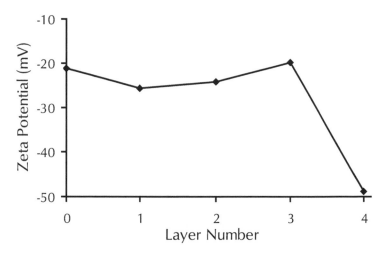

FIGURE 2 (a) Zeta Potential of ZMP nanocontainer in water. Layer number 0: initial Zn2PMoO7, 1: Myristic acid treated ZMP, 2: Zn2PMoO7/PANI, 3: Zn2PMoO7/PANI/ Benzotriozole, 4: Zn2PMoO7/PANI/ Benzotriozole/PAA.

Benzotriazole loaded ZMP nanoparticles encapsulated with PAA layer, after encapsulation with PAA layer negative value of zeta potential value suddenly increases up to -48.9 mV. This sudden increase in value of zeta potential is due to adsorption of negatively charged PAA chains. Charged species are adsorbed on bare ZMP nanoparticles are confirmed by zeta potential and which also shows intra-particle interaction in the ZMP nanocontainer. From Figure 2b it is observed that after addition of each layer, the average particle size goes on increasing gradually. From the results of particle size distribution analysis, it is confirmed that there is formation of multilayered structure of ZMP nanocontainer. The particle size of bare ZMP nanoparticle was found to be around 68.43 nm and it is found to be decreased marginally to 65.32 nm after deposition of myristic acid layer. The adsorption of PANI layer by the emulsion polymerization onto ZMP nanoparticles results into increase in the particle size to an average value of 544.95 nm. The thickness of PANI layer is found to be 98.05 nm. After deposition of benzotriazole that is corrosion inhibitor, the particle size increases to 863.69 nm with layer thickness 159.37 nm, from above it is observed that there is more adsorption of benzotriazole molecules on PANI loaded ZMP nanoparticles. Finally deposition of PAA on benzotriazole layer takes place by emulsion polymerization method which results in increase in average particle size up to 1492.23 nm with an average layer thickness of 314.26 nm. Increase in particle size after deposition of each layer signifies that successful formation of ZMP nanocontainer.

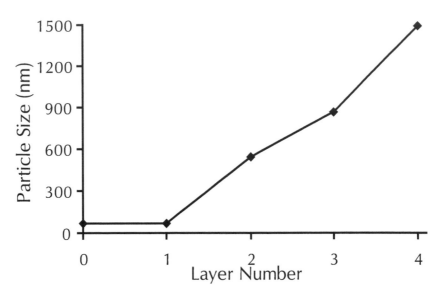

FIGURE 2 (b) Growth in particle size of ZMP nanocontainer. Layer number 0: initial Zn2PMoO7, 1: Zn2PMoO7/PANI, 2: Zn2PMoO7/PANI/Benzotriozole, 3: Zn2PMoO7/PANI/Benzotriozole/PAA.

MORPHOLOGY ANALYSIS OF ZINC PHOSPHOMOLYBDATENANOCONTAINER

A transmission electron microscopy image of zinc phosphomolybdatenanocontainer containing zinc phosphomolybdatenanoparticles at core along with polyelectrolyte material surrounding the zinc phosphomolybdatenanoparticles is depicted in Figure 3. The TEM images indicate the distinct formation of zinc phosphomolybdatenanocontainer distorted spheres. Further, the presence of bright intensity layered structure around the dark zinc phosphomolybdatenanoparticles in TEM image has confirmed the layer formation of PANI, benzotriazole and PAA. The presence of benzotriazole and polyelectrolytes around zinc phosphomolybdatenanoparticles results in an increase in the average particles size of zinc phosphomolybdatenanocontainer, which is around 1250 nm. These results are consistent with the average particle size of zinc phosphomolybdatenanocontainer obtained from the particle size analyser (as reported in section 3.1). Change in the scattering of light around each particles show that multilayer assembly (PANI/benzotriozole/PAA layers) is established on the zinc phosphomolybdatenanoparticles.

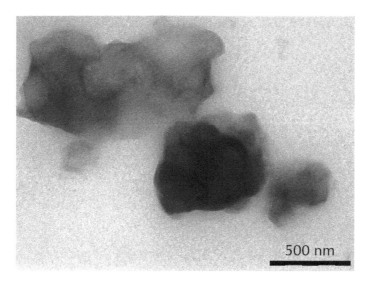

FIGURE 3   TEM image of ZMP nanocontainers

FTIR ANALYSIS OF ZINC PHOSPHOMOLYBDATENANOCONTAINER

Figure 4 signifies the FTIR spectrum of pure Zn2PMoO7 (pattern A), Modified Zn2PMoO7 (pattern B), Zn2PMoO7 loaded with PANI (pattern C), Zn2PMoO7 loaded with PANI and Benzotriazole (pattern D) and Zn2PMoO7 loaded with PANI-Benzotriazole-Polyacrylic acid (pattern E). Figure 4 (Pattern A) shows FT-IR spectra of ZMP nanoparticles. Very strong Mo–O stretching vibration in $[MoO4]2-$ was detected at 825–936 cm-1 and weak Mo–O bendingvibration was found at 437, 494 cm-1 [11,25]. Further, modification of ZMP nanoparticles was done by ultrasound assisted method by using myristic acid (MA). Figure 4 (B) shows an FTIR spectrum of myristic acid modified ZMP nanoparticles. Due to addition of myristic acid to synthesized ZMP nanoparticles characteristics peaks are observed on the bare ZMP nanoparticles at 2924, 2853 and 1541 cm−1 shows stretching vibration of the C–H which came from the –CH3 and –CH2 in the myristic acid respectively. Bending of –OH bond attributed by the characteristic peak at 1541 cm−1. Figure 4 (C) shows the characteristic peak due to addition of PANI layer to the modified ZMP nanoparticles which are at 1230 cm-1 is due to (C–N) stretching mode of the amine group and the peak at 1433 cm-1 reflects C=C stretching mode of the quinoid rings and C=C stretching of benzenoid rings respectively [36]. Secondary =N–H bending showed by the characteristics peak at 1526 cm-1. Above mentioned peaks reflects the formation of polyaniline layer on the modified ZMP nanoparticles. Figure 4 (D) reflects the FTIR spectrum of benzotriazole loaded PANI loaded modified myristic acid ZMP nanoparticles. Benzotriazole layer of PANI coated ZMP nanoparticles shows the characteristic peaks at 1520, 1260, and 750 cm-1 which reflects the effective formation benzotriazole layer. The bands which are close to 750 cm-1 are typical of the benzene ring vibration and the band near to 1520 cm-1 is characteristic of the aromatic and the triazole rings stretching vibration [21]. Finally benzotriazole layer loaded myristic acid modified ZMP nanoparticles coated with PAA layer the FTIR spectrum of which is depicted in figure 4 (E). The characteristic peak at 1732 cm-1 reflects the adsorption of PAA layer which is attributed to carbonyl C=O stretching in PAA [18].

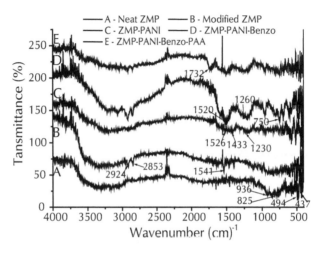

FIGURE 4   FTIR spectra of: (A) Neat ZMP (B) Modified ZMP (C) ZMP loaded with PANI, (D) ZMP loaded with PANI and benzotriazole, (E) ZMP loaded with PANI, benzotriazole and PAA (Polyacrylic acid).

XRD ANALYSIS OF ZINC PHOSPHOMOLYBDATENANOCONTAINER

Figure 5 depicts the XRD patterns of ZMP nanoparticle and ZMP nanocontainers. Figure 5 (pattern A) signifies the XRD pattern of Zn2PMoO7 nanoparticles. It is observed that the synthesized ZMP nanoparticles are crystalline in nature [19]and observed phase of ZMP nanoparticles isscheelite phase [30]. In above XRD pattern the diffraction peaks at 2θ value of 25.2, 27.9, 29.1, 31.7, 34.2, 40.3, 51.8 and 52.8o are corresponds to the planes (112), (004), (114), (211), (200), (220), (312) and (224) [27]. The peaks correspond to 2θ values at 31.7, 34.2, 36.1, 56.1, 62.4 and 68.2o shows the characteristics peaks of phosphate addition. The crystallite size of ZMPnanoparticle was estimated as equal to 74.15 nm at 2θ = 25.2o by Debye Scherrer's formula. The particle size of ZMP nanoparticles is consistent with PSD data reported above. Afterwards due to uncalcined state of zinc phosphomolybdate some impurities peaks of Na2SO4 and KNO3 even after number of hot water washing cycles is observed. Further in XRD pattern of ZMP nanocontainer (Figure 5B), it is found that particles are completely covered with adsorbed layers of polyelectrolyte and corrosion inhibitors. The presence of XRD peaks at 25.2, 36.1, 52.8o clearly shows the presence of zinc phosphomolybdate at the core of the nanocontainer.

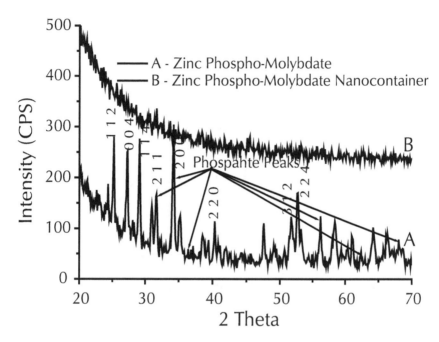

FIGURE 5   XRD pattern of (A) Zn2PMoO7nanoparticles and (B) Zn2PMoO7 nanocontainers.

TGA AND DTA OF ZINC PHOSPHOMOLYBDATENANOCONTAINER

TGA plot of zinc phosphomolybdate nanocontainers shown in Figure 6 (a) which shows weight loss in ZMP nanoparticle at four different steps. Initially the weight loss of 23 wt. % is observed in the range of 40 – 270 oC (Section I) due to desorption of physically adsorbed as well as hydrated water in the ZMP nanocontainer. Polyacrylic acid is adsorbed on the surface of benzotriazole loaded ZMP nanoparticles, due to burning of PAA layer second weight loss (21 wt. %) is observed in the range of 271 – 440 oC (Section II). The third loss in weight of benzotriazole i.e. about (14 wt. %) between 441 and 525 oC (Section III) is due to oxidative degradation of the corrosion inhibitor (benzotriazole) that is enclosed into the two layers of polyelectrolytes in the ZMP nanocontainers. As weight loss is observed in the section III it does confirms that more adsorption of corrosion inhibitor in the nanocontainer. From above figure it is observed that the adsorbed PANI layer is burned off after 525 oC (Section IV) and weight loss observed about 22 wt. %. The overall weight loss is observed from 271 to

725 oC is due to oxidative degradation of hydrocarbon moieties present in the poly-mer. From above thermo gravimetric analysis it is observed that ZMP nanoparticles are more stable at high temperature.

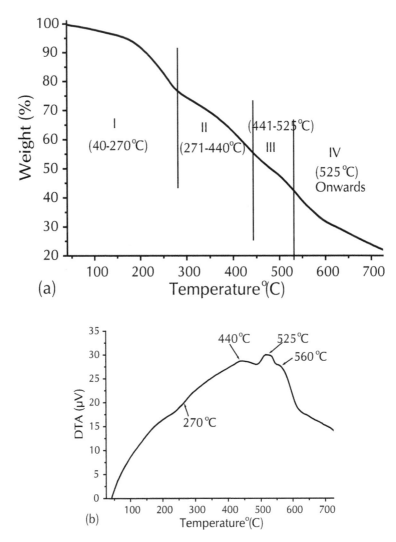

FIGURE 6   (a) TGA and (b) DTA analysis of Zn2PMoO7 nanocontainer.

Figure 6 (b) depicts the DTA plot of ZMP nanocontainers. ZMP nanocontainer DTA plot shows endothermic peak at 270 o C which is due to the removal of physi-cally adsorbed water (In the range from 200-300 0C the water molecules interacting with ZMP nanocontainer surface such as hydroxyls groups releases from the ZMP nanocontainer structure) [38,42]. The exothermic peaks at 400-500 o C are due to multistage decomposition of PAA, PANI and Benzotriazole (corrosion inhibitor) which is also confirmed by TGA analysis. A intense peak in the range of 500 to 550 o C corresponds to oxidative degradation of PAA, PANI and benzotriazole [15]. The endothermic peak in the range of 550-600 o C is due to loss in weight of ZMP nanocontainer after deposition of benzotriazole layer due to excess removal of moisture. The endothermic peak at 650 700 observed which signifies loss in weight due to removal of moisture from PAA layer.

RELEASE STUDY OF CORROSION INHIBITOR FROM ZINC PHOSPHOMOLYBDATENANOCONTAINER

Figure 7 depicts the release rate and release flux of benzotriazole. It is well known that the benzotriazole is a good inhibitor for ferrous metal under acidic environment [20,26,32] also in neutral environment [6,28].Cao et al. put-forth thatbenzotriazole forms the compact passive layer by the adsorption of benzotriazole in its molecular or protonated form. The purpose of this study is to observe the effect of pH of the aqueous medium (2, 4, 7 and 9) on the release rate of the benzotriazole trough ZMP nanocontainer. Final layer of nanocontainer i.e. PAA is pH dependent, due to this it is found that release rate as well as release flux of benzotriazole is goes on increasing gradually with respect to time initially but as time passes it goes on decreasing as the concentration of corrosion inhibitor i.e. benzotriazole increases in the surrounding medium. As the concentration of benzotriazole goes on increasing in the surrounding pH solution which results into decrease in the diffusion rate of benzotriazole due to saturation stage with respect to time. It is clearly observed that release rate and release flux of benzotriazole get decreased with respect to time and increasing pH value. The release rate gets decreases from 0.0065 to 0.085 mgL-1/g of ZMP nanocontainer.min for increase in the pH from 2 to 9 at the end of 10 min. As well as release flux is found to be decreased from 0.00375 to 0.0007 mg.L-1/g of ZMP Nanocontainer.cm2.min for increase in the pH from 2 to 9 at the end of 10 min. From above figure it is observed that more release is observed at more acidic pH i.e. at pH 2 due to ability of PAA to expand its matrix structure at more acidic environment. Due to above ability of benzotriazole it forms passive layer on the surface of ferrous metal leads to more corrosion inhibition in acidic medium.

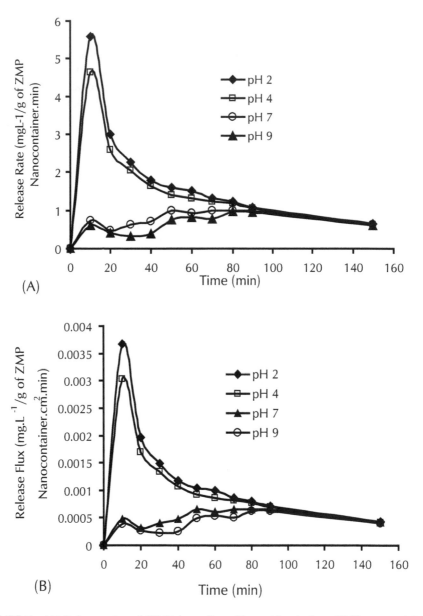

FIGURE 7 (A) Release rate and (B) Release flux of benzotriazole from ZMP nanocontainers at different pH values.

The diffusive ability of the benzotriazole in pH environment was estimated by the equation as follows

$$J_A = -D_{AB}\frac{dC_A}{dr}$$

(2)

Whereas JA = Release flux of benzotriazole, DAB = Diffusivity of benzotriazole, CA = Concentration of benzotriazole and r = radial position in spherical nanocontainer.

From Figure 8 it is observed that the diffusivity of benzotriazole gets decreased from 2.78 × 10-14 to 1.5 × 10-14 cm2/min with respect to increase in the pH value from 2 to 9. Due to ability of PAA to expand its matrix structure in case of highly acidic pH i.e. at pH 2, benzotriazole shows more sustained release at acidic pH as compare to basic pH.

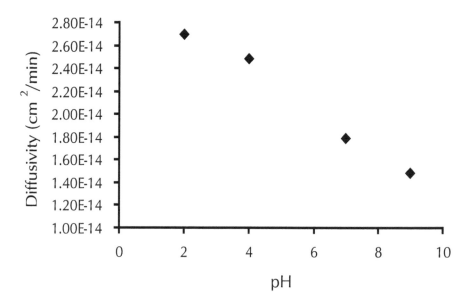

FIGURE 8 Diffusivity of benzotriazole from nanocontainers versus pH.

CORROSION RATE ANALYSIS OF ZINC MOLYBDATENANOCONTAINER/ ALKYD COATINGS

Figure 9 indicates the corrosion rate per year at different wt % loading of ZMP nano-containers in alkyd resin coatings. At first ZMP nanocontainers were incorporated in the alkyd resin with the help of pigment Muller and formulation was coated on MS plates for study corrosion rate analysis. Experimentation was carried out by dipping coated MS in HCl, NaCl and NaOH (5 wt % each) solutions for a period of 750 h. The loss in weight of each species was measured by gravimetric analysis for the analysis of corrosion rate through dip test. Corrosion rate for 0 % loading of ZMP nanocontainer is 0.0765 cm/yr, 0.0849 cm/yr and 0.1207 cm/yr in case of 5% HCl, NaOH and NaCl solution respectively. From above figure, it can be concluded that the corrosion rate was maximum for NaCl solution and minimum for HCl solution. The corrosion rate was changing predominantly at varying percent loading of nanocontainer in all cases. Increase in percentage loading of ZMP nanocontainer from 2 to 5 wt %, decrease in the corrosion rate takes place in all cases of HCl, NaCl and NaOH solution which results into minimum corrosion rate per year observed. From above it can be predicted that corrosion resisting ability of ZMP nanocontainer is maximum in HCl solution compared to NaCl and NaOH solution, which attributes to more passivation in acidic medium resulted from more release of benzotriazole from the ZMP nanocontainer.

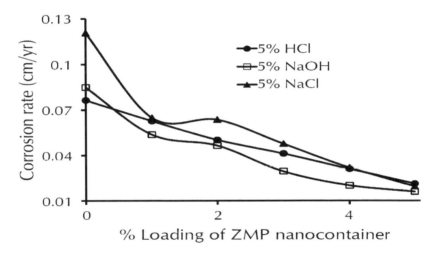

FIGURE 9   Effect of loading of nanocontainer on the corrosion rate after loading into the alkyd resin.

ELECTROCHEMICAL CHARACTERIZATION OF ZINC PHOSPHOMOLYBDATENANOCONTAINER/ALKYD COATINGS

Figure 10 signifies the electrochemical analysis (Tafel plot) of MS panels coated with neat alkyd resin, and coated with 2 and 4 wt % loading of ZMP nanocontainer incorporated in alkyd resin. Above analysis was carried out in 5 wt % aqueous NaCl solution at room temperature. The Tafel plot is plotted as log (current density) as a function of applied potential. In Tafel plot analysis current density is measured in corrosion process for simultaneous redox reactions occurs at the surface of cathode and anode of MS plate. Icorr i.e. corrosion current density and Ecorr i.e. corrosion potential, values were found from the Tafel plot analysis. It is observed that corrosion current density was decreased from 0.027 for pure alkyd resin to 0.012 A/cm2 for 2 wt% ZMP nanocontainers incorporated in pure alkyd resin coatings and it gets decreases upto value of 0.01 A/cm2 when percentage loading of ZMP nanocontainers is increased to 4 wt %. Ecorr values goes on increasing with increase in percentage loading of ZMP nanocontainer and which is found to be shifted to positive side from -7.42 V to -1.034 V with the addition of 4 wt % of ZMP nanocontainers in alkyd resin.

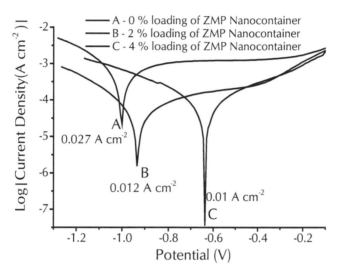

FIGURE 10   Tafel plots of MS panels coated with alkyd/ZMP nanocontainer coatings at different loading of ZMP nanocontainer in alkyd resin carried out in 5 % NaCl solution as an electrolyte at room temperature (25 oC).

From above results it is observed that release rate, corrosion rate, and electrochemical (Tafel plot) analysis shows that the percentage loading of 4 wt % ZMP nanocontainers in alkyd resin shows a considerable improvement in the anticorrosive properties compare to pure alkyd resin.

13.4.4 CONCLUSION

The present work confirmed the layer by layer assembly of ZMP nanocontainers and the release mechanism of corrosion inhibitor. FTIR and TEM study confirms the successful formation of ZMP nanocontainer as a layer by layer system with the aid of ultrasonic irradiation. Zeta potential and particle size analysis also shows the formation of layers and shows appropriate change in the surface charge, which could be responsible for the release mechanism initiated by the change in pH. Release study and corrosion results from Tafel plot and corrosion rate analysis showed significant improvement in the anticorrosion properties of coatings due to the optimum loading of the ZMP nanocontainers.

REFERENCES-CASE STUDY III

1. Aramaki K. Self-healing mechanism of an organosiloxane polymer film containing sodium silicate and cerium (III) nitrate for corrosion of scratched zinc surface in 0.5 M NaCl. Corrosion Science, 44(7), 1621–1632 (2002).
2. Barkade, S. S., Naik, J. B., and Sonawane, S. H. Ultrasound assisted miniemulsion synthesis of polyaniline/Ag nanocomposite and its application for ethanol vapor sensing. Colloids and Surfaces A: Physicochemical and Engineering Aspects, 378(1–3), 94–98 (2011).
3. Benito, S., Graff, A., Stoenescu, R., Broz, P., Saw, C., Heider, H., Marsch, S., Hunziker, P., and Meier, W. Polymer nanocontainers for biomedical applications. European Cells and Materials,6(1), 21–22 (2003).
4. Bhanvase, B. A. and Sonawane, S. H. New approach for simultaneous enhancement of anticorrosive and mechanical properties of coatings: application of water repellent nano CaCO3–PANI emulsion nanocomposite in alkyd resin. Chemical Engineering Journal, 156(1), 177–183(2010).
5. Borisova, D., Mohwald, H., and Shchukin, D. G. Mesoporous silica nanoparticles for active corrosion protection. ACS Nano, 5 (3), 1939–1946 (2011).
6. Cao, P. G., Yao, J. L., Zheng, J. W., Gu, R. A., and Tian, Z. Q. Comparative study of inhibition effects of benzotriazole for metals in neutral solutions as observed with surface enhanced Raman spectroscopy. Langmuir,18(1), 100–104 (2002).
7. Chico, B., Simancas, J., Vega, J., Granizo, N., Diaz, I., De la Fuente, D., and Morcillo, M. Anticorrosive behaviour of alkyd paints formulated with ion-exchange pigments. Progress in Organic Coatings, 61(2–4), 283–290 (2008).
8. Evaggelos, M., Ioannis, K., George, P., and George, K. Release studies of corrosion inhibitors from cerium titanium oxide nanocontainers. Journal of Nanoparticle Research, 13(2), 541–554 (2011).
9. George, C., Dorfs, D., Bertoni, G., Falqui, A., Genovese, A., Pellegrino, T., Roig, A., Quarta, A., Comparelli, R., Curri, M. L., Cingolani, R., and Manna, L. A cast-mold approach to iron oxide and Pt/iron oxide nanocontainers and nanoparticles with a reactive concave surface. Journal of American Chemical Society, 133(7), 2205–2217 (2011).
10. Hu, Y., Chen, Y., Chen, Q., Zhang, L., Jiang, X., and Yang, C. Synthesis and stimuli-responsive properties of chitosan/poly(acrylic acid) hollow nanospheres. Polymer, 46(26), 12703–12710 (2005).
11. Isac, J. and Ittyachen, M. A. Growth and characterization of rare-earth mixed single crystals of samarium barium molybdate. Bulletin of Materials Science,15(4), 349–353(1992).
12. Jose, E., Pereira, S., Susana, I., Cordoba, T., and Roberto, M. T. Polyaniline acrylic coatings for corrosion inhibition: the role played by counter-ions. Corrosion Science, 47(3), 811–822 (2005).
13. Kalendova, A. and Vesely, D. Study of the anticorrosive efficiency of zincite and periclase-based core-shell pigments in organic coatings. Progress in Organic Coatings, 4(1), 5–19 (2009).
14. Kartsonakis, I. A. and Kordas, G. Synthesis and characterization of cerium molybdate nanocontainers and their inhibitor complexes. Journal of the American Ceramic Society, 93(1), 65–73 (2009).
15. Kartsonakis, I., Daniilidis, I., and Kordas, G. Encapsulation of the corrosion inhibitor 8-hydroxyquinoline into ceria nanocontainers. Journal of Sol-Gel Science and Technology, 48(1–2) 24–31 (2008).
16. Kim, B. J., Oh, S. G., and Im, S. S. Investigation on the solubilization locus of aniline – HCl salt in SDS micelles with 1H–NMR spectroscopy. Langmuir, 17, 565–266 (2001).

17. Kumar, A., Stephenson, L. D., and Murray, J. N. Self-healing coatings for steel. Progress in Organic Coatings, 55(3), 244–253 (2006).

18. Lu, X., Yu, Y., Chen, L., Mao, H., Wang, L., Zhang, W., and Wei, Y. Poly(acrylic acid)-guided synthesis of helical polyaniline microwires. Polymer, 46(14), 5329–5333 (2005).

19. Marques, A. P. A., Melo, D. M. A., Paskocimas, C. A., Pizani, P. S., Joya, M. R., Leite, E. R., and Longo, E. Photoluminescent BaMoO4 nanopowders prepared by complex polymerization method (CPM). Journal of Solid State Chemistry, 179(3), 658–678 (2006).

20. Matheswaran, P. and Ramasamy, A. K. Influence of benzotriazole on corrosion inhibition of mild steel in citric acid medium. E-Journal of Chemistry,7(3), 1090–1094 (2010).

21. Mennucci, M. M., Banczek, E. P., Rodrigues, P. R. P., and Costa, I. Evaluation of benzotriazole as corrosion inhibitor for carbon steel in simulated pore solution. Cement and Concrete Composites, 31(6) 418–424 (2009).

22. Murphy, E. B. and Wudl, F. The world of smart healable materials. Progress in Polymer Science,35(1–2), 223–251 (2010).

23. Nesterova, T., Dam-Johansen, K., and Kiil, S. Synthesis of durable microcapsules for self-healing anticorrosive coatings: a comparison of selected methods. Progress in Organic Coatings, 70(4), 342–352 (2011).

24. Paliwoda-Porebska, G., Stratmann, M., Rohwerder, M., Potje-Kamloth, K., Lu, Y., and Pich, A. Z. On the development of polypyrrole coatings with self-healing properties for iron corrosion protection. Corrosion Science, 47(12), 3216–3233 (2005).

25. Phuruangrat, A., Thongtem, T., and Thongtem, S. Synthesis of nanocrystalline metal molybdates using cyclic microwave radiation. Materials Science-Poland,28(2), 557–563 (2010).

26. Popova, A. and Christov, M. Evaluation of impedance measurements on mild steel corrosion in acid media in the presence of heterocyclic compounds. Corrosion Science, 48(10), 3208–3221 (2006).

27. Raj, A. M. E. S., Mallika, C., Swaminathan, K., Sreedharan, O. M., and Nagaraja, K. S. Zinc (II) oxide-zinc (II) molybdate composite humidity sensor. Sensors and Actuators B: Chemical, 81(2–3), 229–236 (2002).

28. Ramesh, S. and Rajeswari, S. Corrosion inhibition of mild steel in neutral aqueous solution by new triazole derivatives. Electrochimica Acta, 49(5), 811–820 (2004).

29. Revie, R. W. Corrosion and corrosion control, 4th ed., John Wiley & Sons, New Jersey (2008).

30. Ryu, J. H., Yoon, J. W., Lim, C. S., Oh, W. C., and Shim, K. B. Microwave-assisted synthesis of camoo4nano-powders by a citrate complex method and its photoluminescence property.Journal of Alloys and Compound, 390(1–2) 245–249 (2005).

31. Schweitzer, P.A. Paint and coatings: applications and corrosion Resistance, CRC Press (2006).

32. Selvi, S. T., Raman, V., and Rajendran, N. Corrosion inhibition of mild steel by benzotriazole derivatives in acidic medium. Journal of Applied Electrochemistry, 33(12), 1175–1182 (2003).

33. Shchukin, D. G. and Möhwald, H. Surface-engineered nanocontainers for entrapment of corrosion inhibitors. Advanced Functional Materials, 17(9), 1451–1458 (2007).

34. Shchukin, D. G., Lamaka, S. V., Yasakau, K. A., Zheludkevich, M. L., Ferreira, M. G. S., and Mohwald, H. Active anticorrosion coatings with halloysite nanocontainers. The Journal of Physical Chemistry C, 112(4) 958–964 (2008).

35. Shchukin, D. G., Zheludkevich, M., Yasakau, K., Lamaka, S., Ferreira, M. G. S., and Mowald, H. Layer-by-layer assembled nanocontainers for self healing corrosion protection. Advanced Materials, 18(13), 1672–1678 (2006).

36. Sun, Y., Macdiarmid, A. G., and Epstein, A. J. Polyaniline: synthesis and characterization of pernigraniline base. Journal of the Chemical Society, Chemical Communications, 7, 529–531 (1990).
37. Suryanarayana, C., Rao, K. C., and Kumar, D. Preparation and characterization of microcapsules containing linseed oil and its use in self-healing coatings. Progress in Organic Coatings, 63 (1), 72–78 (2008).
38. Takeuchi, M., Martra, G., Coluccia, S., and Anpo, M. Investigations of the structure of H2O clusters adsorbed on TiO2 surfaces by near-infrared absorption spectroscopy. TheJournalof PhysicalChemistryB, 109(15), 7387–7391 (2005).
39. Tedim, J., Poznyak, S. K., Kuznetsova, A., Raps, D., Hack, T., Zheludkevich, M. L., and Ferreira, M. G. S., Enhancement of active corrosion protection via combination of inhibitor-loaded nanocontainers. Applied Materials and Interfaces, 2(5), 1528–1535 (2010).
40. Wu, D. Y., Meure, S., and Solomon, D. Self-healing polymeric materials: a review of recent developments.Progress in Polymer Science,33(5), 479–522 (2008).
41. Yang, J., Lee, J., Kang, J., Lee, K., Suh, J., Yoon, H., Huh, Y., and Haam, S. Hollow silica nanocontainers as drug delivery vehicles. Langmuir, 24(7), 3417–3421 (2008).
42. Yoshino, K., Fukushima, T.,and Yoneta, M. Structural, optical and electrical characterization on ZnO film grown by a spray pyrolysis method. Journal of Material Science: Materials in Electronics, 16(7), 403–408 (2005).
43. Zheludkevich, M. L., Serra, R., Montemor, M. F., and Ferreira, M. G. S. Oxide nanoparticle reservoirs for storage and prolonged release of corrosion inhibitors. Electrochemistry Communications, 7(8), 836–840 (2005).
44. Zheludkevich, M. L., Shchukin, D. G., Yasakau, K. A., Möhwald, H., and Mario, G. S. Anticorrosion coatings with self-healing effect based on nanocontainers impregnated with corrosion inhibitor. Chemistry of Materials, 19(3), 402–411 (2007).

13.5 CASE STUDY IV- APPLICATION OF GOLD NANOPARTICLES (AUNPS) IN BIOSENSORS

13.5.1 INTRODUCTION

This chapter deals mainly with application of gold nanoparticles (AuNPs) in biosensors. Synthetic approaches, based on physical and on chemical procedures, as well as structural and optical features, are also described. Owing to the large surface-to-volume ratio, high surface reaction activity, high catalytic efficiency, strong adsorption ability and their biocompatibility, numerous biosensors have been constructed using gold nanoparticles with higher selectivity, better stability and a lower detection limits. Not only optical properties, electrochemical properties of AuNPs have also been harnessed. Still, development of new protocols for preparing functionalized gold nanoparticles and using them for biosensing is an active research area. The synthesis and application of nanoparticles (NP) have fueled the growth of nanotechnology, foundation of which is based on their size and shape. Owing to their nanometer size, high surface-to-volume ratio, and ability to couple with surface plasmons of neighboring metal particles or with electromagnetic wave, they have a variety of interesting spectroscopic, electronic, and chemical properties different than their bulk counterparts [37, 138, 156]. Particular interest has been focused on the noble metal nanoparticles as technologically they play vital role in important fields such as catalysis [102], optoelectronic devices[66] and surface-enhanced Raman scattering (SERS) [19,75]. As their size scale is similar to that of biological molecules (e.g., proteins, DNA) and structures (e.g., viruses and bacteria), they are enormously exploited for various biomedical applications also like biosensing [67,136], imaging [65], gene and drug delivery [71,132,162]. Among various noble metal NPs, colloidal gold (CG) NPs have gained much more attention and are being extensively studied because of different synthetic approaches, variable size, good biocompatibility, relatively large surface, and their variable optical behavior and catalytic properties [76,127,139]. Their optical and electrical properties are known to be dramatically affected by their size, shape, and surrounding surface environments [84,105,170], which in turn depend on the methods used for the synthesis, molar ratio of reductant to gold ion, type of stabilizer [10], reaction time, and temperature [121], pH [129] and refractive index [115,119,138].

13.5.2 HISTORY OF GOLD NANOPARTICLES:

Gold is the first metal discovered by humans, and is the subject of one of the most ancient themes of investigation in science. The extraction of gold started in the 5 th millennium B.C. near Varna (Bulgaria) and reached 10 tons per year in Egypt around 1200–1300 B.C. First data on CG can be found in treatises by Chinese, Arabic and Indian scientists, who prepared CG and used it, in particular, for medical purposes as early as 5 ± 4th centuries B. C. Colloidal gold was used to make ruby glass and for coloring ceramics, and these applications are still continuing now. The most famous example is the Lycurgus Cup that was manufactured in the 5th to 4th century B.C, using silver and gold NPs in approximate ratio of 7:3. The presence of these metal NPs gives special color display for the glass. When viewed in reflected light, for example in daylight, it appears green. However, when a light is shone into the cup and transmitted through the glass, it appears red. This glass can still be seen in British museum.

The beginning of scientific research on CG dates back to the mid-19th century, when Michael Faraday published an article on synthesis and properties of CG [43]. In this article, Faraday described, for the first time, aggregation of CG in the presence of electrolytes, the protective effect of gelatin and other high-molecular-mass compounds and the properties of thin films of CG. Colloidal gold solutions prepared by Faraday are still stored in the Royal Institution of Great Britain in London. Richard Zsig-Mondy [174] was the first to describe the methods of synthesis of CG with different particle sizes using different reducing agents. Zsigmondy used colloidal gold as the main experimental object when inventing (in collaboration with Siedentopf) an ultra-microscope. In 1925, Zsigmondy was awarded the Nobel Prize in 'Chemistry for his demonstration of the heterogeneous nature of colloid solutions and for the methods he used, which have since became fundamental in modern colloid chemistry'112]. Studies by the Nobel Prize laureate Theodor Svedberg on the preparation, analysis of mechanisms of colloidal gold formation and their sedimentation properties (with the use of the ultra-centrifuge he had invented) are among important studies [133]. Now days, CG is used by scientists as a perfect model for studies of optical properties of metal particles, the mechanisms of aggregation and stabilization of colloids. As a result, it has led to the exponential increase in number of research efforts for applications in almost every sphere such as biological detection, controlled drug delivery, low-threshold laser, optical filters, and also sensors, among others [4,16,72]. Various researchers have reviewed the synthesis, properties and role of NPs in various fields [9,37,48,92,114,117].

13.5.3 SYNTHESIS METHODS FOR GOLD NANOPARTICLES (AUNPS)

The methods used for the synthesis of metal NPss in colloidal solution are very important as they control the size and shape of NPs, which in turn affects their properties. Moreover, successful utilization of NPs in biological assays relies on the availability of nanomaterials in desired size, their morphology, water solubility and surface functionality. Several reviews on the synthesis of nanoparticles are available [93]. Some reviews dedicatedly covered the synthesis of gold nanoparticles [109,127,140]. Synthesis methods of CG (and other metal colloids) can arbitrarily be divided into following two major categories:

Dispersion methods (metal dispersion) It has been checked and is correct.

Condensation methods (reduction of the corresponding metal salts).

- Dispersion methods are based on destruction of the crystal lattice of metallic gold in high-voltage electric field [28,30]. The yield and shape of AuNPs formed under electric current depend not only on the voltage between electrodes and the current strength, but also on the presence of electrolytes in the solution. The addition of even very small amount of alkalis or chlorides and the use of high-frequency alternating current for dispersion can substantially improve the quality of gold hydrosols.

- Condensation methods are more commonly employed than dispersion methods. CG is most often prepared by reduction of gold halides (for example, of HAuCl4) with the use of chemical reducing agents and/or irradiation. The systematic adjustment of the reaction parameters, such as reaction time, temperature, concentration, and the selection of reagents and surfactants can be used to control size, shape, and quality of NPs. The condensation methods can further be classified into:

Citrate Reduction Method

The Brust–Schiffrin Method: Two-Phase Synthesis and Stabilization by Thiols

Micro-emulsion, Reversed Micelles, Surfactants, Membranes, and Poly-electrolytes

Seeding-Growth method

Physical Methods (ultrasonic, UV, IR or ionizing radiation or laser photolysis)

CITRATE REDUCTION METHOD:

The simplest and most commonly used method for the preparation for AuNPs is the aqueous reduction of HAuCl4 by sodium citrate at boiling point [142,159]. Particles synthesized by this method are nearly mono-dispersed having controlled size which in turn depends on the initial reagent concentrations [57,109] and can be easily characterized by their plasmon absorbance band at about 520 nm for 10-15 nm sized particles. NPs from other noble metals may also be prepared by citrate reduction, such as silver particles from AgNO3, palladium from H2[PdCl4], and platinum from H2[PtCl6] [22,80116]. The similarities in the preparation of these different metal colloids allow the synthesis of mixed-metal particles, which may have functionality different from each individual metal. Although sodium citrate is the most common reducing agent, metal NPs can also be synthesized by the use of both strong and weak reducing agents including borohydride, alcohols, hydrogen gas, gallic acid, glutamic acid, polyvinyl pyrollidone, etc. Reduction can also be achieved via photochemical reduction, thermal reduction, sonochemical reduction and electrochemical reduction, and other reducing agents[95,157]. Table 1 shows different kinds of reducing and stabilizing agents used for the synthesis of gold nanoparticles.

TABLE 1   References to common reducing agents and stabilizers used in the synthesis of NPs in colloidal solution

Important Factors for NP synthesis	Example
Reducing Agents/Reduction Methods	Alcohols [78,101]
	Hydrogen Gas [48,102,148]
	Sodium Borohydride [110,135]
	Sodium Citrate [83,173]
	Thermal Reduction [46]
	Photochemical Reduction [107,108]
	Sono-chemical Reduction [167]
	Metal Vapor Condensation[70,111,134]
	Electrochemical Reduction [70]
Stabilizers	Surfactants [128,165,168]
	Polymers [34,39,49]
	Dendrimers [104]
	Block copolymers[113,152,166]
	Other Ligands [53,96,155]

THE BRUST–SCHIFFRIN METHOD: TWO-PHASE SYNTHESIS AND STABILIZATION BY THIOLS:

Another procedure that has become extremely popular for smaller AuNPs synthesis in organic solvent is two phase reduction method developed by Brust et. al. This synthesis technique is inspired by Faraday's two-phase system and uses thiol ligands that strongly bind gold due to the soft character of both Au and S [11]. The stabilization of AuNPs with alkanethiols was first reported in 1993 by Mulvaney and Giersig, who showed the possibility of using thiols of different chain lengths for stabilization [50]. The method has had a considerable impact on the overall field in less than a decade. NPs prepared by this method can be repeatedly isolated and re-dissolved in common organic solvents without irreversible aggregation or decomposition. Subsequently, the Brust route was explored for the synthesis of wide range of monolayer protected clusters [7,20,124,126].

MICRO-EMULSION, REVERSED MICELLES, SURFACTANTS, MEMBRANES, AND POLYELECTROLYTES

The AuNPs can also be prepared by the two-phase micro-emulsion method in which, first the metal-containing reagent is transferred from an aqueous to an organic phase. After the addition of a surfactant solution to this system, a micro-emulsion, i.e., a dispersion of two immiscible liquids, is formed. In micro-emulsion methods, alkane thiols are often added to the reaction solution, and these additives form dense self-assembled monolayers on the gold surface. This method was employed for the preparation of self-assembled two- and three-dimensional ensembles of AuNPs [12,13,50].

SEEDING-GROWTH METHOD

The seeding-growth procedure is another popular technique that has been used for a century. Recent studies have successfully led to the controlled size distribution (typically 10-15%) in the range of 5-40 nm, where the sizes can be manipulated by varying the ratio of seed to metal salt [64,97,119]. The step-by-step particle enlargement is more effective than a one-step seeding method to avoid secondary nucleation [21]. Gold nanorods have been conveniently fabricated using the seeding-growth method [15].

PHYSICAL METHODS

Physical methods based on ultrasonic, UV, IR or ionising radiation or laser photolysis [91,146] and electrochemical methods [89] of reduction are much less commonly employed than chemical methods. The advantage of the former methods is that impurities of chemical compounds are absent in the resulting sols (on the metal particle surface) [41].

13.5.4 PROPERTIES OF GOLD NANOPARTICLES

Gold nanoparticles (AuNPs) possess many physical and chemical properties owing to their easy synthesis, unique optical properties, high surface to volume ratio, biocompatibility, and tunable shape and size. All these properties make them excellent scaffolds for the fabrication of chemical and biological sensors [8,14,117,165]. Some reviews are available which have discussed the properties of nanomaterials in detail [36,154]. Figure 1 shows different types of properties of AuNPs, which are discussed in detail in following sections.

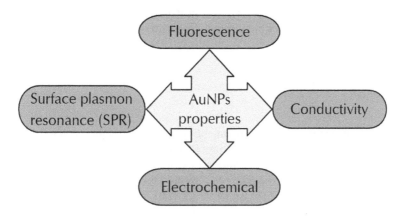

FIGURE 1   Properties of gold nanoparticles (AuNPs).

OPTICAL PROPERTIE

SURFACE PLASMON BAND (SPB)

The SPB is due to the collective oscillations of the electron gas at the surface of NPs (6s electrons of the conduction band for AuNPs) that is correlated with the electromagnetic field of incoming light, i.e., the excitation of the coherent oscillation of the conduction band. The nature of SPB was rationalized in a master publication authored by Mie in 1908 [99]. According to Mie theory, the total cross section composed of the SP absorption and scattering is given as a summation over all electric and magnetic oscillations. Mie theory attributes the plasmon band of spherical particles to the dipole oscillations of free electrons in conduction band occupying the energy states immediately above the Fermi energy level.

The main characteristics of the SPB are (i) its position around 520 nm; (ii) its sharp decrease with decreasing core size for AuNPs due to the onset of quantum size effects. The decrease of SPB intensity with decrease in particle size is accompanied by broadening of the Plasmon bandwidth and (iii) step like spectral structures indicating transitions to the discrete unoccupied levels of the conduction band. The SPB maximum and bandwidth are however influenced by the particle shape, medium dielectric constant, and temperature. The refractive index of the solvent has been shown to induce a shift of the SPB, as predicted by Mie theory. But since all AuNPs need some kind of stabilizing ligands or polymer, the band energy is rarely exactly as predicted by Mie theory. Applications of the sensitivity of the position of the SPB are known, especially in the fields of sensors and biology. A shift of the SPB of AuNPs has been measured upon adsorption of gelatin, and quantitative yield measurements of the adsorbed amount were obtained[24]. Phase transfer of dodecylamine-capped AuNPs dispersed in an organic solvent into water containing the surfactant cetyltrimethylammonium bromide (CTAB) was monitored by color changes initiated upon shaking [141]. Surface interaction of AuNPs with functional organic molecules was probed by monitoring the shifts of SPB position.

FLUORESCENT PROPERTIES

AuNPs have attracted increasingly attention for the unique nano-optics properties. These fascinating optical properties, including those of surface plasmon band (SPB), surface-enhanced Raman scattering (SERS) and Raleighresonance scattering (RRS), have been well documented. In contrast, the studies on photoluminescence from AuNPs are very limited [58,60,61,77,82,169]. Shen et. al., [125] used fluorescent AuNPs for the detection of 6-mercaptopurine. The fluorescent AuNPs with mean diameter of ~15 nm were synthesized in aqueous solution and significant enhancement in fluorescence emission was observed upon AuNPs self-assembly with 6MP. A new fluorescent method for sensitive detection of biological thiols in human plasma was developed using a near-infrared (NIR) fluorescent dye, FR 730 [122]. The sensing approach was based on the strong affinity of thiols to gold and highly efficient fluorescent quenching ability of AuNPs. In the presence of thiols, NIR fluorescence would enhance dramatically due to desorption of FR 730 from the surfaces of AuNPs, which allowed the analysis of thiol-containing amino acids in a very simple approach. The size of Au NPs was found to affect the fluorescent assay and the best response for cysteine detection was achieved when using AuNPs of diameter 24 nm. Visible luminescence has been reported for water-soluble AuNPs, for which a hypothetical mechanism involving 5 d10 f6 (sp)1 inter-band transition has been suggested[58,100].

ELECTROCHEMICAL PROPERTIES

Just as ionic space charges, the electrical double layer exist at all electrified metal/electrolyte solution interfaces, nanoparticles in solutions (colloids, metal sols, regardless of the metal, and semiconductor nanoparticles) have double layers with ionic surface excesses on the solution side that reflect any net electronic charge residing on metal NP surface (or its capping ligand shell). In this light, one can say that all metal-like nanoparticles are intrinsically electroactive and act as electron donor/ acceptors to the quantitative extent of their double-layer capacitances. The electron charge storage capacity per nanoparticle however depends on the nanoparticle size (surface area), nanoparticle double-layer capacitance (CDL), and potential (relative to nanoparticle zero charge). This capacity can be quite substantial; for example, a 10-nm-diameter nanoparticle with CDL = 120 aF (equivalent to 40 μF/cm2) can store ~750 e/V. This capacity is capable, as a "colloidal microelectrode", of driving electrochemical reactions such as proton reduction toH2. The quantitative demonstrations of this property by Henglein [52] represented the beginning of modern understanding of the electrochemistry of metal NPs. In the earliest experiments, NPs were charged by chemical reactions. The transition to electrochemical control by Ung et. al., [144] was made by showing that solutions of Ag NPs capped with poly(acrylic acid) could also be charged at (macroscopic) working electrodes, diffusing to undergo electron transfer at electrode/electrolyte interfaces. Murray thoroughly reviewed the electrochemical properties of metal nanoclusters, such as Ag nanoclusters and gold nanoclusters [100]. However, no report so far is available about the quantitative characterization of AuNP population by their redox current. The reason is probably that the faradic current is intrinsically small at the sub-micro ampere or nano-amper level, which yields a very poor sensitivity and limit of detection. When gold clusters are prepared and used for different applications, their populations are usually very low. When the nano-clusters are small enough, they have molecular-like electrochemical voltammograms and it has been observed that the electrochemical current is determined by the diffusion of the molecular-like clusters [98]. The diffusion coefficients of these nanoclusters are around or smaller than ~10−6 cm2/s, which is close to or smaller than "usual" inorganic ions in their water solutions. Therefore, both the Cottrell equation and Levich equation apply to relate the Faradic current and population of gold clusters. An amperometric analysis can be applied because the faradic current is intrinsically proportional to the gold cluster population. When relatively larger gold clusters (14–28 kDa) were examined, quantized double layer charging/discharging voltamograms showed distinguishable peaks of 1e- transfers. In these dispersions, the gold nanoclusters behave as quantum capacitors, instead of electroactive species. When the AuNPs are larger than 5 nm, they are not considered like molecules. Each AuNP consists of a number of gold atoms. For example, assuming the AuNPs as perfect spheres and their densities are identical to bulk gold, AuNP of 5 nm diameter, 10 nmdiameter, or 20 nm diameter contains about 3.8 × 103, 3.1 × 104, 2.5 × 105 gold atoms, respectively. Therefore, when the average diameter of the AuNPs is a known value, the number of AuNP can be estimated by analyzing the quantity of element Au.

13.5.5   *BIOSENSORS AND NANOMATERIALS*

BIOSENSORS

Analytical chemistry plays an important role in our everyday life because almost every sector of industry and public service relies on quality control. Majority of chemical analysis methods are time-consuming and heavily employ expensive reagents and equipment in order to achieve high selectivity and low detection limits. Biosensors emerge as upbeat technology to face this challenge.

A biosensor is generally defined as an analytical device which converts a biological response into a quantifiable and processable signal. Figure 2 shows schematically the parts comprising a typical biosensor: a) bio-receptors that specifically bind to the analyte; b) an interface architecture where a specific biological event takes place and gives rise to a signal picked up by c) the transducer element; the transducer signal (which could be anything from the in-coupling angle of a laser beam to the current produced at an electrode) is converted to an electronic signal and amplified by a detector circuit using the appropriate reference and sent for processing.

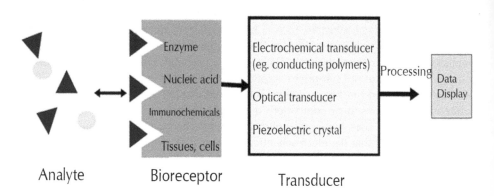

FIGURE 2   Components of a biosensor.

Successful biosensor should meet the following conditions:
- The biocatalyst must be highly specific for the purpose of the analysis.
-     Be stable under normal storage conditions and show a low variation between assays.
- The reaction should be independent of physical parameters such as stirring, pH and temperature.
- The response should be accurate, precise, reproducible and linear over the concentration range of interest, without dilution or concentration.
- It should also be free from electrical or other transducer induced noise.

- If the biosensor is to be used for invasive monitoring in clinical situations, the probe must be tiny and biocompatible, having no toxic or antigenic effects.
- For rapid measurements of analytes from human samples, biosensor should provide real-time analysis.
- The complete biosensor should be cheap, small, portable and capable of being used by semi-skilled operators.

Typical recognition elements used in biosensors are: enzymes, nucleic acids, antibodies, whole cells, and receptors. Of these, enzymes are among the most common. [40] Taking into account the biomolecule that recognizes the target analyte, biosensors can be named as (i) affinity sensors, when the bioreceptor uses non-covalent interactions like antibody-antigen reactions or DNA strand hybridization, and (ii) catalytic or enzyme sensors, when the analyte is the enzyme substrate, or it can be detected by measuring the signal produced by one substrate or product of the enzymatic reaction involving the analyte.

Affinity biosensors make use of the specific capabilities of an analyte to bind to a biorecognition element. This group can be further divided into immunosensors (which rely on specific interactions between an antibody and an antigen), nucleic acid biosensors (which make use of the affinity between complementary oligonucleotides), and biosensors based on interactions between an analyte (ligand) and a biological receptor. Some whole-cell biosensors act as recognition elements responding to (trigger) substances by expressing a specific gene. Catalytic biosensors make use of biocomponents capable of recognizing (bio) chemical species and transforming them into a product through a chemical reaction. This type of biosensor is represented mostly by enzymatic biosensors, which make use of specific enzymes or their combinations. Many whole cell biosensors also rely on biocatalytic reactions.

Biosensors are also classified according to the parameter that is measured by the physicochemical transducer of the biological event. Thus, classically biosensors are grouped into optical, electrochemical, acoustic and thermal ones. Optical transducers of most common enzyme biosensors are based on optical techniques such as absorption, reflectance, luminescence, chemi-luminescence, evanescent wave, surface plasmon resonance, and interferometry.

The early era of biosensing research and development was first sparked with the defining paper by Clark [32] and his invention of the oxygen electrode in 1955/56 [31]. The subsequent modification of the oxygen electrode led up to another publication in 1962 [33], which reported the development of the first glucose sensor and the enhancement of electrochemical sensors (e.g. polarographic, potentiometric and conductometric) with enzyme-based transducers. Soon after Clark's proposal, Updike et. al., [144] introduced modifications to this first approach to avoid oxygen concentration dependence. Clark's work and the subsequent transfer of his technology to Yellow Spring Instrument Company (Ohio, USA) led to the successful commercial launch of the first dedicated glucose biosensor in 1975. Since then, many serious players in the field of medical diagnostics, such as Bayer, Boehringer Mannheim, Eli Lilly, Lifescan, DKK Corporation etc., invested in the development and the mass scale production of biosensors, which are utilized in health care [3,143], environmental monitoring [38,143], food and drink [74], the process industries [143], defense and security.

AUNP BASED BIOSENSOR

Biosensors are quickly becoming prevalent in modern society and have many applications, including detecting biological hazards and diagnosing certain diseases. With the recent advances in nanotechnology, nanomaterials have received great interests in the field of biosensors due to their exquisite sensitivity in chemical and biological sensing [63]. Nanoparticles, with their unique, size-dependent properties, are an extremely promising technology for biosensor creation. They offer key advantages through increased biocompatibility and a method of simple visual recognition of sensing, although this takes place at the cost of some of the resolution found in other types of biosensors. Owing to the unique properties of nanomaterials, direct electrochemistry and catalytic activity of many proteins have been observed at electrodes modified with various nanomaterials, semiconductor nanoparticles, and metal nanoparticles [23,88,120,146,158]. Various nanostructures have been examined as hosts for protein immobilization via approaches including protein adsorption, covalent attachment, protein encapsulation, and sophisticated combinations of methods. Studies have shown that nanomaterials can not only provide a friendly platform for the assembly of protein molecules but also enhance the electron-transfer process between protein molecules and the electrode.

Metal nanoparticles have many unique properties like large surface-to-volume ratio, high surface reaction activity, high catalytic efficiency, and strong adsorption ability. The conductivity properties of nanoparticles at nanoscale dimensions allow the electrical contact of redox-centers in proteins with electrode surfaces[54,58].These nanoparticles have been used to facilitate the electron transfer in nanoelectronic devices owing to the roughening of conductive sensing interface, catalytic properties of the nanoparticles. Among the nanomaterials used as component in biosensors, gold nanoparticles (AuNPs) have received greatest interests because they have several kinds of intriguing properties [148,149]. Gold colloid has many advantages for biosensor applications like gold nanoparticles can provide a stable surface for the immobilization of biomolecules, such that the molecules retain their biological activities. Modification of an electrode surface with gold nanoparticles provides a microenvironment similar to that of the redox proteins in native systems and offers the protein molecules more freedom in orientation, which can weaken the insulating property of the protein shell for the direct electron transfer and facilitate the electron transfer through the conducting tunnels of colloidal gold [86]. Gold nanoparticles can form conducting electrodes and are the site of electron transfer when anchored to the substrate surface, allowing direct electron transfer between redox proteins and electrode surfaces with no mediators required [161]. They can act as an electron-conducting pathway between prosthetic groups and electrode surface. Moreover, AuNPs have an ability to permit fast and direct electron transfer between a wide range of electroactive species and electrode materials. In addition, the light-scattering properties and extremely large enhancement ability of the local electromagnetic field enables AuNPs to be used as signal amplification tags in diverse biosensors. The roles that AuNPs have played in the biosensing process and the mechanism of AuNPs for improving the analytical performances are discussed in upcoming sections.

AUNPS BASED OPTICAL BIOSENSORS:

Optical biosensors generally measure changes in light or photon output. For optical biosensing utilizing AuNPs, the optical properties provide a wide range of opportunities, all of which ultimately arise from the collective oscillations of conduction band electrons ("plasmons") in response to external electromagnetic radiation [17]. There are several optical sensing modalities for AuNPs, and the Surface Plasmon Resonance (SPR) is the one that attracted most intensive research. SPR, which is an optical phenomenon arising from the interaction between an electromagnetic wave and the conduction electrons in a metal, is used for probing and characterizing physicochemical changes of thin films on metal surface (Figure 3). This resonance is a coherent oscillation of the surface conduction electrons excited by electromagnetic radiation.

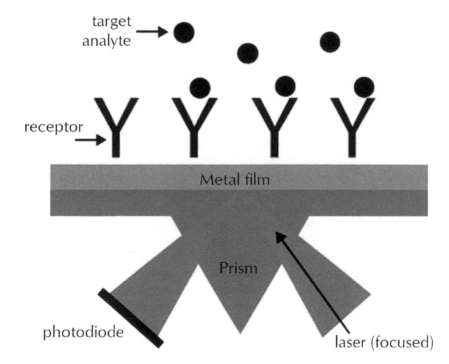

FIGURE 3   Surface plasmon resonance detection unit [36].

Due to their unique size, gold nanoparticles selectively absorb and reflect certain wavelengths in the visible range of light. This range depends on the size of the particles, with roughly spherical gold nanoparticles less than ~40 nm in diameter appearing red and shifting in color from pink to purple as the size of the particles increases. The same color-shifting effect can be achieved by bringing two smaller gold nanoparticles together so that their absorbance properties behave as if the smaller particles were a larger single particle. This effect lasts only as long as the particles are in sufficient proximity to each other, enabling the creation of a sensing mechanism [73,147]. This sensing mechanism has previously been demonstrated by Guarise et. al., [51] by first stabilizing 12 nm gold nanoparticles with a monothiol and then binding them together with a dithiol cleavable by hydrazine. The bound particles shift in absorbance from near 520 nm (appearing red) to a new peak near 600 nm (appearing purple). A colorimetric adenosine biosensor based on the aptazyme-directed assembly of Au NPs is reported in literature [85]. Very recently, gold nanoparticles have been used for the detection of sequence-specific DNA binding using colorimetric biosensing strategy [106]. Gold nanoparticle (AuNP)-based highly sensitive and colorimetric detection of the temporal evolution of superoxide dismutase (SOD1) aggregates have been carried out by Hong et. al.[55]. Combination of unlabeled DNAzyme and AuNPs have been employed for colorimetric detection of lead (Pb+2) [152]. The binding of specific molecules onto the surface of metallic films can induce a variation in the dielectric constant, which can cause a change in the reflection of laser light from a metal-liquid surface and have been studied intensively in SPR to provide better analytical characteristics. The signal amplification mechanism of AuNPs can be generally summarized into two points: (i) the electronic coupling between the localized surface plasmon of AuNPs and the propagating plasmon on the SPR gold surface and (ii) the high density and high molecular weight of AuNPs increase the apparent mass of the analytes immobilized on them.

Gold nanoparticles have also been used for optical virus detection using spatial arrangement of Au nanoparticles on the virus-like particles (VLP) surface [103]. This structure produces a red shift in the absorption spectrum due to Plasmon coupling between adjacent Au particles, leading to the construction of an optical virus detection system. A fiber-based biosensor has been developed for organophosphorous pesticide determination using LSPR effect of AuNPs [81]. An acetylcholinesterase (AChE) layer has been self-assembled by covalent coupling onto the GNP layer. When suitable pesticides presented, the activity of AChE to hydrolyze acetylcholine chloride would be inhibited leading to the change of the light attenuation due to a local increase of the refractive index. The comparative study of the fiber sensor with and without AuNPs suggested that the AuNPs coated on optical fiber can substantially enhance the sensitivity of the sensor. Since GNP-based SPR biosensors can be fabricated into an array format as well, Matsui et. al.[95] demonstrated a SPR sensor chip for detection of dopamine using dopamine-imprinted polymer gel with embedded gold nanoparticles. It was observed in the studies that SPR signal of the chip is much higher than a chip immobilizing a lower density of AuNPs or no AuNPs. Chang et. Al [25] used gold nanoparticles to develop DNA based biosensor for the detection of mercury using an amplified surface plasmon resonance as "turn-on" indicator. Inorganic mercury ion

(Hg2+) has been shown to coordinate to DNA duplexes that feature thymine–thymine (T–T) base pair mismatches. The general concept used in this approach is that the "turn-on" reaction of a hairpin probe via coordination of Hg2+ by the T–T base pair results in a substantial increase in the SPR response, followed by specific hybridization with a gold nanoparticle probe to amplify the sensor performance.

Surface enhanced Raman scattering (SERS) is another optic transduction mode that can greatly benefit from the use of GNPs Cao et al. reported a multiplexed detection method of oligonucleotide targets using oligonucleotides functionalized nanoparticles and Raman-active dyes [56]. The gold nanoparticles facilitated the formation of a silver coating that acted as a promoter for the Raman scattering of the dyes. High sensitivity down to the 20 fM DNA level has been reported. Gold nanoparticles can also be used to enhance the fluorescence signal of a labeled antibody. Simonian et. al.[130] used gold nanoparticles for organophorous analytes using fluorophore (7-hydroxy-9H-(1, 3-dichloro-9,9-dimethylacridin-2-one) phosphate, which binds weekly to the active site of an enzyme, that in turn binds covalently to the gold nanoparticles. The fluorescence intensity of the fluorophore is significantly enhanced through the strong local electric field of the gold nanoparticle. The sensor showed linear response in μm when exposed to paraoxon solutions. Sun et al [132] used gold nanoparticles for fabrication of biosensor for BSA using self- assembly of butyl rhodamine B fluorescent dye. A recent report presents a method of preparing well-ordered nanoporous gold arrays using a porous silicon (PSi) template to fabricate an optical DNA biosensor [44]. The mechanism of the optical response caused by DNA hybridization on the Au–PSi surface was qualitatively explained by the electromagnetic theory and electrochemical impedance spectroscopy (EIS).

Current gold nanoparticle solution based methods of biosensing are not limited strictly to this one type of nanoparticle but can incorporate other particles as well. Peptide linked gold nanoparticle – quantum dot biosensors have been created by Chang et al., that rely on the ability of the gold nanoparticles to quench the photoluminescence of the quantum dots when in their close proximity bound state [26]. The method of sensing is also considered an "on sensor" since the default state of the particles is "off" (no luminescence), and it is converted to "on" (luminescence) once sensing takes place.

AUNPS BASED ELECTROCHEMICAL BIOSENSOR

In the last few years, electrochemical biosensors created by coupling biological elements with electrochemical transducers based on or modified with gold nanoparticles have played an increasingly important role in biosensor research. Gold nanoparticles can, very usefully, provide a stable surface for the immobilization of biomolecules, such that the molecules retain their biological activities. Modification of an electrode surface with gold nanoparticles provides a microenvironment similar to that of the redox proteins in native systems and offers the protein molecules more freedom in orientation. This results in weakening of insulating property of the protein shell for direct electron transfer and facilitate the electron transfer through the conducting tunnels of colloidal gold[86], [161]. They can act as an electron-conducting pathway between prosthetic groups and electrode surface. Gold nanoparticle-modified electrode surfaces can be prepared in three ways: (a) by binding gold nanoparticles with functional groups of self-assembled monolayers (SAMs); (b) by direct deposition of nanoparticles onto the bulk electrode surface; (c) by incorporating colloidal gold into the electrode by mixing the gold with the other components in the composite electrode matrix.

The effect of gold nanoparticles (50-130 nm in diameter) on the response of tyrosinase based biosensor for phenol detection was investigated [123]. The addition of gold nanoparticles to the biosensor membrane led to improvement in the response time by a reduction of approximately 5 folds to give response times of 5-10 s. The linear response range of the phenol biosensor was also extended from 24 to 90 mM of phenol. Gold nanoparticles (AuNPs) were electrodeposited onto a glassy carbon (GC) electrode to increase the sensitivity of the tyrosinase (TYR) electrode for pesticide detection [69]. The quantitative relationships between the inhibition percentage and pesticide concentration in various water samples were measured at the TYR-AuNP-GC electrode, showing an enhanced performance attributed by the use of AuNPs. Ahirwal and Mitra recently developed an electrochemical immunosensor using gold nanoparticles and an attached antibody for investigating their sensing capabilities with cyclic voltammetry and electrochemical impedance spectroscopy [2]. The resulting data indicated that this sandwich-type immunosensor allowed for antibody stability and adequate sensitivity, suggesting its potential application for immunoassays. Wang et. al., [151] fabricated gold nanostructure modified electrodes by a simple one-step electrodeposition method for electrochemical DNA biosensor. The DNA immobilization and hybridization on gold nano-flower modified electrode was studied with the use of $[Ru(NH_3)_6]^{3+}$ as a hybridization indicator. The double-stranded DNA complex was chemi-absorbed to a gold electrode to produce an electrochemical biosensor for L-histidine (L-histidine-dependent DNAzymes) [77].

Based on surface plasmon resonance of gold nanoparticles (AuNPs), electrochemical biosensor (EC-SPR configuration) has also been demonstrated for hydrogen peroxide (H_2O_2) [172]. One advantage of the EC-SPR configuration is the ability to simultaneously obtain information about the electrochemical and optical properties of films with thicknesses in the nanometer range. Cytochrome c has been stably immobilized onto the Au/TiO_2 film and enhanced analytical performance i.e 4-fold larger than obtained at the Au/TiO_2 film was observed. The enhanced photocurrents generated are from surface plasmon resonance of AuNPs, which was confirmed by the good match between action spectrum for photocurrent changes and ultraviolet-visible light (UV-Vis) adsorption spectrum of AuNPs. Besides this characteristic, the present biosensor for H_2O_2 has also exhibited a low detection limit, a wide dynamic linear range, and good stability.

Composite based electrodes have shown improved selectivity by inhibiting the interference reaction at the electrode. The large surface area, electrochemical properties, catalytic abilities and inherent biocompatibility make composites suitable for use in amperometric biosensors. Composites of gold nanoparticles; carbon based materials, Prussian blue nanoparticles have been utilized for the fabrication of electrochemical biosensors [59,131,163,164,171]. Gold nanoparticle-coated multiwall carbon nano-tube-modified electrodes have been used for electrochemical determination of methyl parathion and hybridization of oligonucleotides [90]. In recent approaches, the utilization of DNA sequences attached on AuNPs and carbon nanotubes (CNTs) have been reported to improve the recognition power of genosensors[29,160]. A simple and sensitive sandwich-type electro-chemiluminescence immunosensor for α-1-fetoprotein (AFP) on a (nano-Au) modified glassy carbon electrode (GCE) has been developed using silica doped Ru(bpy)32+ and nano-Au composite as labels [164]. The prepared Ru-silica@Au composite nanoparticles own the large surface area, good biocompatibility and highly effective electrochemiluminescence properties. The immunosensor performed high sensitivity and wide liner for detection AFP in the range of 0.05–50 ng/mL with limit of detection of 0.03 ng/mL. Meanwhile, Liu et. al., [85,164] have reported a new electrode interface by using l-cysteine–gold particle composite immobilized in the network of a Nafion membrane on a glassy carbon electrode. This HRP biosensor exhibited good response to H2O2, and displayed the remarkable sensitivity and repeatability. An amperometric uric acid biosensor was fabricated using gold nanoparticle (AuNP)/multiwalled carbon nanotube (MWCNT) composite, which exhibited linearity from 0.01–0.8 mM with limit of detection (LOD) of 0.01 mM. The sensor measured uric acid levels in serum of healthy individuals and persons suffering from gout [27]. AuNP–CaCO3 composite has been prepared and applied for HRP biosensor [18]. AuNP–CaCO3 can retain the porous structure and inherits the advantages from its parent materials, such as satisfying biocompatibility and good solubility and dispersibility in water.

Utilization of hybrids formed by AuNPs and dendrimers has been reported as efficient systems to improve charge transfer on electrode surfaces and created the concept of electroactive nanostructured membranes (ENM) [35]. This strategy involved the utilization of dendrimer-polyamido-amine generation 4 (PAMAM G4) containing AuNPs and PVS as polyelectrolyte matrices for bilayer fabrication on ITO (indium tin oxide) electrodes. This modified electrode was utilized with redox mediator (Me) around AuNPs to improve the electrochemical performance on electrode/electrolyte. Composite of gold and poly(propyleneimine) dendrimer have been used for electrochemical DNA biosensor [5]. The DNA probe was effectively wired onto the GCE/PPI-AuNP via Au-S linkage and electrostatic interactions. AuNPs were also assembled on the surface of AgCl@PANI core-shell to fabricate AuNPs–AgCl@PANI hybrid material [45,160]. This hybrid material has been used to develop an amperometric glucose biosensor. The composite could provide a biocompatible surface for high enzyme loading and due to size effect, AuNPs could act as a good catalyst for both oxidation and reduction of H2O2. Feng et. al.,[45] employed a AuNP/PANI nanotube membrane on a glassy carbon electrode for the impedimetric sensing of the immobilization and hybridization of non-labelled DNA, thus obtaining a much wider dynamic detection range and lower detection limit for the DNA analysis.

13.5.6 CONCLUSION

The combination of low toxicity, high surface area, biocompatibility and colloidal stability allow gold nanoparticles to be safely integrated into the biosensors. Many research groups have demonstrated that they can be used for a wide range of sensing applications, ranging from chemical to biological sample. Major advancements in biosensors revolve around immobilization and interface capabilities of biological material with electrode surface. The use of nanomaterials and their composite results not only on stable immobilization matrix but also act as catalyst for many reactions, thus resulting in enhanced signal response. Looking into many newly explored properties of AuNPs, it is believed there will be a tremendous growth in coming era towards the development of AuNPs based bio-sensing devices for therapeutic and diagnostic applications, like in cancer treatment for laser ablation, chemotherapy, and so on.

KEYWORDS

- **Corrosion resistance**
- **Gold nanoparticles (AuNPs)**
- **Hematite nanoparticles**
- **Mechanical alloying**
- **Nanocontainer**
- **Phosphomolybdate (ZMP)**
- **Polysaccharides**

REFERENCES

1. Adlim, M., M., Abu Bakar, et al. "Synthesis of chitosan-stabilized platinum and palladium nanoparticles and their hydrogenation activity."Journal of Molecular Catalysis A: Chemical,212(1–2), 141–149 (2004).
2. Ahirwal, G. K. M. and C. K. "Gold nanoparticles based sandwich electrochemical immunosensor."Biosensors and Bioelectronics,25(9), 2016–2020 (2010).
3. Alcock, S. and Turner, A."Continuous analyte monitoring to aid clinical practice."Engineering in Medicine and Biology Magazine, IEEE,13(3), 319–325 (1994).
4. Andrievskii, R. A. "Directions in Current Nanoparticle Research."Powder Metallurgy and Metal Ceramics,42(11), 624–629 (2003).
5. Arotiba, O., Owino, J., et al. "An Electrochemical DNA Biosensor Developed on a Nanocomposite Platform of Gold and Poly (propyleneimine) Dendrimer."Sensors,8(11), 6791–6809 (2008).
6. Bagotsky, V. S. Fundamentals of Electrochemistry. John Wiley & Sons, Hoboken, New Jersey (2006).
7. Bhat, S. and Maitra, U."Facially amphiphilic thiol capped gold and silver nanoparticles."Journal of Chemical Sciencesi,120(6), 507–513 (2008).
8. Boisselier, E., Diallo, A. K., et al. "Encapsulation and Stabilization of Gold Nanoparticles with "Click" Polyethyleneglycol Dendrimers."Journal of the American Chemical Society,132(8), 2729–2742 (2010).

9. Bönnemann, H. and Richards, R. M."Nanoscopic Metal Particles – Synthetic Methods and Potential Applications."European Journal of Inorganic Chemistry,2001(10), 2455–2480 (2001).

10. Boopathi, S., Senthilkumar, S., et al. "Facile and One Pot Synthesis of Gold Nanoparticles Using Tetraphenylborate and Polyvinylpyrrolidone for Selective Colorimetric Detection of Mercury Ions in Aqueous Medium."Journal of Analytical Methods in Chemistry,2012(2012).

11. Brust, M., Fink, J., et al. "Synthesis and reactions of functionalised gold nanoparticles."Journal of the Chemical Society, Chemical Communications, (16), 1655–1656 (1995).

12. Brust, M. and Kiely, C. J."Some recent advances in nanostructure preparation from gold and silver particles: a short topical review."Colloids and Surfaces A: Physicochemical and Engineering Aspects,202(2–3), 175–186 (2002).

13. Brust, M., Walker, M., et al. "Synthesis of thiol-derivatised gold nanoparticles in a two-phase Liquid-Liquid system."Journal of the Chemical Society, Chemical Communications, (7), 801–802 (1994).

14. Bunz, U. H. F. and Rotello, V. M."Gold Nanoparticle–Fluorophore Complexes: Sensitive and Discerning "Noses" for Biosystems Sensing."Angewandte Chemie International Edition,49(19), 3268–3279, (2010).

15. Busbee, B. D., Obare, S. O., et al. "An Improved Synthesis of High-Aspect-Ratio Gold Nanorods."Advanced Materials,15(5), 414–416 (2003).

16. Bohren,C F and D. R. H. A. Absorption of Light by Small Particles, Wiley, New York (1983).

17. Murphy, C. J.,Gole,A. M., Hunyadi, S. E., Stone, J. W., Sisco,P. N., Alkilany,A., Kinard, B.E., and Hankins, P."Chemical sensing and imaging with metallic nanorods."Chem Commun, (Camb)5(2008).

18. Cai, W. Y., Xu,Q., et al. "Porous gold-nanoparticle-CaCO3 hybrid material: Preparation, characterization, and application for horseradish peroxidase assembly and direct electrochemistry."Chemistry of materials,18(2), 279–284 (2006).

19. Campion, A. and Kambhampati, P."Surface-enhanced Raman scattering."Chem. Soc. Rev.,27(4), 241–250 (1998).

20. Carotenuto, G. and Nicolais, L."Size-controlled synthesis of thiol-derivatized gold clusters."Journal of Materials Chemistry,13(5), 1038–1041 (2003).

21. Carrot, G., Valmalette,J. C., et al. "Gold nanoparticle synthesis in graft copolymer micelles." Colloid & Polymer Science,276(10), 853–859 (1998).

22. Cassagneau, T. and Fendler, J. H.,"Preparation and Layer-by-Layer Self-Assembly of Silver Nanoparticles Capped by Graphite Oxide Nanosheets."The Journal of Physical Chemistry B,103(11), 1789–1793 (1999).

23. Chai, F., Wang, C.,et al. "Colorimetric detection of Pb2+ using glutathione functionalized gold nanoparticles."ACS Applied Materials & Interfaces,2(5), 1466–1470 (2010).

24. Chandrasekharan, N., Kamat,P. V., et al. "Dye-Capped Gold Nanoclusters, Photoinduced Morphological Changes in Gold/Rhodamine 6G Nanoassemblies."The Journal of Physical Chemistry B,104(47), 11103–11109 (2000).

25. Chang, C. C., Lin,S., et al. "An amplified surface plasmon resonance "turn-on" sensor for mercury ion using gold nanoparticles."Biosensors and Bioelectronics,30(1), 235–240 (2011).

26. Chang, E., Miller,J. S., et al. "Protease-activated quantum dot probes."Biochemical and Biophysical Research Communications,334(4), 1317–1321 (2005).

27. Chauhan, N. and Pundir, C. S.,"An amperometric uric acid biosensor based on multi-walled carbon nanotube–gold nanoparticle composite."Analytical Biochemistry,413(2), 97–103(2011).

28. Choi, W. K., Liew,T. H., et al. "A Combined Top-Down and Bottom-Up Approach for Precise Placement of Metal Nanoparticles on Silicon."Small,4(3), 330–333 (2008).

29. Chu, H., Yan, J.,et al. "Electrochemiluminescent detection of the hybridization of oligo-nucleotides using an electrode modified with nanocomposite of carbon nanotubes and gold nanoparticles."Microchimica Acta,175(3), 209–216 (2011).

30. Cigang, X., Harm van, Z., et al. "A combined top-down bottom-up approach for introduc-ing nanoparticle networks into nanoelectrode gaps."Nanotechnology,17(14), 3333 (2006).

31. Clark, L. C. "Monitor and control of blood and tissue oxygen tensions."Transactions American, Society for Artificial Internal Organs,2(1956).

32. Clark, L. C. and Clark,E. W."A personalized history of the Clark oxygen electrode."International Anesthesiology Clinics,25(3), 1 (1987).

33. Clark, L. C. and Lyons, C."Electrode Systems For Continuous Monitoring In Cardiovas-cular Surgery."Annals of the New York Academy of Sciences,102(1), 29–45 (1962).

34. Corbierre, M. K., Cameron,N. S., et al. "Polymer-Stabilized Gold Nanoparticles with High Grafting Densities."Langmuir,20(7), 2867–2873 (2004).

35. Crespilho, F. N., Emilia Ghica,M., et al. "A strategy for enzyme immobilization on lay-er-by-layer dendrimer-gold nanoparticle electrocatalytic membrane incorporating redox mediator."Electrochemistry communications,8(10), 1665–1670 (2006).

36. Daniel, M. C. and Astruc, D."Gold Nanoparticles: Assembly, Supramolecular Chemistry, Quantum-Size-Related Properties, and Applications Toward Biology, Catalysis, and Nano technology."ChemInform,35(16), no–no (2004).

37. Daniel, M. C. and Astruc, D."Gold nanoparticles: assembly, supramolecular chemis-try, quantum-size-related properties, and applications toward biology, catalysis, and nanotechnology."Chemical reviews,104(1), 293–346 (2004).

38. Dennison, M. and Turner, A."Biosensors for environmental monitoring."Biotechnology advances,13(1), 1–12 (1995).

39. Ding, Y., Hu,Y., et al. "Polymer-assisted nanoparticulate contrast-enhancing materials."ScienceChina Chemistry,53(3), 479–486 (2010).

40. Eggins, B. R. Chemical sensors and biosensors, Wiley(2002).

41. Ershov, B. "Metal Nanoparticles in Aqueous Solu tions: Electronic, Optical, and Catalytic Properties." Ross. Khim. Zh,45(3), 20 (2001).

42. Faraday, M. "The Bakerian Lecture: Experimental Relations of Gold (and Other Metals) to Light."Philosophical Transactions of the Royal Society of London,147, 145–181 (1857).

43. Feng, J., Zhao,W., et al. "A label-free optical sensor based on nanoporous gold arrays for the detection of oligodeoxynucleotides."Biosensors and Bioelectronics,30(1), 21–27 (2011).

44. Feng, Y., Yang,T., et al. "Enhanced sensitivity for deoxyribonucleic acid electrochemical impedance sensor: Gold nanoparticle/polyaniline nanotube membranes."Analytica chi-mica acta,616(2), 144–151 (2008).

45. Fleming, D. A. and Williams, M. E."Size-Controlled Synthesis of Gold Nanoparticles via High-Temperature Reduction."Langmuir,20(8), 3021–3023 (2004).

46. Freestone, I., Meeks,N., et al. "The Lycurgus Cup — A Roman nanotechnology."Gold Bulletin,40(4), 270–277 (2007).

47. Fu, X., Wang,Y., et al. "Shape-Selective Preparation and Properties of Oxalate-Stabilized Pt Colloid."Langmuir,18(12), 4619–4624 (2002).

48. Gibson, M. I., Danial,M., et al. "Sequentially Modified, Polymer-Stabilized Gold Nanoparticle Libraries: Convergent Synthesis and Aggregation Behavior."ACS Combinatorial Science,13(3), 286–297 (2011).

49. Giersig, M. and Mulvaney, P."Preparation of ordered colloid monolayers by electrophoretic deposition."Langmuir,9(12), 3408–3413 (1993).

50. Guarise, C., Pasquato,L., et al. "Reversible aggregation/deaggregation of gold nanoparticles induced by a cleavable dithiol linker."Langmuir,21(12), 5537–5541 (2005).

51. Henglein, A. "Reactions of organic free radicals at colloidal silver in aqueous solution. Electron pool effect and water decomposition."The Journal of Physical Chemistry,83(17), 2209–2216 (1979).

52. Hermes, J. P., Sander,F., et al. "Nanoparticles to Hybrid Organic-Inorganic Superstructures."CHIMIA International Journal for Chemistry,65(4), 219–222 (2011).

53. Hernández-Santos, D., González-García, M. B., et al. "Metal-nanoparticles based electroa nalysis."Electroanalysis,14(18), 1225–1235 (2002).

54. Hong, S., Choi,I., et al. "Sensitive and Colorimetric Detection of the Structural Evolution of Superoxide Dismutase with Gold Nanoparticles."Analytical Chemistry,81(4), 1378–1382 (2009).

55. Hossain, M. K., Huang, G. G., et al. "Characteristics of surface-enhanced Raman scattering and surface-enhanced fluorescence using a single and a double layer gold nanostructure."Physical Chemistry Chemical Physics,11(34), 7484–7490 (2009).

56. Hostetler, M. J., Wingate, J. E., et al. "Alkanethiolate Gold Cluster Molecules with Core Diameters from 1.5 to 5.2 nm: Core and Monolayer Properties as a Function of Core Size."Langmuir,14(1), 17–30 (1998).

57. Huang, T. and Murray, R. W."Visible Luminescence of Water-Soluble Monolayer-Protected Gold Clusters."The Journal of Physical Chemistry B,105(50), 12498–12502 (2001).

58. Huo, Z., Zhou, Y., et al. "Sensitive simultaneous determination of catechol and hydroquinone using a gold electrode modified with carbon nanofibers and gold nanoparticles."Microchimica Acta,173(1), 119–125 (2011).

59. Hwang, Y. N., Jeong, D. H., et al. "Femtosecond Emission Studies on Gold Nanoparticles."The Journal of Physical Chemistry B,106(31), 7581–7584(2002).

60. Wilcoxon, J. P.,Martin,J. E., Parsapour, F., Wiedenman, B., and Kelley, D. F."Photoluminescence from nanosize gold clusters." J. Chem. Phys.,108, 9137(1998).

61. Jain, K. "Current status of molecular biosensors."Medical device technology,14(4), 10 (2003).

62. Jana, N. R., Gearheart,L., et al. "Evidence for Seed-Mediated Nucleation in the Chemical Reduction of Gold Salts to Gold Nanoparticles." Chemistry of materials,13(7), 2313–2322 (2001).

63. Jiang, W., Papa,E., et al. "Semiconductor quantum dots as contrast agents for whole animal imaging."Trends in Biotechnology,22(12), 607–609 (2004).

64. Kamat, P. V. "Photophysical, Photochemical and Photocatalytic Aspects of Metal Nanoparticles."The Journal of Physical Chemistry B,106(32), 7729–7744 (2002).

65. Karhanek, M., Kemp, J. T., et al. "Single DNA Molecule Detection Using Nanopipettes and Nanoparticles."Nano Letters,5(2), 403–407 (2005).

66. Katz, E., Willner,I., et al. "Electroanalytical and bioelectroanalytical systems based on metal and semiconductor nanoparticles."Electroanalysis,16(1–2), 19–44 (2004).

67. Kim, G. Y., Shim, J.,et al. "Optimized coverage of gold nanoparticles at tyrosinase electrode for measurement of a pesticide in various water samples."Journal of Hazardous Materials,156(1–3), 141–147 (2008).

68. Klabunde, K. J. and Cardenas-Trivino, C.In Active Metals: Preparation,Characterization, Applications. VCH, New York (1996).

69. Kohler, N., Sun,C., et al."Methotrexate-Modified Superparamagnetic Nanoparticles and Their Intracellular Uptake into Human Cancer Cells."Langmuir,21(19), 8858–8864 (2005).

70. Kostoff, R., Koytcheff, R., et al. "Structure of the nanoscience and nanotechnology applications literature."The Journal of Technology Transfer,33(5), 472–484 (2008).

71. Kreibig, U. and Genzel, L."Optical absorption of small metallic particles."Surface Science,156, Part 2(0), 678–700 (1985).

72. Kress-Rogers, E. Handbook of Biosensors and Electronic Noses: Medicine, Food and the Environment. CRC Press, Boca Raton, USA (1996).

73. Kwon, K., Lee,K. Y., et al. "Controlled Synthesis of Icosahedral Gold Nanoparticles and Their Surface-Enhanced Raman Scattering Property."The Journal of Physical Chemistry C,111(3), 1161–1165 (2006).

74. Lahav, M., Shipway, A. N., et al. "An enlarged bis-bipyridinium cyclophane-Au nanoparticle superstructure for selective electrochemical sensing applications."Journal of Electroanalytical Chemistry,482(2), 217–221 (2000).

75. Li, L. D., Chen,Z. B., et al. "Electrochemical real-time detection of l-histidine via self-cleavage of DNAzymes."Biosensors and Bioelectronics,26(5), 2781–2785 (2011).

76. Li, Y., Boone,E., et al. "Size Effects of PVP−Pd Nanoparticles on the Catalytic Suzuki Reactions in Aqueous Solution."Langmuir,18(12), 4921–4925 (2002).

77. Li, Y., Schluesener,H., et al. "Gold nanoparticle-based biosensors."Gold Bulletin,43(1), 29–41 (2010).

78. Lin, C. S., Khan,M. R., et al. "Platinum states in citrate sols by EXAFS."Journal of Colloid and Interface Science,287(1), 366–369 (2005).

79. Lin, T. J., Huang,K. T., et al. "Determination of organophosphorous pesticides by a novel biosensor based on localized surface plasmon resonance."Biosensors and Bioelectronics,22(4), 513–518 (2006).

80. Link, S., Beeby, A., et al. "Visible to Infrared Luminescence from a 28-Atom Gold Cluster."The Journal of Physical Chemistry B,106(13), 3410–3415 (2002).

81. Link, S. and El-Sayed, M. A."Size and Temperature Dependence of the Plasmon Absorption of Colloidal Gold Nanoparticles."The Journal of Physical Chemistry B,103(21), 4212–4217 (1999).

82. Link, S. and El-Sayed, M. A."Shape and size dependence of radiative, non-radiative and photothermal properties of gold nanocrystals."International Reviews in Physical Chemistry,19(3), 409–453 (2000).

83. Liu, J. and Lu, Y."Fast colorimetric sensing of adenosine and cocaine based on a general sensor design involving aptamers and nanoparticles."Angewandte Chemie,118(1), 96–100 (2006).

84. Liu, S., Leech,D., et al. "Application of colloidal gold in protein immobilization, electron transfer, and biosensing."Analytical letters,36(1), 1–19 (2003).

85. Liu, Y., Yuan, R., et al. "Direct electrochemistry of horseradish peroxidase immobilized on gold colloid/cysteine/nafion-modified platinum disk electrode."Sensors and Actuators B: Chemical,115(1), 109–115 (2006).

86. Luo, X. L., Xu,J. J., et al. "A novel glucose ENFET based on the special reactivity of MnO2 nanoparticles."Biosensors and Bioelectronics,19(10), 1295–1300 (2004).

87. Ma, H., Yin,B., et al. "Synthesis of Silver and Gold Nanoparticles by a Novel Electrochemical Method."ChemPhysChem,5(1), 68–75 (2004).

88. Ma, J. C. and Zhang, W. D."Gold nanoparticle-coated multiwall carbon nanotube-modified electrode for electrochemical determination of methyl parathion."Microchimica Acta,175(3), 309–314 (2011).

89. Mallick, K., Witcomb, M. J., et al. "Polymer-stabilized colloidal gold: a convenient method for the synthesis of nanoparticles by a UV-irradiation approach."Applied Physics A: Materials Science & Processing,80(2), 395–398 (2005).

90. Martin, C. R. "Membrane–Based Synthesis of Nanomaterials."Chemistry of materials,8(8), 1739–1746 (1996).

91. Masala, O. and Seshadri, R."Synthesis Routes For Large Volumes Of Nanoparticles."Annual Review of Materials Research,34(1), 41–81 (2004).

92. Matsui, J., Akamatsu,K., et al. "SPR Sensor Chip for Detection of Small Molecules Using Molecularly Imprinted Polymer with Embedded Gold Nanoparticles."Analytical Chemistry,77(13), 4282–4285 (2005).

93. Mayer, A. B. R. and Mark, J. E."Colloidal gold nanoparticles protected by water-soluble homopolymers and random copolymers."European Polymer Journal,34(1), 103–108 (1998).

94. Meli, L. and Green, P. F."Aggregation and Coarsening of LigandStabilized Gold Nanoparticles in Poly(methyl methacrylate) Thin Films."ACS Nano,2(6), 1305–1312 (2008).

95. Meltzer, S., Resch,R., et al. "Fabrication of Nanostructures by Hydroxylamine Seeding of Gold Nanoparticle Templates."Langmuir,17(5), 1713–1718(2001).

96. Menard, L. D., Gao,S. P., et al. "Sub-Nanometer Au Monolayer-Protected Clusters Exhibiting Molecule-like Electronic Behavior: Quantitative High-Angle Annular Dark-Field Scanning Transmission Electron Microscopy and Electrochemical Characterization of Clusters with Precise Atomic Stoichiometry."The Journal of Physical Chemistry B,110(26), 12874–12883 (2006).

97. Mie, G. "Beiträge zur Optik trüber Medien, speziell kolloidaler Metallösungen."Annalen der Physik,330(3), 377–445 (1908).

98. Murray, R. W. "Nanoelectrochemistry: Metal Nanoparticles, Nanoelectrodes, and Nanopores."Chemical reviews,108(7), 2688–2720 (2008).

99. Narayanan, R. and El-Sayed, M. A."Effect of Catalytic Activity on the Metallic Nanoparticle Size Distribution: Electron-Transfer Reaction between Fe(CN)6 and Thiosulfate Ions Catalyzed by PVP–Platinum Nanoparticles."The Journal of Physical Chemistry B,107(45), 12416–12424 (2003).

100. Narayanan, R. and El-Sayed, M. A."Shape-Dependent Catalytic Activity of Platinum Nanoparticles in Colloidal Solution."Nano Letters,4(7), 1343–1348 (2004).

101. Niikura, K., Nagakawa,K., et al. "Gold Nanoparticle Arrangement on Viral Particles through Carbohydrate Recognition: A Non-Cross-Linking Approach to Optical Virus Detection."Bioconjugate Chemistry,20(10), 1848–1852 (2009).

102. Nijhuis, C. A., Oncel, N., et al. "Room-Temperature Single-Electron Tunneling in Dendrimer-Stabilized Gold Nanoparticles Anchored at a Molecular Printboard."Small,2(12), 1422–1426 (2006).

103. Norman, T. J., Grant,C. D., et al. "Near Infrared Optical Absorption of Gold Nanoparticle Aggregates."The Journal of Physical Chemistry B,106(28), 7005–7012 (2002).

104. Ou, L. J., Jin,P. Y., et al. "Sensitive and Visual Detection of Sequence-Specific DNA-Binding Protein via a Gold Nanoparticle-Based Colorimetric Biosensor."Analytical Chemistry,82(14), 6015–6024 (2010).

105. Pal, A., Esumi,K., et al. "Preparation of nanosized gold particles in a biopolymer using UV photoactivation."Journal of Colloid and Interface Science,288(2), 396–401(2005).

106. Park, J. E., Atobe,M., et al. "Synthesis of multiple shapes of gold nanoparticles with controlled sizes in aqueous solution using ultrasound."Ultrasonics Sonochemistry,13(3), 237–241 (2006).

107. Pillai, Z. S. and Kamat, P. V."What Factors Control the Size and Shape of Silver Nanoparticles in the Citrate Ion Reduction Method?"The Journal of Physical Chemistry B,108(3), 945–951 (2003).

108. Pittelkow, M., Moth-Poulsen, K.,et al. "Poly(amidoamine)-Dendrimer-Stabilized Pd(0) Nanoparticles as a Catalyst for the Suzuki Reaction."Langmuir,19(18), 7682–7684 (2003).

109. Prasad, B. L. V., Stoeva,S. I., et al. "Gold Nanoparticles as Catalysts for Polymerization of Alkylsilanes to Siloxane Nanowires, Filaments, and Tubes."Journal of the American Chemical Society,125(35), 10488–10489 (2003).

110. Zsigmondy,R. and Akademishe Verlags-gesellschaft, M.B.H. 44,45,46. "Das Kolloide Gold." (1925).

111. Rahme, K., Vicendo, P., et al. "A Simple Protocol to Stabilize Gold Nanoparticles using Amphiphilic Block Copolymers: Stability Studies and Viable Cellular Uptake."Chemistry – A European Journal,15(42), 11151–11159 (2009).

112. Raveendran, P., Fu,J., et al. "Completely "Green" Synthesis and Stabilization of Metal Nanoparticles."Journal of the American Chemical Society,125(46), 13940–13941 (2003).

113. Richardson, M. J., Johnston,J. H., et al. "Monomeric and Polymeric Amines as Dual Reductants/Stabilisers for the Synthesis of Gold Nanocrystals: A Mechanistic Study."European Journal of Inorganic Chemistry,2006(13), 2618–2623 (2006).

114. Richter, J., Seidel,R., et al. "Nanoscale palladium metallization of DNA."Advanced Materials, 12(7), 507–507 (2000).

115. Saha, K., Agasti,S. S., et al. "Gold Nanoparticles in Chemical and Biological Sensing."Chemical reviews,112(5), 2739–2779 (2012).

116. Sau, T. K. and Murphy, C. J."Room Temperature, High-Yield Synthesis of Multiple Shapes of Gold Nanoparticles in Aqueous Solution."Journal of the American Chemical Society,126(28), 8648–8649 (2004).

117. Sau, T. K., Pal,A., et al. "Size Controlled Synthesis of Gold Nanoparticles using Photochemically Prepared Seed Particles."Journal of Nanoparticle Research,3(4), 257–261 (2001).

118. Schierhorn, M., Lee,S. J., et al. "Metal–Silica Hybrid Nanostructures for Surface-Enhanced Raman Spectroscopy."Advanced Materials,18(21), 2829–2832 (2006).

119. Schmid, G. and Corain, B."Nanoparticulated Gold: Syntheses, Structures, Electronics, and Reactivities."ChemInform,34(44), no–no (2003).

120. Shang, L., Yin, J.,et al. "Gold nanoparticle-based near-infrared fluorescent detection of biological thiols in human plasma."Biosensors and Bioelectronics,25(2), 269–274 (2009).

121. Sharina, A., Lee,Y., et al. "Effects of Gold Nanoparticles on the Response of Phenol Biosensor Containing Photocurable Membrane with Tyrosinase."Sensors,8(10), 6407–6416 (2008).

122. Sharma, J., Mahima,S., et al. "Solvent-Assisted One-Pot Synthesis and Self-Assembly of 4-Aminothiophenol-Capped Gold Nanoparticles."The Journal of Physical Chemistry B,108(35), 13280–13286 (2004).

123. Shen, X.C., Jiang,L. F., et al. "Determination of 6-mercaptopurine based on the fluorescence enhancement of Au nanoparticles."Talanta,69(2), 456–462 (2006).

124. Shimizu, T., Teranishi,T., et al. "Size Evolution of Alkanethiol-Protected Gold Nanoparticles by Heat Treatment in the Solid State."The Journal of Physical Chemistry B,107(12), 2719–2724 (2003).

125. Shipway, A. N., Lahav,M., et al. "Nanostructured Gold Colloid Electrodes."Advanced Materials,12(13), 993–998 (2000).

126. Shon, Y. S., Gross,S. M., et al. "Alkanethiolate-Protected Gold Clusters Generated from Sodium S-Dodecylthiosulfate (Bunte Salts)."Langmuir,16(16), 6555–6561 (2000).

127. Shou, Q., Guo,C., et al. "Effect of pH on the single-step synthesis of gold nanoparticles using PEO–PPO–PEO triblock copolymers in aqueous media."Journal of Colloid and Interface Science,363(2), 481–489 (2011).

128. Simonian, A. L., Good,T. A., et al. "Nanoparticle-based optical biosensors for the direct detection of organophosphate chemical warfare agents and pesticides."Analytica chimica acta,534(1), 69–77 (2005).

129. Song, Z., Yuan, R.,et al. "Multilayer structured amperometric immunosensor based on gold nanoparticles and Prussian blue nanoparticles/nanocomposite functionalized interface."Electrochimica Acta,55(5), 1778–1784 (2010).

130. Sun, X., Liu,B., et al. "A novel biosensor for bovine serum albumin based on fluorescent self-assembled sandwich bilayers."Luminescence,24(1), 62–66 (2009).

131. Svedberg, T. Colloid Chemistry ACS Monography,New York, Chem. Catalog. Co, 16,(1924).

132. Swihart, M. T. "Vapor-phase synthesis of nanoparticles."Current Opinion in Colloid &amp,Interface Science,8(1), 127–133 (2003).

133. Tabuani, D., Monticelli,O., et al. "Palladium Nanoparticles Supported on Hyperbranched Aramids: Synthesis, Characterization, and Some Applications in the Hydrogenation of Unsaturated Substrates."Macromolecules,36(12), 4294–4301 (2003).

134. Taton, T. A., Mirkin,C. A., et al. "Scanometric DNA Array Detection with Nanoparticle Probes."Science,289(5485), 1757–1760 (2000).

135. Templeton, A. C., Pietron,J. J., et al. "Solvent Refractive Index and Core Charge Influences on the Surface Plasmon Absorbance of Alkanethiolate Monolayer-Protected Gold Clusters."The Journal of Physical Chemistry B,104(3), 564–570 (2000).

136. Templeton,A. C., Wuelfing,W. P., et al. "Monolayer-Protected Cluster Molecules."Accounts of Chemical Research,33(1), 27–36 (1999).

137. Thanh, N. T. K. and Rosenzweig, Z."Development of an Aggregation-Based Immunoassay for Anti-Protein A Using Gold Nanoparticles."Analytical Chemistry,74(7), 1624–1628 (2002).

138. Thomas, K. G. and Kamat, P. V."Chromophore-Functionalized Gold Nanoparticles."Accounts of Chemical Research,36(12), 888–898 (2003).

139. Thomas, K. G., Zajicek,J., et al. "Surface Binding Properties of Tetraoctylammonium Bromide-Capped Gold Nanoparticles."Langmuir,18(9), 3722–3727(2002).

140. Turkevich, J., Stevenson,P. C., et al. "The Formation of Colloidal Gold."The Journal of Physical Chemistry,57(7), 670–673 (1953).

141. Turner, A. P. F. "Biosensors: Past, present and future."(1996)(Last accessed October 6, 2005).

142. Ung, T., Giersig,M., et al. "Spectroelectrochemistry of Colloidal Silver."Langmuir,13(6), 1773–1782 (1997).

143. Updike, S. J. and Hicks,G. P."The enzyme electrode."Nature,214(5092), 986–988 (1967).

144. Fabrikanos, V.A.,Athanassiou,S., andLieser, K. H."Dastelung stabiler hydrosole von gold und silber durch reduktion mit athylendiamintetraessingsaure." Z. Naturforsch. B,18, 612 (1963).

145. Yang,W.H.,Schatz,G. C., and Van Duyne, R. P."Discrete dipole approximation for calculating extinction and Raman intensities for small particles with arbitrary shapes."J. Chem. Phys.,103, 7 (1995).

146. Wang, J., Liu,G., et al. "Electrochemical stripping detection of DNA hybridization based on cadmium sulfide nanoparticle tags."Electrochemistry communications,4(9), 722–726 (2002).

147. Wang, J., Polsky, R.,et al. "Silver-enhanced colloidal gold electrochemical stripping detection of DNA hybridization."Langmuir,17(19), 5739–5741 (2001).

148. Wang, J., Xu,D., et al. "Magnetically-induced solid-state electrochemical detection of DNA hybridization."Journal of the American Chemical Society,124(16), 4208–4209 (2002).

149. Wang, L., Chen, X., Wang, X., Han, X., Liu, S., and Zhao, C."Electrochemical synthesis of gold nanostructure modified electrode and its development in electrochemical DNA biosensor."Biosensors and Bioelectronics,30(1), 151–157 (2011).

150. Wang, X., Kawanami,H., et al. "Amphiphilic block copolymer-stabilized gold nanoparticles for aerobic oxidation of alcohols in aqueous solution."Chemical Communications, (37), 4442–4444 (2008).

151. Wang, Z., Lee,J. H., et al. "Label-Free Colorimetric Detection of Lead Ions with a Nanomolar Detection Limit and Tunable Dynamic Range by using Gold Nanoparticles and DNAzyme."Advanced Materials,20(17), 3263–3267 (2008).

152. Wang, Z. L. "Functional Oxide Nanobelts: Materials, Properties and Potential Applications in Nanosystems and Biotechnology."Annual Review of Physical Chemistry,55(1), 159–196(2004).

153. Warner, M. G., Reed,S. M., et al. "Small, Water-Soluble, Ligand-Stabilized Gold Nanoparticles Synthesized by Interfacial Ligand Exchange Reactions."Chemistry of materials,12(11), 3316–3320 (2000).

154. Whetten, R. L., Shafigullin,M. N., et al. "Crystal Structures of Molecular Gold Nanocrystal Arrays."Accounts of Chemical Research,32(5), 397–406 (1999).

155. Wilson, O. M., Hu,X., et al. "Colloidal metal particles as probes of nanoscale thermal transport in fluids."Physical Review B,66(22), 224301 (2002).

156. Xiao, Y., Patolsky,F., et al. "Plugging into Enzymes": Nanowiring of Redox Enzymes by a Gold Nanoparticle."Science,299(5614), 1877 (2003).

157. Xiulan, S., Xiaolian,Z., et al. "Preparation of gold-labeled antibody probe and its use in immunochromatography assay for detection of aflatoxin B1."International Journal of Food Microbiology,99(2), 185–194 (2005).

158. Yan, W., Feng,X., et al. "A super highly sensitive glucose biosensor based on Au nanoparticles-AgCl@ polyaniline hybrid material."Biosensors and Bioelectronics,23(7), 925–931 (2008).

159. Yáñez-Sedeño, P. and Pingarrón, J. M."Gold nanoparticle-based electrochemical biosensors."Analytical and Bioanalytical Chemistry,382(4), 884–886 (2005).

160. Yang, P. H., Sun,X., et al. "Transferrin-Mediated Gold Nanoparticle Cellular Uptake."Bioconjugate Chemistry,16(3), 494–496 (2005).

161. Yang, S., Qu,L., et al. "Gold nanoparticles/ethylenediamine/carbon nanotube modified glassy carbon electrode as the voltammetric sensor for selective determination of rutin in the presence of ascorbic acid."Journal of Electroanalytical Chemistry,645(2), 115–122 (2010).

162. Yuan, S., Yuan, R., et al. "Sandwich-type electrochemiluminescence immunosensor based on Ru-silica@Au composite nanoparticles labeled anti-AFP."Talanta,82(4), 1468–1471 (2010).

163. Zeng, S., Yong,K. T., et al. "A Review on Functionalized Gold Nanoparticles for Biosensing Applications."Plasmonics,6(3), 491–506 (2011).

164. Zhai, S., Hong,H., et al. "Synthesis of cationic hyperbranched multiarm copolymer and its application in self-reducing and stabilizing gold nanoparticles."Science China Chemistry,53(5), 1114–1121 (2010).

165. Zhang, J., Du,J., et al. "Sonochemical Formation of Single–Crystalline Gold Nanobelts."Angewandte Chemie,118(7), 1134–1137 (2006).

166. Zhang, Y. X. and Zeng, H. C."Surfactant–Mediated Self-Assembly of Au Nanoparticles and Their Related Conversion to Complex Mesoporous Structures." Langmuir,24(8), 3740–3746 (2008).

167. Zheng, J., Petty,J. T., et al. "High Quantum Yield Blue Emission from Water-Soluble Au8 Nanodots."Journal of the American Chemical Society,125(26), 7780–7781 (2003).

168. Zhong, Z., Patskovskyy, S., et al. (2004). "The Surface Chemistry of Au Colloids and Their Interactions with Functional Amino Acids."The Journal of Physical Chemistry B,108(13), 4046–4052 (2004).

169. Zhong, Z., Wu,W., et al. "Nanogold-enwrapped graphene nanocomposites as trace labels for sensitivity enhancement of electrochemical immunosensors in clinical immunoassays: Carcinoembryonic antigen as a model."Biosensors and Bioelectronics,25(10), 2379–2383 (2010).

170. Zhu, A., Luo,Y., et al. "Plasmon-Induced Enhancement in Analytical Performance Based on Gold Nanoparticles Deposited on TiO2 Film."Analytical chemistry,81(17), 7243–7247 (2009).

171. Zhu, T., Vasilev, K., et al. "Surface Modification of Citrate-Reduced Colloidal Gold Nanoparticles with 2-Mercaptosuccinic Acid."Langmuir,19(22), 9518–9525 (2003).

172. Zsigmondy, R. "Die chemische Natur des Cassiusschen Goldpurpurs."Justus Liebigs Annalen der Chemie,301(2–3), 361–387 (1898).

CHAPTER 14

MICRO-STRUCTURAL INVESTIGATION OF GRAPHITE POWDERS FOR SENSIBLE REACTIVITY COMPARISON

HEINRICH BADENHORST, BRIAN RAND, and WALTER FOCKE

CONTENTS

14.1 INTRODUCTION

Synthetic graphite is an important industrial material and is used in many high temperature applications, ranging from structural and moderator components in nuclear reactors to electrodes for arc furnaces in the steel production industry. Thus, knowledge of the high temperature oxidative behavior of graphite is essential for understanding any structural changes which may occur due to oxidation during accidental air-ingress situations or degradation during normal operation.

Synthetic graphite is produced *via* a multi-step, re-impregnation process resulting in very complex microstructures and porosity [8]. In general though, graphite is considered to be a fairly simple and well understood allotrope of carbon. It is assumed to be comprised of layered planes of hexagonally bonded carbon atoms with crystallites of varying sizes and thickness. However, the closest approximation to this ideal structure can only be found in mined natural graphite flakes [9]. Despite the complexity and large property variations found in graphitic materials, kinetic investigations are routinely conducted on graphite samples [2-5,7,11,12] from different sources and origins, without any examination of the material microstructure.

New developments in the field of scanning electron microscopy (SEM) allow very high resolution imaging with excellent surface definition [1,6]. The use of high-brightness field-emission guns and in-lens detectors allow the use of very low (~ 1kV) acceleration voltages. This limits electron penetration into the sample and significantly enhances the surface detail which can be resolved, making this technique ideal for examining the morphology of graphite materials.

The aim of this Chapter is to demonstrate the necessity of visually inspecting the microstructure of graphite materials both before and during oxidation. The microstructure not only demonstrates the complex development of the surface area during oxidation but also highlights the presence of trace impurities and exposes the underlying crystallinity. All of these factors are critical when attempting to compare the oxidative reactivity of graphite samples from different sources and origins. Especially in cases where the exact history of the material is not known.

Simply analyzing the purity or ash content is not enough due to the considerable impact extremely low levels of impurities have on the oxidation rate. Furthermore, the use of X-ray diffraction to determine the crystallinity is shown to be an inadequate representation of the observed behavior.

Graphite samples from different origins which have been exposed to different pretreatments are often subjected to an oxidative reactivity comparison without any consideration being given to the underlying sample microstructures. Four samples from natural and synthetic origins were thoroughly examined before and after being partially oxidized using a high resolution FEGSEM at low acceleration voltages. Despite the fact that all four are considered to be high purity, highly crystalline samples, based on ash content and XRD results, they exhibit widely varying reactivities. Based on the observed microstructures and the influences of trace impurities, it is possible to qualitatively explain the measured differences in reaction rate. This demonstrates the clear need to first examine and classify the microstructure, crystallinity and impurity effects present in a given graphite material before comparing its reactivity to other samples.

14.2 MATERIALS AND METHODS

Four powdered graphite samples were examined during this investigation. The first two are proprietary nuclear grade graphite samples, one from a natural source (NNG) and one synthetically produced material (NSG). Both samples were intended for use in the nuclear industry and were subjected to high levels of purification including halogen treatment. The ash contents of these samples were very low, with the carbon content being >99.9 mass %. The exact history of both materials are not known. The third graphite (RFL) was obtained from a commercial source (Graphit Kropfmühl AG Germany). This is a large flake, natural graphite powder and was purified by the supplier with an acid treatment and a high temperature soda ash burn up to a purity of 99.91 mass %. A fourth sample was produced for comparative purposes by heating the RFL sample to 2700°C for 6 hours in a TTI furnace (Model: 1000–2560–FP20). This sample was designated PRFL since the treatment was expected to further purify the material. All thermal oxidation was conducted in a TA Instruments SDT Q600 thermogravimetric analyzer (TGA). The samples were all oxidized to a burn-off of around 30%, at which point the oxidizing atmosphere was rapidly changed to inert. SEM images were obtained using an ultra-high resolution field-emission microscope (Zeiss Ultra Plus 55 FEGSEM) equipped with an in-lens detection system. The powder X-ray diffraction (PXRD) spectra of the graphite samples were obtained using a PANalytical X-pert Pro powder diffractometer with variable divergence and receiving slits, and an X'celerator detector using iron-filtered cobalt Kα radiation.

14.3 DISCUSSION AND RESULTS

14.3.1 FEGSEM RESOLUTION

Initially the RFL sample was only purified up to a temperature of 2400 °C. When this sample was subsequently oxidized, the purification was found to have been only partially effective. It was still possible to detect the effects of trace levels of catalytic impurities, as can be seen from Figure 1.

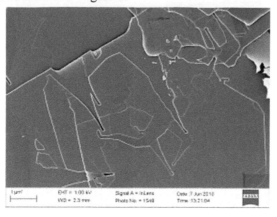

FIGURE 1 FEGSEM image of partially purified RFL (30k x magnification)

Since the catalyst particles tend to trace channels into the graphite, as seen in Figure 2, the consequences of their presence can be easily detected. As a result it is possible to detect a single, minute catalyst particle which is active on a large graphite flake. This effectively results in the ability to detect impurities present at extremely low levels.

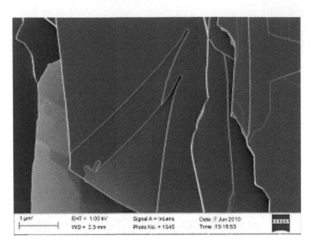

FIGURE 2   Channeling catalyst particles (40k x magnification)

When the tips of these channels are examined, the ability of the high resolution FEGSEM, operating at low voltages, to resolve surface detail and the presence of catalytic particles is further substantiated. As can be seen from Figure 3, the machine is capable of resolving the catalyst particle responsible for the channeling.

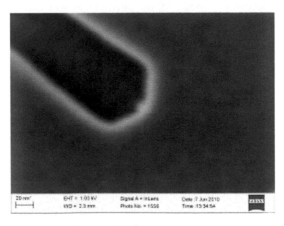

FIGURE 3   Individual catalyst particle (1000k x magnification)

In this case the particle in the image has a diameter of around ten nanometres. This demonstrates the powerful capability of the instrument and serves to validate future assertions regarding the presence or absence of impurities. These figures also serve to demonstrate the massive influence on graphite reactivity which is possible from exceptionally low levels of impurities. In Figure 1 and Figure 2, nanometer-sized particles are shown to create several micron wide channels in the graphite. This greatly increases the exposed surface area of the graphite and hence its oxidation rate. The extent of this influence is to a certain degree dependent on the exact type of catalytic behavior exhibited by the relevant impurity.

14.3.2   AS-RECEIVED MATERIAL

The as-received NNG material is shown in Figure 4. A wide particle distribution is evident.

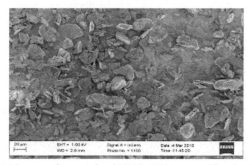

FIGURE 4   FEGSEM image of as-received NNG (750 x magnification)

When the particles are examined more closely, they are found to be highly agglomerated, as shown in Figure 5.

FIGURE 5   Close-up of as-received NNG (3k x magnification)

All samples were subsequently wet-sieved in ethanol to break-up the agglomerates. For NNG and NSG, the size fractions between 25 and 45 μm were retained, whilst for RFL and PRFL the size fractions between 200 and 250 μm were retained. When the sieved NNG flakes are examined they are free of extraneous flakes but still appear to be composites with rounded edges, as can be seen in Figure 6.

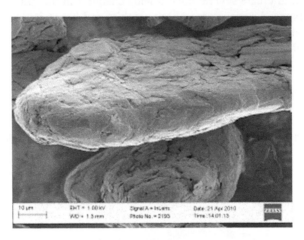

FIGURE 6   Close-up of sieved NNG (3k x magnification)

In an effort to reduce the composites to single particles, the sample was sonicated in ethanol for 5 minutes. This had an adverse effect on some of the particles as shown in Figure 7.

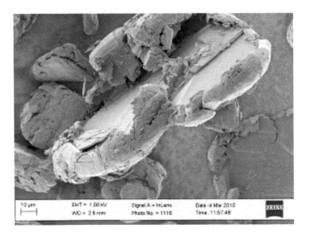

FIGURE 7   Sonicated NNG (2k x magnification)

Despite the short duration and low intensity of the treatment, the flake edges are folded and curled inward. This demonstrates the malleability of the graphite and the ease with which the flakes may be damaged and the structure modified.

FIGURE 8   Potato shaped NNG particle (4k x magnification)

It was concluded that the sieved particles were in fact not composites but particles which were severely damaged during jet-milling. This is a common practice to obtain the so-called potato shaped graphite particles as the one shown in Figure 8. In contrast, the sieved RFL flakes exhibit the morphology expected for crystalline graphite flakes, as can be seen in Figure 9.

FIGURE 9   Sieved RFL graphite flakes (175 x magnification)

The layered edge structure is immediately visible on some flakes, as seen in Figure 10. Some flakes do however show damaged edges similar to Figure 7, but the extent is limited to the edge and the flat flake structure remains intact and visible.

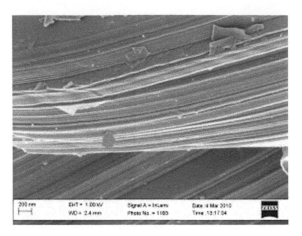

FIGURE 10   RFL edge (72k x magnification)

When examined closely, the basal plane appears intact, with some minor folding and roughening as is visible in Figure 11.

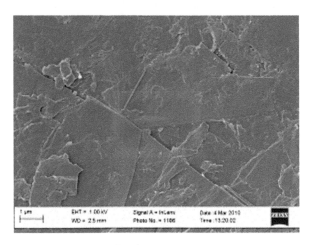

FIGURE 11   RFL edge (24k x magnification)

The synthetic graphite on the other hand has a completely different structure from the two natural graphite samples. The expected flat, layered flake structure is not readily visible, as can be seen in Figure 12.

FIGURE 12   NSG (500 x magnification)

In this sample two distinct particle types are distinguishable. The first are long and comparatively thin, needle-like particles. These particles tend to have a fairly flat surface with folds or creases stretching along the particle, as shown in Figure 13.

FIGURE 13   NSG needle particle (4k x magnification)

When examined edge-on in Figure 14, it becomes apparent that these are folds in the basal plane. The structure is comparable to that of a stack of papers which have been repeatedly folded lengthwise. Thus it is clear that the basal plane is orientated along the lateral surface of these particles and seems to form a continuous plane along this direction. Despite this folding the basal plane is readily distinguishable and at first glance it appears that these particles may have similar crystallinity to the natural particles, based solely on the size, smoothness and layered nature of the observed basal planes.

FIGURE 14   NSG needle particle edge (23k x magnification)

During fabrication of synthetic graphite a filler material known as needle coke is produced. The particles in Figure 13 and 14 are most likely derived from the needle coke with its characteristic elongated, needle-like shape. This filler is mixed with a binder which can be either coal tar or petroleum derived pitch. The pitch is in a molten state when added and the mixture is then either extruded or molded. The resulting artefact can then be re-impregnated with pitch if a high density material is required. The second group of particles are most likely derived from this molten pitch. They are highly disordered with a characteristic mosaic texture, as can be seen in Figure 15, probably derived from the flow phenomena during impregnation.

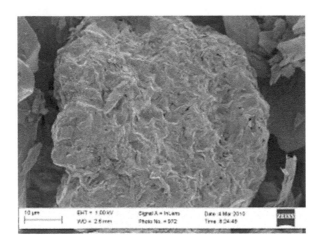

FIGURE 15   NSG pitch particle (4k x magnification)

When examined more closely as in Figure 16, some layered regions are discernable, but for these particles it is difficult to define a clear basal orientation. At this point it is difficult to discern to what extent crystallinity even exists.

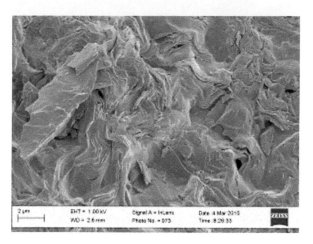

FIGURE 16   NSG pitch particle close-up (14k x magnification)

This classification is not absolute and some particles may be found which have characteristics that are a combination of both needle coke and pitch. As expected, when the heat treated RFL sample, PRFL, is examined in Figure 17, a structure very similar to that of the original RFL graphite is found.

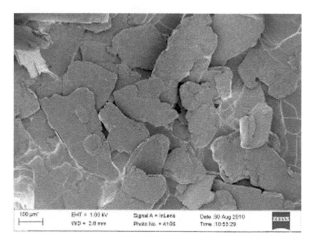

FIGURE 17   PRFL sample (250 x magnification)

However, when the basal planes are examined more closely as in Figure 18, a difference becomes evident.

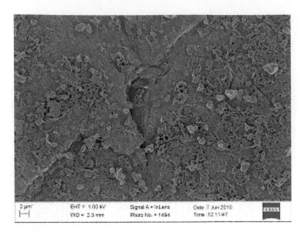

FIGURE 18   PRFL basal surface (5k x magnification)

For some flakes the surface appears to be covered in fine structures. When these are inspected in Figure 19, they appear to be voids within the larger crystal structure.

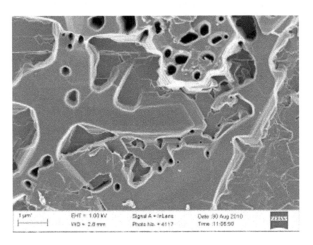

FIGURE 19   PRFL crystal voids (30k x magnification)

When the edges of these structures are examined the expected graphitic layering is clearly visible in Figure 20.

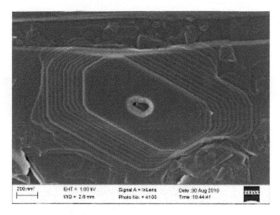

FIGURE 20   PRFL basal structures (120k x magnification)

It was concluded that these voids are the remnants of impurity inclusions which were subsequently evaporated by the high temperature heat treatment. Since, the SEM cannot detect compositional differences but only topography, these inclusions are not visible in the as-received RFL material. Since, these are natural graphite flakes it is highly likely that these inclusions developed during the original formation of the graphite crystals.

14.3.3 OXIDIZED MICROSTRUCTURE

When the oxidized NNG microstructure is examined in Figure 21, fairly complex and irregular structures are found. This is consistent with the expectation of potato graphite, that is extensively damaged and crumpled particles. As the outer roughness is removed by oxidation, the multifaceted nature of the particle interior is revealed.

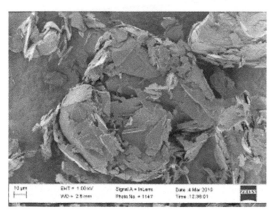

FIGURE 21   Oxidized NNG (1k x magnification)

However when the particles are inspected closely, the basal plane is clearly notice-able as in Figure 22. Furthermore it is largely intact with some minor layering and steps, compared to the size of the edge.

FIGURE 22   Oxidized NNG basal plane (9k x magnification)

When viewed from above the edges are erratic and in some cases have a charac-teristic spiked or saw-tooth nature as seen for example in Figure 23. Both of these fea-tures are caused by impurities, which will be discussed in subsequent sections. Despite their shape, the edges themselves are fairly smooth, in some cases spanning several microns. This is due to the level of crystalline perfection and which is apparent when compared to the synthetic material.

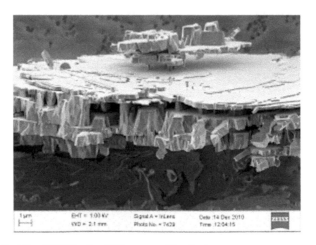

FIGURE 23   Oxidized NNG edge (13k x magnification)

Oxidized RFL flakes still largely exhibit the expected flake-like nature expected for highly crystalline graphite as can be seen in Figure 24. With the overall flake shape remaining easily identifiable.

FIGURE 24   Oxidized RFL flakes (150 x magnification)

However, some flakes show the presence of pits growing in the flake whilst others are in the process of disintegrating completely, presumably due to imperfections in the original flake crystal structure, as shown in Figure 25 A and B respectively. It may be noted that the pits are highly erratic and have no clear structure.

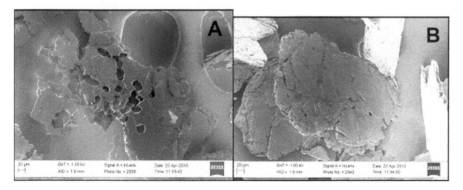

FIGURE 25   Degraded RFL flake structures (500 x magnification)

When the basal plane of the intact flakes are examined, the surface is very similar to that of the NNG graphite, as can be seen from Figure 26. It is fairly undamaged with some minor steps whilst the activities of impurity particles are also visible.

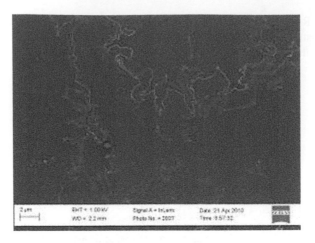

FIGURE 26   Oxidized RFL basal plane (10k x magnification)

The edges of these flakes are also smooth with the same erratic, spiked nature as NNG particles, with widths of a few micron as seen in Figure 27.

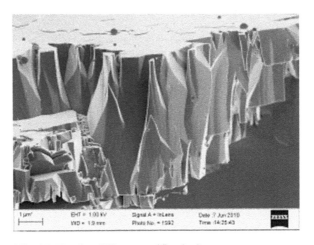

FIGURE 27   Oxidized RFL edge (25k x magnification)

The situation is remarkably different in the case of the synthetic graphite particles. It still remains possible to clearly discern between needle particles and pitch particles as is evident in Figure 28.

FIGURE 28 Oxidized NSG (400 x magnification)

The needle particles bear some resemblance to the natural graphite flakes, with the basal plane still readily identifiable in Figure 29.

FIGURE 29 Oxidized NSG needle particle (4k x magnification)

However, when the basal plane is examined closely in Figure 30 there is a stark contrast with the natural graphite basal plane. The basal surface is severely degraded, with attack possible virtually anywhere.

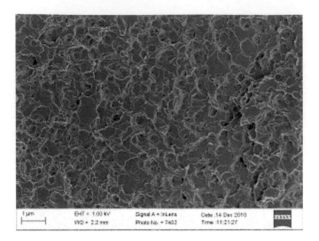

FIGURE 30   Oxidized NSG needle particle basal plane (25k x magnification)

The cavities were extensively investigated and no traces of impurities were found to be present. Instead the oxidation hollow has the characteristic corkscrew like shape of a screw dislocation as can be seen from Figure 31. The pits tend to have a vaguely hexagonal shape.

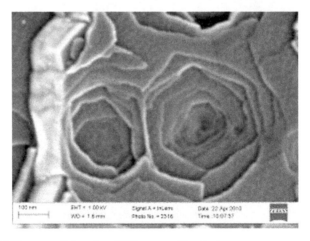

FIGURE 31   NSG screw dislocation (320k x magnification)

It is also interesting to note that in some regions the defect density is not as high as in others, as can be seen for the different horizontal bands in Figure 32 A and also the different regions visible in Figure 32 B. This may imply different levels of crystalline perfection in these regions.

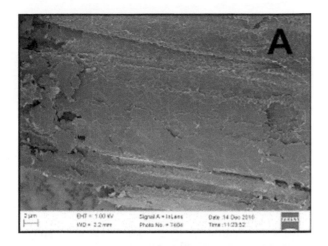

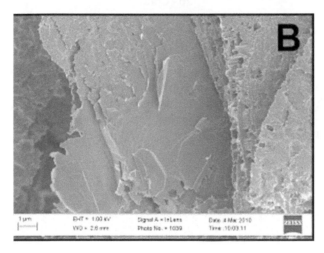

FIGURE 32   NSG crystallinity differences (7k and 19k x magnification)

These stark differences between the natural and synthetic materials are not expected if one examines the XRD spectra of the samples which are very similar, these are shown in Figure 33.

XRD Spectra

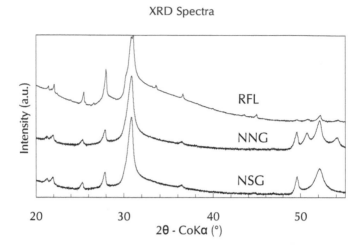

FIGURE 33   XRD spectra

Using the empirical Scherrer equation [10] it is possible to calculate the domain sizes for the three samples (the heat treatment is not thought to affect the natural graphite crystallinity and it has an identical spectra to the RFL sample) from their XRD spectra to get a quantitative idea of the differences between their respective levels of crystallinity. These are given in Table 1.

TABLE 1   Crystallite sizes

	L_A (nm)
NNG	38
NSG	26
RFL	26

The differences in the basal attack would indicate far larger differences in crystallinity than these estimates indicate, demonstrating that XRD alone cannot be taken as an indication of the level of crystalline perfection. When examined edge on as in Figure 34, it can be seen that the needle particles retained their original structure, however any gaps or fissures in the folds have grown in size. This implies the development of complex slit-like porosity which would not have occurred to the same extent if direct basal attack was not possible to a large degree.

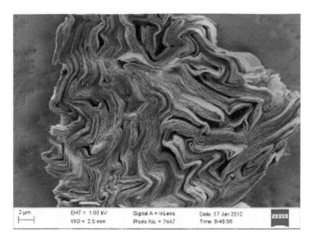

FIGURE 34 Slit-like pore development in NSG (8k x magnification)

When the particles edges are examined more closely in Figure 35, the low level of crystalline perfection is further substantiated. The maximum, continuous edge widths are no more than a few hundred nanometres, far less than the several micron observed in the natural samples.

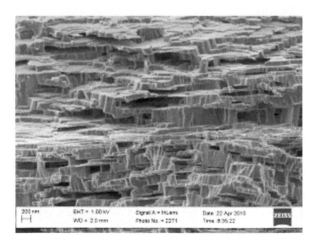

FIGURE 35 Degraded edge structure of NSG (50k x magnification)

The complex microstructural development characteristic of this sample is even more pronounced in the pitch particles, as can be seen from Figure 36.

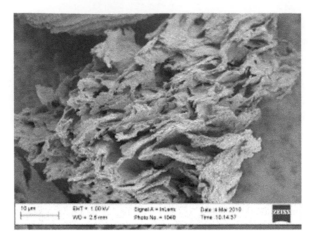

FIGURE 36   Oxidized pitch particle (4k x magnification)

These particles lack any long range order, however when their limited basal like areas area examined more closely as in Figure 37, a texture very similar to the basal plane of needle particles is found, indicating similar levels of crystalline perfection.

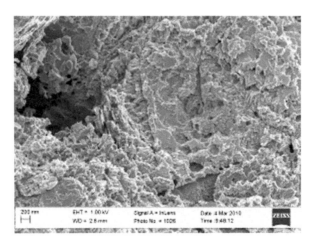

FIGURE 37   Oxidized pitch particle surface (49k x magnification)

At first glance the oxidized PRFL particles look very similar to the RFL sample, with the flake structure still clearly visible, as can be seen in Figure 38.

FIGURE 38 Oxidized PRFL (200 x magnification)

However, when examined more closely as in Figure 39, splendid, pristine edges are noticeable, consistent with the 120° angles expected of perfectly crystalline graphite.

FIGURE 39 Oxidized PRFL edge viewed from above (15k x magnification)

When viewed edge-on the edges are smooth and continuous over several microns with no random or erratic behavior, although edge width varies significantly, as shown in Figure 40.

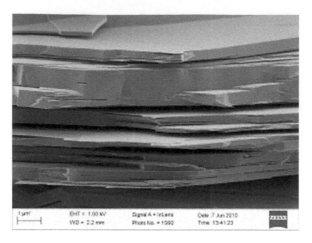

FIGURE 40   Oxidized PRFL edge (25k x magnification)

Despite the variations in size and imperfections in the overall structure, in some cases the edges are very homogenous and have very large widths, as can be seen in Figure 41.

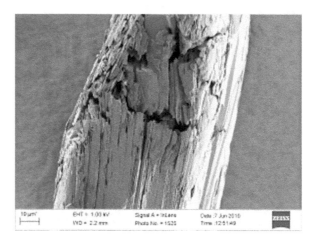

FIGURE 41   Oxidized PRFL with large edge (2k x magnification)

When the basal planes of these flakes are examined they are found to be virtually untouched with some minor irregularities visible in Figure 42.

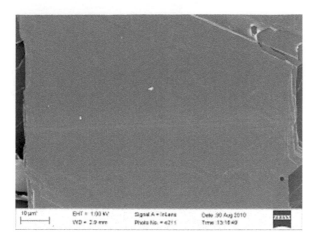

FIGURE 42 Oxidized PRFL basal plane (100 x magnification)

In some regions the remnants of the surface irregularities mentioned in the previous section are still visible, as shown in Figure 43. These are regions with a comparatively high proportion of exposed edges and as such may be assumed to oxidize rapidly.

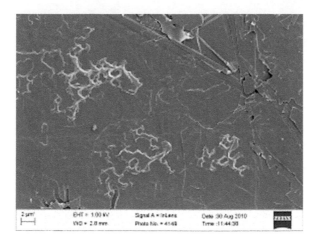

FIGURE 43 Oxidized PRFL fine surface structures (7k x magnification)

Some flakes still have pit growth within the structure as seen in Figure 44. However, in this case the pits are always hexagonal and tend to be found along twinning boundaries.

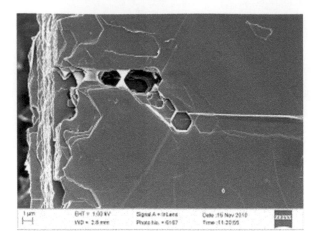

FIGURE 44   Hexagonal pits in oxidized PRFL (10k x magnification)

When under developed pits are examined closely they are found to be free of impurities and structured in the characteristic corkscrew manner expected for screw dislocations, as shown in Figure 45.

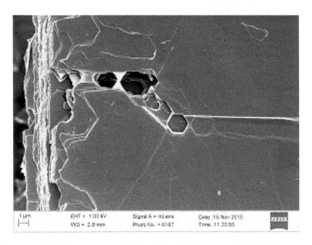

FIGURE 45   Hexagonal pits on screw dislocation (30k x magnification)

No evidence could be found for the presence of any impurities in the PRFL sample. Overall, the PRFL graphite clearly demonstrates the expected behavior for highly crystalline graphite flakes, free of the effects of impurities.

14.3.4 CATALYTIC IMPURITIES

When examined in detail, the highly erratic, almost fractal like nature of the NNG graphite is noticeable in Figure 46.

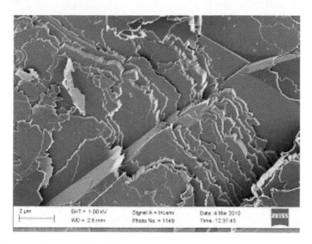

FIGURE 46   Erratic edge of NNG (18k x magnification)

A similar edge structure is found in the RFL graphite as can be seen from Figure 47.

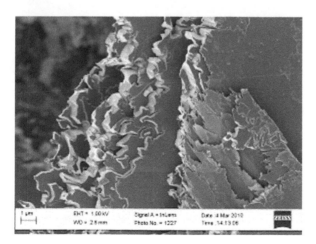

FIGURE 47   Erratic edge of RFL (18k x magnification)

When the edges are scrutinized more closely, as in Figure 48, the reason for these edge formations becomes clear. They are caused by minute impurities which randomly trace channels into the graphite.

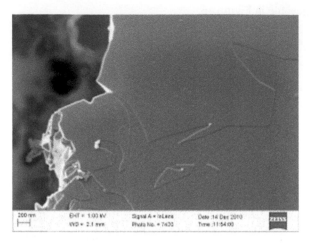

FIGURE 48   Catalyst activity (65k x magnification)

In certain cases, the activity is very difficult to detect, requiring the use of excessive contrast before they become noticeable as shown in Figure 49 A and B.

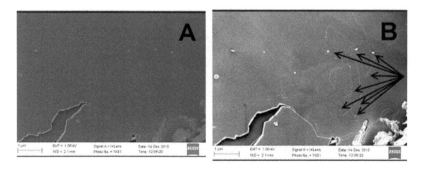

FIGURE 49   Contrast detection of catalyst activity (38k x magnification)

A very wide variety of catalytic behaviors were found. Broadly, these could be arranged into three categories. The first, show in Figure 50, are small, roughly spherical catalyst particles. Channels resulting from these particles are in most cases triangular in nature. In general it was found that these particles tend to follow preferred channeling directions, frequently executing turns at precise, repeatable angles, as demonstrated in Figure 50B. However, exceptions to these observed behaviors were also found, as illustrated in Fig 50C.

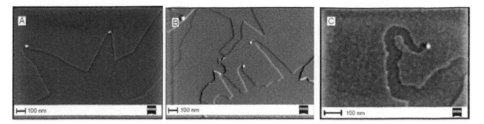

FIGURE 50   Small, spherical catalyst particles

The second group contained larger, erratically shaped particles, some examples are shown in Figure 51.

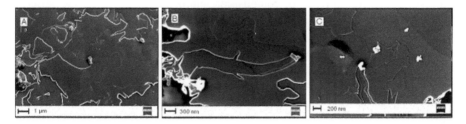

FIGURE 51   Small, spherical catalyst particles

These particles exhibited random, erratic channeling. Where it is likely that the previous group may have been in the liquid phase during oxidation, this is not true for this group, since the particles are clearly capable of catalyzing channels on two distinct levels simultaneously, as can be seen in Figure 51 B and C. The final group contains behaviors which could not be easily placed into the previous two categories, of which examples are shown in Figure 52.

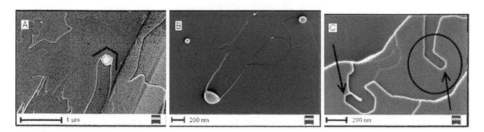

FIGURE 52   Small, spherical catalyst particles

The fairly large particle in Figure 52A cannot be clearly distinguished as having been in the liquid phase during oxidation, yet the tip of the channel is clearly faceted with 120° angles. The particle in Figure 52B was clearly molten during oxidation as it has deposited material on the channel walls. It is curious to note that since the channel walls have expanded a negligible amount compared to the channel depth, the activity of the catalyst deposited on the wall is significantly less than that of the original particle. Finally the startling behavior found in the partially purified RFL graphite, where a small catalyst particle is found at the tip of a straight channel, ending in a 120 ° tip (clearly noticeable in Figure 3), where the width of the channel is roughly an order of magnitude larger than the particle itself. In this case channeling was always found to proceed along preferred crystallographic directions.

Such a wide variety of catalytic behavior is not unexpected for the natural graphite samples under consideration. Despite being highly purified, the purification treatments are unlikely to penetrate the graphite particles completely. As already discussed, the graphite contains inclusions which were most likely trapped during formation and as such may in certain cases be completely shielded. These inclusions have virtually limitless possibilities in terms of composition and hence lead to the wide variety of observed behaviors.

Despite extensive investigation, no evidence of catalytic activity was found in the synthetic sample, NSG. However, the erratic and highly damaged nature of the synthetic graphite basal plane, makes detection of minute impurities very difficult. As mentioned in the previous section, pits found in the as-received natural samples, are erratically shaped, as shown in Figure 53.

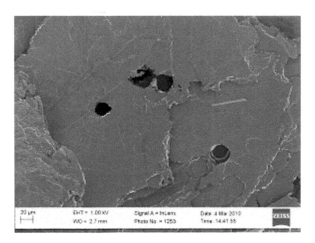

FIGURE 53   Pitting in RFL graphite (650 x magnification)

Underdeveloped pits are often associated with erratically shaped impurities, as shown in Figure 54A and B.

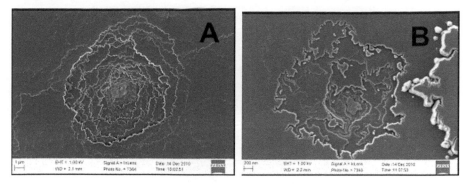

FIGURE 54 Impurity particles associated with pitting (13k and 60k x magnification)

The myriad of different catalytic behaviors found in these purified natural graphite samples coupled with the enormous impact catalyst activity has on reactivity, demonstrates the folly of simply checking the impurity levels or ash content as a basis for reactivity comparison.

14.3.5 INHIBITING IMPURITIES

The last morphological characteristic which has not been accounted for is the presence of spiked or saw-tooth like edge formations found in the natural graphite samples, as can be seen in Figure 55.

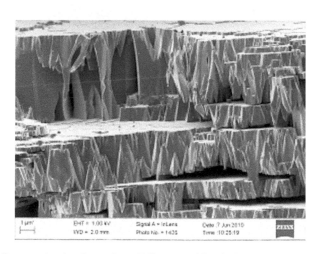

FIGURE 55 Saw-tooth edge formations (15k x magnification)

The closer inspection reveals that invariably the pinnacle of these structures is covered by a particle, as seen in Figure 56.

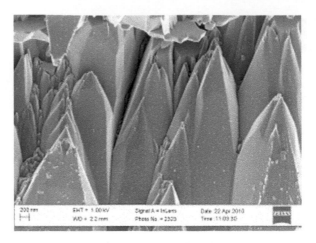

FIGURE 56   Close-up of saw-tooth structures (50k x magnification)

Thus these formations are caused by inhibiting particles which shield the underlying graphite from attack. These layers shield subsequent layers leading to the formation of pyramid like structures topped by a single particle. In some cases as in the left hand side of Figure 57, these start off as individual structures, but then as oxidation proceeds around them, the particles are progressively forced closer together to form inhibition ridges, as can be observed in the right hand side of Figure 57.

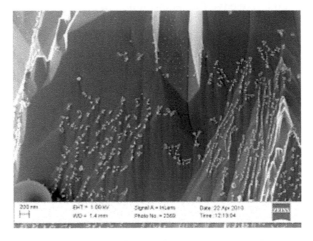

FIGURE 57   Inhibiting particles stacked along ridges (50k x magnification)

In extreme cases these particles may remain atop a structure until it is virtually completely reduced, for example resulting in the nano-pyramid shown in Figure 58.

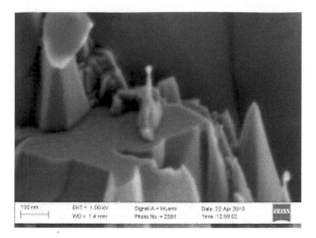

FIGURE 58   Nano-pyramid (300k x magnification)

In some cases particles are found which appear to neither catalyze nor inhibit the reaction, such as the spherical particles seen in Figure 59.

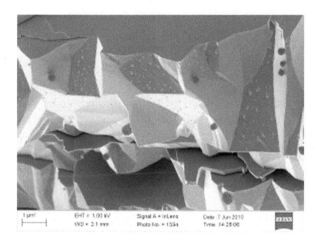

FIGURE 59   Spherical edge particles (25k x magnification)

These may be catalyst particles which agglomerate and deactivate due to their size. The graphite is oxidized away around them, until they are left at an edge, as seen in Figure 60.

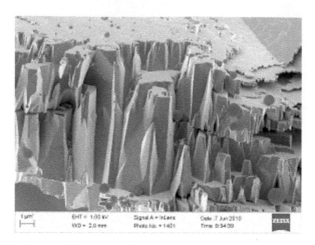

FIGURE 60   Spherical particles accumulating at edge (15k x magnification)

The accumulation of inhibiting particles at the graphite edge will inevitably lead to a reduction in oxidation rate as the area covered by these particles reaches a significant proportion of the total surface area. This may appreciably affect the shape of the observed conversion function, especially at high conversions.

14.3.6   REACTIVITY

As a comparative indication of reactivity the samples were subjected to oxidation in pure oxygen under a temperature program of 4 °C/min in the TGA. The measured reaction rate as a function temperature is shown in Figure 61.

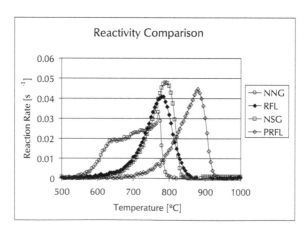

FIGURE 61   Reactivity comparison

As a semi-quantitative indication of relative reactivity the onset temperatures were calculated and are shown in Table 2. It is clear from Table 2 and Figure 61 that the NNG sample has the highest reactivity. The NSG and as-received RFL samples have similar intermediate reactivity, although the NSG sample does exhibit a higher peak reactivity. The sample with the lowest reactivity is the purified PRFL sample. Given the microstructure and impurities found in the respective samples these results are not unexpected. The NNG and NSG samples have comparably complex microstructures which would both have relatively high surface areas. This is further increased by their small particle size compared to RFL and PRFL. Despite the higher crystalline perfection of the NNG sample the presence of impurities increases its reactivity significantly above that of the NSG.

TABLE 2   Onset temperatures

	Temp (°C)
NNG	572
RFL	696
NSG	704
PRFL	760

The RFL material also has excellent crystallinity and a more ideal flake structure, with significantly lower edge surface area. However, the RFL sample still contains considerable amounts of catalytically active impurities, thus increasing its reactivity. As expected the sample with the lowest reactivity is the PRFL sample which has a low surface area combined with zero impurities.

The RFL sample shows a more gradual decrease in reactivity after reaching the peak reaction rate, than for example NSG. This may be due to the presence of inhibiting impurities accumulating at the edges but further investigation is required. Current research is focused on directly linking the microstructures and influence of impurities to the observed conversion behaviors.

14.5   CONCLUSION

Four graphite samples with similar, high levels of purity were investigated, three from a natural and one from a synthetic origin. The as-received and oxidized microstructures as well as the influence of impurities on these structures were thoroughly examined using a high resolution FEGSEM. One of the natural graphite samples was found to be comprised of highly damaged and distorted particles, most likely as a result of jet-milling. When oxidized these samples exhibited very complex, multifaceted microstructures.

The second natural graphite was found to be far closer to the ideal, flake-like structure expected for highly crystalline natural graphite and when oxidized these structures remained intact to a large extent. Both natural graphite samples showed significant levels of catalytic impurities which were active during oxidation revealing a myriad of different behaviours. Despite having very low ash contents, >99.9% purity, preliminary reactivity results indicate that these impurities significantly increase the sample reactivity despite increased crystalline perfection.

The synthetic graphite exhibited some very complex microstructures with a clear distinction between needle coke and pitch derived particles being possible. When oxidized both particles exhibited extensive damage to the basal areas and edges, indicating very low levels of crystalline perfection with oxidative attack being possible anywhere on the surface. This was unexpected since the NSG had an XRD spectra very similar to the natural graphite samples, indicating that this is perhaps not a clear indication of crystalline perfection. The needle structures remained intact during oxidation with slit-like pore development, but the pitch particles degraded into very complex and intricate structures. Despite showing no traces of catalytic impurities the synthetic sample exhibited intermediate reactivity due to the low crystalline perfection and complex structures with relatively high surface areas.

The final natural graphite sample was a purified version of the more ideal, flake-like natural graphite and showed no traces of catalytic activity. As expected this sample exhibited the lowest reactivity and the oxidized microstructure revealed the hexagonal edges and unreactive basal plane expected for highly crystalline graphite flakes.

Despite the fact that all four are considered to be high purity, highly crystalline samples, based on ash content and XRD results, they exhibit widely varying reactivities. Based on the observed microstructures and the influences of trace impurities, it is possible to qualitatively explain the measured differences in reaction rate. This demonstrates the clear need to first examine and classify the microstructure, crystallinity and impurity effects present in a given graphite material before comparing its reactivity to other samples. Is it hoped that in future work will allow these observed features to be directly linked with sample reactivity and conversion behavior.

KEYWORDS

- **Graphite**
- **Influence of impurities**
- **Microstructure**
- **Natural**
- **Oxidation**
- **Synthetic**

ACKNOWLEDGMENTS

This work is based upon research supported by the Skye foundation and the South African Research Chairs Initiative of the Dept. of Science and Technology and the National Research Foundation. Any opinion, findings and conclusions or recommendations expressed in this material are those of the authors and therefore Skye, the NRF and DST do not accept any liability with regard thereto.

REFERENCES

1   Cazaux, J. From the physics of secondary electron emission to image contrast in scanning electron microscopy. *J. of Elec. Micros.* **61**, 261-284 (2012).

2   Fuller, E. L. and Okoh, J. M. Kinetics and mechanisms of reaction of air with nuclear grade graphites: IG-110. *J. of Nucl. Mat.* **240**, 241-250 (1997).

3   Guo, W., Xiao, H., and Guo, W. Modelling of TG curves of isothermal oxidation of graphite. *Mat. Sci. and Eng.* **474**, 197-200 (2008).

4   Hinssen, H.-K., Kühn, K., Moormann, R., Schlögl, B., Fechter, M., and Mitchell, M. Oxidation experiments and theoretical examinations on graphite materials relevant for the PBMR. *Nucl. Eng. and Design.* **238**, 3018-3025 (2008).

5   Kim, E. S., Lee, K. W., and No, H. C. Analysis of geometrical effects on graphite oxidation through measurement of internal surface area. *J. of Nucl. Mat.* **348**, 174-180 (2006).

6   Liu, J. The versatile FEG-SEM: From ultra-high resolution to ultra-high surface sensitivity. *Micros. and Microanal.* **9**, 144-145 (2003).

7   Moormann, R., Hinssen, H.-K., and Kühn, K. Oxidation behaviour of an HTR fuel element matrix graphite in oxygen compared to a standard nuclear graphite. *Nucl. Eng. and Design* **227**, 281-284 (2004).

8   Pierson, H. O. *Handbook of carbon, graphite, diamond and fullerenes. Properties, processing and applications.* New Jersey, USA: Noyes Publications (1993).

9   Wissler, M. Graphite and carbon powders for electrochemical applications. *J. Power Sources* **156**, 142-150.

10 Warren, B. E. X-ray diffraction in random layer lattices. *The Phys. Rev.* **59**, 693-698 (1941).

11 Xiaowei, L., Jean-Charles, R., and Suyuan, Y. Effect of temperature on graphite oxidation behaviour. *Nucl. Eng. and Design* **227**, 273-280 (2004).

12 Zaghib, K., Song X., and Kinoshita, K. Thermal analysis of the oxidation of natural graphite: isothermal kinetic studies. *Thermochimica Acta*, **371**, 57-64 (2001).

CHAPTER 15

A NANOFILLER PARTICLES AGGREGATION IN ELASTOMERIC NANO-STRUCTURED COMPOSITES— THE IRREVERSIBLE AGGREGATION MODEL

YU. G. YANOVSKY, G. V. KOZLOV, and G. E. ZAIKOV

CONTENTS

15.1 INTRODUCTION

The aggregation of the initial nanofiller disperse particles is more or less large particles aggregates always occurs in the course of technological process of making particulate-filled polymer composites in general [1] and elastomer composites in particular [2]. The aggregation process acts on composites (nanostructured composites) macroscopic properties [1, 3]. For composites a process of nanofiller aggregation gains a special significance, since its intensity can be the one, that nanofiller particles aggregates size exceeds 100 nm – the value, which assumes (although and conditionally enough [4]) as an upper nanoscale limit for nanoparticle. In other words, the aggregation process can result to the situation, when primordially supposed nanostructured composite ceases to be the one. Therefore, at present several methods exist, which allowed to suppress nanoparticles aggregation process [2, 5]. Proceeding from this, in the present paper theoretical treatment of disperse nanofiller aggregation process in butadiene-styrene rubber matrix within the frameworks of irreversible aggregation models was carried out.

Disperse nano sized particles aggregation process in elastomer matrix has been studied. The modified model of irreversible aggregation particle-cluster was used for theoretical analysis of this process. The necessity of a modification is defined by simultaneous formation of a large number of nanoparticles aggregates. The offered approach allows us to predict some final parameters of nanoparticles aggregates as a function of the initial particles size, their contents and other factors number.

15.2 EXPERIMENTAL

The objects of the analysis were elastomer composites based on butadiene-styrene rubber (BSR). BSR of industrial production, mark SKS-30 ARK was used, which contains 7.0-12.3% cis– and 71.8-72.0% trans links with density of 920-930kg/m^3. The rubber is completely amorphous one. Mineral shungite (Zazhoginskii deposit of III-th variety) makes up ~30% globular amorphous metastable carbon and ~60% high-disperse silicates (see Table 1.). Its structure is fullerene-like one. Nano- and micro-dimensional disperse particles of schungite were obtained at the Institute of Applied Mechanics of the Russian Academy of Sciences from the industrially mined mineral by unique technology (Russian Federation patent No.2442657 of 20.02.2012) on a planetary ball mill Retsch PM 100 (Germany).

TABLE 1 The chemical composition of mineral shungite

SiO$_2$	TiO$_2$	Al$_2$O$_3$	FeO	MgO	CaO	Na$_2$O	K$_2$O	S	C	H$_2$O
57.0	0.2	4.0	2.5	1.2	0.3	0.2	1.5	1.2	29.0	4.2

The process of the preparation of samples for tests consisted of a few stages. A mixture of a polymeric matrix and fillers in equal volume ratios and of other ingredients was prepared in mixer. Thereafter for the mixtures obtained the optimum time of vulcanization was determined and the very process of vulcanization was carried out in special moulds. As a result, plates from the elastomer material of size 15×15cm and thickness of 2mm were obtained. Then from these plates the specimens of size 1×1cm were cut and, to decrease roughness of the specimen surface, micro cuts with the aid of a microtome were made. The resulting specimens were investigated by the AFM. For macro tests standard specimens were cut from the 15 × 15cm 2mm thick plates investigated in accordance with the State Standards 270-75.

The analysis of the received in milling process shungite particles were monitored with the aid of analytical disk centrifuge (CPS Instruments, Inc., USA), allowing to determine with high precision the size and distribution by sizes within the range from 2 nm up to 50 mcm.

Nanostructure was studied on scanning probe microscopes Nano-DST (Pacific Nanotechnology, USA) and Easy Scan DFM (Nanosurf, Switzerland) by semi-contact method in the force modulation regime. Scanning probe microscopy results were processed with the aid of specialized software package SPIP (Scanning Probe Image Processor, Denmark). SPIP is a powerful program package for processing of images, obtained on scanning probe microscopy (SPM), atomic forced microscopy (AFM), scanning tunneling microscopy (STM), scanning electron microscopes, transmission electron microscopes, interferometers, confocal microscopes, profilometers, optical microscopes and so on. The given package possesses the whole functions number, which are necessary at images precise analysis, in the number of which the following are included:

- The possibility of three-dimensional reflected objects obtaining, distortions automatized leveling, including Z-error mistakes removal for examination of separate elements and so on;
- Quantitative analysis of particles or grains, more than 40 parameters can be calculated for each found particle or pore: Area, perimeter, average diameter, the ratio of linear sizes of grain width to its height distance between grains, coordinates of grain center of mass can be presented in a diagram form or in a histogram form.

15.3   DISCUSSION AND RESULTS

For theoretical treatment of the processes of nanofiller particles aggregates growth and final sizes the traditional irreversible aggregation models are inapplicable, since, it is obvious, that in composites aggregates a large number of simultaneous growth sites takes place. Therefore the model of multiple growth sites, offered in paper [6], was used for a description of nanofiller aggregation.

In Figure 1 the images of the studied composites, obtained in the force modulation AFM regime, and corresponding to them the distributions of a nanoparticles aggregates fractal dimension d_f are adduced [7]. As it follows from the values d_f (d_f=2.40-2.48), nanofiller particles aggregates in the composites under consideration are formed by a mechanism of a particle-cluster (P-Cl), i.e. they are Witten-Sander clusters [8]. The variant «a», of above model [6] was chosen for the modeling. According to this variant the mobile particles are added to the lattice, consisting of a large number of

"seeds" with density of c_0 at the modeling beginning [6]. Such model generates structures, which have fractal geometry on short scales of length with value $d_f=2.5$ (see Fig.
1) and homogeneous structure on large scales of length. A relatively high particles
concentration c is required in the model for formation of uninterrupted network [6].

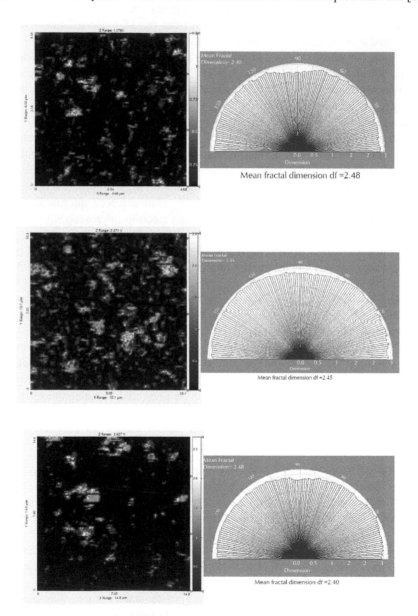

Mean fractal dimension df =2.48

Mean fractal dimension df =2.45

Mean fractal dimension df =2.40

FIGURE 1 The images, obtained in the force modulation regime, for composites, filled with
technical carbon (a), nanoshungite (b), microshungite (c), and corresponding to them fractal
dimensions d_f.

In case of "seeds" high concentration c_0 for the variant «a» of the model the following relationship was obtained [6]:

$$R_{max}^{d_f} = N = c/c_0,$$ (1)

where R_{max} is a maximal radius of nanoparticles cluster (aggregate), N is a number of nanoparticles per one aggregate, c is a concentration of nanoparticles, c_0 is "seeds" number, which is equal to nanoparticles clusters (aggregates) number.

The value N can be estimated according to the following equation [9]:

$$2R_{max} = \left(\frac{S_n N}{\pi q}\right)^{1/2},$$ (2)

where S_n is cross-sectional area of nanoparticles, from which aggregate consists, q is packing coefficient, equal to 0.74.

The experimentally obtained value for nanoparticles and theirs aggregates was taken from [7]. A diameter $2R_{agr}$ for nanoparticles aggregate was accepted as $2R_{max}$ (Table 2). The value S_n was also calculated according to the experimental values of nanoparticles radius r_n (Table 2). In Table 2 the values N for the studied nanofillers, obtained according to the above indicated method, were adduced. It is significant, that the value N is a maximum one for nanoshungite despite of the larger values r_n in comparison with technical carbon.

Further, the equation (1) allows to estimate the maximal radius R_{max}^T of nanoparticles aggregate within the frameworks of the aggregation model [6]. These values R_{max}^T are adduced in Table 2, from which their reduction in a sequence of technical carbon-nanoshungite-microshungite, that fully contradicts to the experimental data, i.e. to R_{agr} change (Table 2). However, we must not neglect the fact, that the equation (1) was obtained within the frameworks of computer simulation, where the initial aggregating particles sizes are the same in all cases [6]. For real composites the values r_n can be distinguished essentially (Table 2). It is expected, that the value R_{agr} or R_{max}^T will be the higher, than the larger radius of nanoparticles, forming aggregate.

Then theoretical value of a radius R_{agr}^T of nanofiller particles cluster (aggregate) can be determined as follows:

$$R_{agr}^T = k_n r_n N^{1/d_f},$$ (3)

where k_n is proportionality coefficient, in the present work accepted empirically equal to 0.9.

The comparison of experimental R_{agr} and calculated according to the equation (3) R_{agr}^T values of the studied nanofillers particles aggregates radius shows their good correspondence (the average discrepancy of R_{agr} and R_{agr}^T makes up 11.4%). Therefore, the theoretical model [6] gives a good correspondence to the experiment only in case of consideration of aggregating particles real characteristics and, in the first place, their size.

Let us consider, two more important aspects of nanofiller particles aggregation within the frameworks of the model [6]. Some features of the indicated process are defined by nanoparticles diffusion at composites processing. Specifically, length scale, connected with diffusible nanoparticle, is correlation length ξ of diffusion. By definition, the growth phenomena in sites, remote more than ξ, are statistically independent. Such definition allows to connect the value ξ with the mean distance between nanofiller particles aggregates L_n. The value ξ can be calculated according to the equation [6] as

$$\xi^2 \approx \tilde{n}^{-1} R_{agr}^{d_f - d + 2},$$
(4)

where c is a concentration of nanoparticles, d is dimension of Euclidean space, in which a fractal is considered (it is obvious, that in our case $d=3$). The value c should be accepted equal to nanofiller volume contents φ_n, which is calculated [10] as follows:

$$\varphi_n = \frac{W_n}{\rho_n}.$$
(5)

Here W_n is nanofiller mass contents, ρ_n is its density, determined according to the equation [3]:

$$\rho_n = 0.188(2r_n)^{1/3}.$$
(6)

The values r_n and R_{agr} were determined experimentally (see graph of Fig. 2 [7]). In Fig. 3 the relation between L_n and ξ is adduced, which, as it is expected, proves to be linear and passing through coordinates origin. This means, that the distance between nanofiller particles aggregates is limited by mean displacement of statistical walks, by which nanoparticles are simulated. The relationship between L_n and ξ can be expressed analytically as follows:

$$L_n = 9.6\xi, nm$$
(7)

The second important aspect of the model [6], in reference to simulation of nanofiller particles aggregation, is a finite nonzero initial particles concentration c or φ_n effect, which takes place in any real systems. This effect is realized at the condition ξ

$\approx R_{agr}$, that occurs at the critical value $R_{agr}(R_c)$, determined according to the relationship [6]:

$$c \sim R_c^{d_f - d} . \tag{8}$$

The relationship (8) right side represents cluster (particles aggregate) mean density. This equation establishes, that fractal growth continues only, until cluster density reduces up to medium density, in which it grows. The calculated according to the relationship (8) values R_c for the considered nanoparticles are adduced in Table 2, from which it follows, that they give reasonable correspondence with the experimental values R_{agr} (the average discrepancy of R_c and R_{agr} makes up 24 %).

TABLE 2   The parameters of irreversible aggregation model of nanofiller particles aggregates growth

Filler	Experimental radius of nanofiller aggregate R_{agr}, nm	Radius of nanofiller particle r_n, nm	Number of particles in one aggregate N	Radius of nanofiller aggregate R_{max}^T, the equation (1), nm	Radius of nanofiller aggregate R_{agr}^T, the equation (3), nm	Radius of nanofiller aggregate Rc, the equation (8), nm
Technical carbon	34.6	10	35.4	34.7	34.7	33.9
Nanoshungite	83.6	20	51.8	45.0	90.0	71.0
Microshungite	117.1	100	4.1	15.8	158.0	255.0

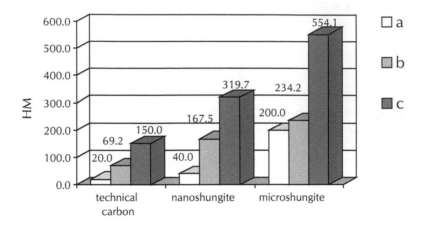

FIGURE 2 The initial particles diameter (a), their aggregates size in composite (b) and distance between nanoparticles aggregates (c) for composites, filled with technical carbon, nano- and microshungite.

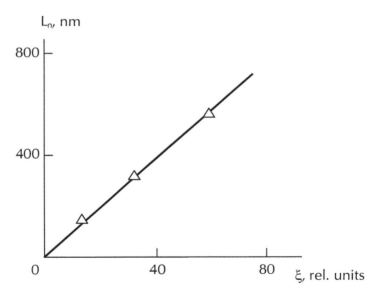

FIGURE 3 The relation between diffusion correlation length ξ and distance between nanoparticles aggregates L_n for studied composites.

Since, the treatment [6] was obtained within the frameworks of a more general model of diffusion-limited aggregation, then its correspondence to the experimental data indicated unequivocally, that aggregation processes in these systems were controlled by diffusion. Therefore let us consider briefly nanofiller particles diffusion. Statistical walkers diffusion constant ξ can be determined with the aid of the relationship [6]:

$$\xi \approx (\zeta t)^{1/2},$$ (9)

where t is walk duration.

The equation (9) supposes (at t=const) ξ increase in a number technical carbon-nanoshungite-microshungite as 196-1069-3434 relative units, i.e. diffusion intensification at diffusible particles size growth. At the same time diffusivity D for these particles can be described by the well-known Einstein's relationship [11]:

$$D = \frac{kT}{6\pi\eta_c r_n \alpha},$$ (10)

where k is Boltzmann constant, T is temperature, η_c is a composite medium viscosity, α is numerical coefficient, which further is accepted equal to 1.

In its turn, the value η can be estimated according to the equation [12]:

$$\frac{\eta_{\tilde{n}}}{\eta_0} = 1 + \frac{2.5\varphi_n}{1 - \varphi_n},$$ (11)

where η_0 and η_c are initial polymer and its mixture with nanofiller viscosities, accordingly, φ_n is nanofiller volume contents.

The calculation according to the equations (10) and (11) shows, that within the indicated above nanofillers number the value D changes as 1.32-1.14-0.44 relative units, i.e. reduces in three times, that was expected. This apparent contradiction is due to the choice of the condition t=const (where t is composite production duration) in the equation (9). In real conditions the value t is restricted by nanoparticle contact with growing aggregate and then instead of t the value t/c_0 should be used, where c_0 is seeds concentration, determined according to the equation (1). In this case the value ξ for the indicated nanofillers changes as 0.288-0.118-0.086, i.e. it reduces in 3.3 times that corresponds fully to the calculation according to the Einstein's relationship Equation (10). This means, that nanoparticles diffusion in polymer matrix obeys classical laws of Newtonian rheology [11].

15.4 CONCLUSIONS

Aggregation of disperse nanofiller particles in elastomeric matrix can be described theoretically within the frameworks of a modified model of irreversible aggregation particle-cluster. The obligatory consideration of nanofiller initial particles size is a feature of the indicated model application to description of the real systems. The indicated particles diffusion in polymer matrix obeys classical laws of Newtonian liquids hydrodynamics. The offered approach allows to predict nanoparticles aggregates final parameters as a function of the initial particles size, their contents and other factors number.

KEYWORDS

- **Composites aggregation**
- **Diffusion**
- **Elastomer matrix**
- **Nanoparticles**
- **Newtonian liquids hydrodynamics**

REFERENCES

1 Kozlov, G. V. Yanovskii, Yu. G., and Zaikov, G. E. *Structure and Properties of Particulate-Filled Polymer Composites: The Fractal Analysis*. New York, Nova Science Publishers, Inc.,p. 282 2010

2 Edwards, D. C. Polymer-filler interactions in rubber reinforcement. *J. Mater. Sci.*, **25** (12), 4175 (1990).

3 Mikitaev, A. K., Kozlov, G. V., and Zaikov, G. E. *Polymer Nanocomposites: The Variety of Structural Forms and Applications*. New York, Nova Science Publishers, Inc., p. 319 (2008).

4 Buchachenko, A. L. The nanochemistry – direct way to high technologies of new century. *Uspekhi Khimii*, **72**(5), 419 (2003).

5 Kozlov, G. V., Yanovskii, Yu. G., Burya, A. I., and Aphashagova, Z. Kh. Structure and properties of particulate-filled nanocomposites phenylone/aerosol. *Mekhanika Kompozitsionnykh Materialov i Konstruktsii*, **13**(4), 479 (2007).

6 Witten, T. A. and Meakin, P. Diffusion-limited aggregation at multiple growth sites. *Phys. Rev. B*, **28**(10), 5632 (1983).

7 Yanovsky, Yu. G., Valiev, Kh. Kh., Kornev, Yu. V. Karnet, Yu. N., Boiko, O. V. Kosichkina, K. P., Yumashev, O. B. The role of scale factor in estimation of the mechanical properties of composite materials with nanofillers. Nanomechanics Science and Technology: *An International Journal*, **1**, **3**, 187-211 (2010).

8 Witten, T. A. and Sander, L. M. Diffusion-limited aggregation. *Phys. Rev. B*, **27**(9), 5686 (1983).

9 Bobryshev, A. N., Kozomazov, V. N., Babin, L. O., and Solomatov, V. I. Synergetics of Composite Materials. Lipetsk, NPO ORIUS, p. 154 (1994).

10 Sheng, N., Boyce, M. C., Parks, D. M., Rutledge, G. C., Abes, J. I., Cohen, R. E. Multiscale micromechanical modeling of polymer/clay nanocomposites and the effective clay particle. *Polymer*, **45**(2), 487 (2004).

11 Happel, J. and Brenner, G. The Hydrodynamics at Small Reynolds Numbers. Moscow, *Mir*, p. 418 (1976).

12 Mills, N. J. The rheology of filled polymers. *J. Appl. Polymer Sci.*, **15**(11), 2791 (1971).

INDEX